THE ART OF ECOLOGY

The Art of
Ecology

Writings of

G. Evelyn Hutchinson

Edited by
David K. Skelly, David M. Post, and Melinda D. Smith

Foreword by
Thomas E. Lovejoy

Yale UNIVERSITY PRESS
New Haven and London

Frontispiece: G. Evelyn Hutchinson at Woods Hole in the 1930s. Archives of the Yale University Library.

Yale University Press books may be purchased in quantity for educational, business, or promotional use. For information, please e-mail sales.press@yale.edu (U.S. office) or sales@yaleup.co.uk (U.K. office).

Designed by Mary Valencia.
Set in Adobe Garamond type by Technologies 'N Typography, Merrimac, Mass.
Printed in the United States of America.

Library of Congress Control Number: 2010927632
ISBN: 978-0-300-15449-8 (paperback)

A catalogue record for this book is available from the British Library.

This paper meets the requirements of ANSI/NISO Z39.48–1992 (Permanence of Paper).
10 9 8 7 6 5 4 3 2 1

CONTENTS

FOREWORD

In 1962 as a Yale freshman with a passion for biology, I attended the weekly afternoon Zoology Department seminar and endeavored to understand an almost bewildering variety of topics from molecular biology to ecology and evolution. I noted that some of the most interesting questions came from the back of the lecture hall in a strong British accent soon identified as that of G. Evelyn Hutchinson. That was my introduction to one of the most amazing minds I have ever had the privilege to know.

Within weeks I heard him deliver the "green pigeon" talk at the opening of the Bingham Laboratory of Oceanography and Ornithology (cited in the introductory essay to Part Five). Years later I swallowed hard and dared to ask the great polymath to be the advisor for my thesis. That began a rich intellectual adventure which so many shared and which this volume will open to yet many more.

To reread these works, and the marvelous (to use one of Evelyn's favorite but carefully used words) introductory essays to each section, is to be reminded of the astounding breadth of his knowledge and his insights. They ranged from details of lacustrine ecosystems and interactions between two species of water bugs to encompassing theory, biogeochemistry, and even exobiology and the question of whether life somewhere else might be built around silicon instead of carbon.

One of his great strengths—quite obvious in this volume—is the strong link between the empirical and theory. While capable of extraordinary synthesis as in the *Treatise on Limnology*, and soaring leaps of theory as in n-dimensional niche theory, everything was anchored in or related to natural history observation and fact. For some that could have been an impediment, but for Evelyn the natural history observation was fact that theory had to fit—although he was perfectly capable of holding something like the green pigeons in some intellectual anteroom until he found a place where it would fit.

That is why his work is so robust and much of it foundational for modern ecology. When a burst of mathematical ecology was under attack largely because it did not have that fundamental link with real nature, he remarked that his student Robert MacArthur "really knew his warblers."

His scholarship had such richness in part because he diligently set anything new in the context of previous work. This perhaps reached its zenith in *An Introduction to Population Ecology*, a textbook with superabundant footnotes. At the very outset when referring to Ockham's razor he provides a long footnote on William of Ockham. When referring to territory he supplies a footnote that could stand alone as an essay. He set a standard for understanding how current advances in ecology build from previous achievements—an example for some of today's cyber world students who seem to feel any citation more than five years old is irrelevant.

He achieved all of this with only his undergraduate degree from Cambridge, having chosen instead to go to South Africa and from there by serendipity to Yale in 1928. In 1970–71, the secretary of the university sent Yale graduate and Cary Trust trustee Frank Stubbs to see Evelyn because of the trust's interest in establishing what became the Cary Institute for Ecosystem Studies. Evelyn included me in the meeting (I was in the last throes of the dissertation). The gentlemanly Frank Stubbs kept referring to Dr. Hutchinson and Dr. Lovejoy, until finally the great professor intervened: "I think you should know. Neither of us is 'Doctor.' Tom is too young and I am too old."

Those straight-to-the-point statements were an example of how he thought, and—one might believe—how he managed to keep order in a mind so full of knowledge. For example, "Differential calculus is more interesting because it describes how things change." That same directness—pointing out that if ecology hoped to progress like physics it would need to embrace theory—led to reconsideration by the editors of *Ecology* and the publication in 1942 of Raymond Lindeman's *Trophic-Dynamic Aspect of Ecology*, which is in many senses the base of ecosystem ecology.

Less well known is his interest in and concern about environmental issues. As early as the 1940s he was lecturing on the greenhouse effect and the potential climate change from rising greenhouse gases. When receiving the Franklin medal he said he hoped "the various things we are doing to the atmosphere will cancel each other out." In 1967 he contributed a large section, "Ecological Biology in Relation to the Maintenance and Improvement of the Human Environment," in a report to the National Academy of Science.

What might be less apparent from reading this anthology—were it not for the essay at the start of the second section—is his amazing set of students. Some were stellar intellects, and the rest of us clearly did better than we might have otherwise. Both his stimulation and his inspiration inevitably taught us to think differently. We were attracted by the great mind and the possibility of being part of his intellectual atelier. He brought the best out of us simply by treating us with respect and inspiring us to do our utmost. It was also a very diverse group: rainforest ecology, coral reef ecology, butterfly studies, limnology, and plant population biology were just a few of the subjects concentrated on by Hutchinson's students during my time, so any graduate student discussion was greatly enriched.

In his remarks upon receiving the first Tyler Award in 1974, he declined to admit to "paternity" as the "Father of Ecology," saying that went properly to Charles Darwin. He went on to say, "I will, however, proudly admit to being the intellectual father of a great number of ecologists, for an extraordinary succession of incomparable young men and women who have studied with me, and I with them."

Gerardo Budowski was not one of them, but as a student in the Yale School of Forestry took just one of his courses (Ecological Principles). Decades later he said to me: "My mind is still full of him." This anthology makes that possible for many more.

Thomas E. Lovejoy

PREFACE

The aim of this book is to collect, in one place, writings of one of the greatest biologists of the twentieth century. G. Evelyn Hutchinson is often declared the father of modern ecology. Along the way he made foundational contributions to the study of freshwater environments and did much to foster the systems view of the natural world. In fact, Hutchinson thought and wrote about a dizzying spectrum of topics. A close study of his writings makes it clear that Hutchinson's great contributions to science flow directly from the extraordinary breadth of his accumulated knowledge coupled with his keen, synthetic intellect.

We can go further. If a student wished to read the work of just one scientist in order to understand the major patterns and principles of ecology, there is no question that Hutchinson would be that person. While we do not recommend this approach, it is certain that ecologists working in the twenty-first century cannot consider themselves well founded without reading and understanding Hutchinson's writings.

There have been two particular motivations active in the development of this anthology. The first is purely scientific. Hutchinson's work is critical to the continuing development of ecology. Though his publications are, at this point, decades old, he remains highly cited. Nevertheless, these citations center on a very small fraction of his total output. The acknowledged classic papers are included here. But there are many others worth our time and attention. It has been a great challenge to pare this anthology down to its current dimensions. There is much more out there, but we feel that the present effort represents Hutchinson's work well.

We hope that anyone reading this anthology will gain an appreciation of the several facets of his work and why it remains salient and inspiring. We also hope to motivate readers to go beyond the boundaries of this volume to explore Hutchinson's work further. We provide brief excerpts from some of his several books and give some indication of his influences on his students and others. Those who undertake a closer study of this work will be well compensated for their efforts.

The second motivation for the anthology is historical. We believe Hutchinson is worth studying as a scientist. This collection is intended to complement other works, out and forthcoming, documenting the history of ecology or focusing on Hutchinson's life and contribution. In an era in which scientists are encouraged to specialize earlier and more narrowly than ever before, Hutchinson's career reminds us of the spectacular benefits of intellectual breadth. There is little doubt that a tight focus helps us contribute efficiently to the body of scientific knowledge. However, neither is there any question that an ability to use and synthesize information from across the spectrum of knowledge is a signal trait of our greatest scientists.

We have organized peer-reviewed publications, magazine articles, book excerpts, and other documents around several themes. Any organization is imperfect. This is particularly so for someone who was active in so many academic arenas simultaneously. We have divided this work into five parts covering Hutchinson's life and vision, his contributions to the study of freshwater environments, his role as a theorist, and his interest in and inspiration from museums. Introductory essays accompany each part. While the function and length of the essays vary, they are, collectively, intended to place Hutchinson's work in a historical frame and to provide context that will help readers understand the continuing importance of his work and ideas.

Anyone familiar with Hutchinson's publications, and more particularly his writing, will understand the humility we experienced in putting words to pages that will interleave with Hutchinson's own work. Hutchinson remains a model for the communication of scientific understanding. He was a fluid, elegant writer with a dust-dry wit. These trepidations aside, editing this volume has been an immense pleasure. The opportunity to read Hutchinson more broadly and more intensively deepened our own understanding of issues that remain as fundamental as when Hutchinson first wrote about them. We hope this volume will make it easier, now and in the future, to gain insight from one of the great lights of biological thought.

Many people contributed to the realization of this volume. We thank, in particular, Michael Donoghue, Sharon Kingsland, Jane Pickering, David Schindler, and Nancy Slack for taking the time and care that characterize their contributions to the book. We also owe a special debt of gratitude to Stephen Stearns. Steve organized the G. Evelyn Hutchinson Centenary Symposium held at Yale University in October 2003. His interest in honoring one of his undergraduate professors provided the spark leading to the creation of this anthology. Steve Stearns and the other speakers at the symposium, including Deborah Goldberg, Lillian Randall, Oswald Schmitz, Karl Turekian, Peter Vitousek, and Earl Werner, helped us put Hutchinson's work into a modern perspective. Susan Bolden, Jack Borrebach, Jaya Chatterjee, Susan Donoghue, Nathan Havill, and the Manuscripts and Archives staff at the Yale University Library have been extremely helpful. We also thank Jeffrey Park for providing funding support from the Yale Institute of Biospheric Studies. It is certain that the book would not exist without the unflagging support and expert input of Jean Thomson Black at Yale University Press. Finally, we thank our families—Parker, Linda, Cassia, Linnea, Kealoha, Aidan, and Nathaniel—for the many reasons that families need to be thanked when a book is being produced.

Part

1 *Introduction*

The Beauty of the World: Evelyn Hutchinson's Vision of Science

Sharon E. Kingsland

The philosophical analysis of scientific method that most influenced G. Evelyn Hutchinson was the second edition of Harold Jeffreys's *Scientific Inference,* published in 1937.[1] Jeffreys, a geophysicist at Cambridge University, was mainly concerned with problems in physics, but Hutchinson recognized the relevance of Jeffreys's approach to biological problems, especially in evolutionary biology and ecology. But understanding Hutchinson's deep love of science requires that we look beyond such philosophical discussions of method, for his vision also reflected his religious beliefs and sensitivity to the aesthetic dimensions of science. His collection of essays from 1953, *The Itinerant Ivory Tower,* took as its epigraph a statement from Simone Weil, whom Hutchinson recognized as a kindred spirit and to whom he dedicated a later essay.[2] Weil wrote that the true definition of science was that it was the study of the beauty of the world. Science for Hutchinson was far more than a way of investigating nature; it was a

means of enlightenment that, like art, could produce feelings of exaltation that ennobled the human spirit.

Although reflections on the meaning of science are scattered throughout Hutchinson's writing, three essays in particular provide insight into his approach to science. An essay entitled "Methodology and Value in the Natural Sciences in Relation to Certain Religious Concepts" is a personal statement about the value of science.[3] An appreciation of D'Arcy Wentworth Thompson gives us Hutchinson's views about the history of science and what constitutes greatness in science.[4] Finally, a postscript to his textbook, *Introduction to Population Ecology* (1978), discusses his philosophy of science in response to criticism that his theoretical approach was tautological and not properly scientific at all.[5] Hutchinson's framing of the philosophical problems of ecology, although formally influenced by Jeffreys, also reflected his admiration for the treatment of scientific problems that is exemplified in Charles Darwin's *Origin of Species,* from which much of modern ecology springs.[6]

Before considering these texts it is worth recalling one of ecology's grandfathers and reflecting on the ways in which Hutchinson's outlook captured the breadth of vision and moral outlook of scientists from two centuries ago. Lawrence Slobodkin and Nancy Slack, in a biographical essay on Hutchinson, compared him to thinkers such as Benjamin Franklin, Thomas Jefferson, Benjamin Thompson (Count Rumford), J. W. von Goethe, and Alexander von Humboldt.[7] A comparison with Humboldt affords a good vantage point from which to begin an appreciation of Hutchinson's vision of science, for there are illuminating parallels between these two scientists.

Humboldt was born into a family of minor nobility in 1769 and died in 1859, the year that Darwin's *Origin of Species* was published. Darwin revered Humboldt as a naturalist and immersed himself in Humboldt's descriptions of tropical America as he traveled around the world on the famous *Beagle* voyage in the 1830s. Humboldt explored equatorial America and Russia and wrote about a great many subjects in physiology, biogeography, and natural history. Toward the end of his life he wrote a holistic digest of celestial and terrestrial phenomena, *Cosmos,* which made him the most successful author of his generation.[8] Humboldt inspired many of the nineteenth-century European naturalists who became the first generation of ecologists.[9]

Hutchinson, although not self-consciously "Humboldtian," nonetheless approached science in a way that is strikingly similar to that of his great predecessor. In 1808 Humboldt published *Views of Nature, or Contemplations on the Sublime Phenomena of Creation.* The first and second editions were products of his Parisian period, 1804 to 1827, when he worked as an independent scholar in the intellectual hub of Europe.[10] He continued to revise and expand the book into the 1850s. It was both a work of science and a work of poetry: Humboldt wanted people to appreciate the "harmonious cooperation of forces" in nature, so as to understand how the natural world was put together and how it formed a whole.[11] This was part of the broader study of what Humboldt called "terrestrial physics," which included a passion for quantitative study as well as an interest in how the play of forces produced states of equilibrium in nature. In the twentieth century, these themes would be developed through the analysis of biogeochemical cycles, population dynamics, and energy flow through food webs, all fields to which Hutchinson contributed. In Humboldt we see the framing of scientific questions that would constitute the central themes of ecological science in Hutchinson's generation.

For all his enthusiasm for numbers and measurement, Humboldt also wanted people to experience enjoyment at his descriptions of the world. The work of science was at the same time a work of literature, one meant to inspire and to make the reader aware that contemplating nature was a deep source of pleasure, not unlike the pleasure one might experience at seeing a great work of art. In his extensive travels Humboldt had ample opportunity to reflect on the diverse human groups he encountered. Whether looking at the most brutish levels of human society or at the glitter of Parisian high culture, human life, he reflected, was full of woe; everywhere one met the depressing spectacle of man opposed to man. The contemplation of nature, Humboldt believed, was a way to find intellectual repose in this world of strife. When pondering the life of organisms, studying the hidden forces of nature in her sacred sanctuaries, or gazing upward to contemplate the harmony of the heavens as the ancient philosophers did, one found one's mind pulled away from the cares of life toward eternal truths. For Humboldt these meditations were religious in nature.[12]

By Hutchinson's time science had largely divorced itself from the religious context that gave it meaning in an earlier age, but Hutchinson recognized that it was still imperative to question the purpose and value of science. Was the

Portrait of Alexander von Humboldt painted by Friedrich Georg Weitsch, 1806. Oil paint on canvas, 126 × 92.5 cm. The original painting hangs in the Staatliche Museen zu Berlin—Preußischer Kulturbesitz, Nationalgalerie. Photograph reproduced by permission of Bildarchiv Preußischer Kulturbesitz and Art Resource, NY.

value of contemplation fundamentally different than in Humboldt's day? Was science now to be viewed in a purely utilitarian light, as part of our steady march toward the conquest of nature?

Hutchinson was concerned with these questions, and he formulated a response that echoed the Humboldtian vision. In his essay on method and value in science Hutchinson inquired about the value of inductive science.[13] For pragmatic Americans the value of science was often thought to lie in its material benefits. Pure science was thought to have value chiefly because it promised to yield some kind of application down the road. However, Hutchinson cautioned that this was only a secondary end. Improving the world materially was of course all to the good—not, however, as an end in itself, but because it freed the mind and body to experience the value of contemplation. The danger of modern society, he believed, was to think that the conquest of nature was an end and to conclude that contemplative values need not be nurtured. He argued that the only real values we experience are contemplative ones, for through contemplation we experience and appreciate beauty. His idea that we were meant to experience beauty was identical, he believed, to the Christian idea that man was created to praise God. This essay was a modern restatement of the vision that had guided Humboldt in the early nineteenth century.

But what did Hutchinson mean by "beauty"? He explained by relating an anecdote about an experience he had while walking down the drive of his house. On that occasion he spied a brilliant patch of red, which drew his attention and puzzled him: "In a second or two I realized that a pair of scarlet tanagers was mating on a piece of broken root conveniently left by a neighbor's somewhat inconsequential bulldozer; the female was sitting inconspicuously on the root, the male maintaining his position on her by a rapid fluttering of his black and hardly visible wings which tended to vibrate his entire body."[14] He reflected that the sight was strikingly beautiful and that it gave him a sense of pleasure to realize this.

But why would such observations produce an impression of beauty and a sense of pleasure? Thinking about his reaction, Hutchinson realized that the observation was made in the light of a range of associations that came from many prior experiences. The observation caused him to think about birds and their breeding behavior, and to relate this one instance to a general theory of territorial and breeding behavior. That ability to generalize was itself pleasing. But other things entered into the experience: he recalled an amorous and beautiful seventeenth-century song he had recently heard; certain religious and psychoanalytic associations were conjured up in his mind by the sight; and finally the color itself reminded him of specimens of Central American tanagers that he had seen in a museum, which caused him to think about the evolution of these birds. The point of this anecdote was to illustrate how the seemingly simple and direct experience was conceptually enriched by so many kinds of associations that Hutchinson recognized it as "essentially an art form."

That enrichment—which is the result of the integrative action of the nervous system, of the brain's ability to analyze and synthesize—constitutes our experience of beauty, and indeed of love for the object. To Hutchinson this appreciation of beauty was also a motive for doing science. As he wrote, "The purpose of inductive knowledge is to produce conceptual schemes which are found to have beauty and which, therefore, give a certain degree of lovableness to the universe which it did not have before we engaged in the activity."[15] We will value the world, and we will want to preserve and protect the world in its diversity, to the extent that we have this ability to contemplate and to bring to the world a rich set of associations and conceptual schemes, which constitute our love of the world. As Hutchinson pointed out, scientists were reluctant to admit the role of beauty in their work, for their education led them to play down these aesthetic considerations. But to construct a scientific theory, or a conceptual scheme, was to create something comparable to a work of art and which, like art, possessed "beauty." Humboldt, one feels, would have understood perfectly what Hutchinson was talking about and why it was crucial to understand that this aesthetic experience gave value to scientific work, over and above the utilitarian value of science.

Hutchinson's direct inspiration for his framing of these ideas was his friend Rebecca West, the English novelist and essayist. In 1928 she published a personal treatise on aesthetics called *The Strange Necessity,* which was reprinted after her death with an appreciative forward by Hutchinson.[16] West had tried to pin down why art was important to her, what it contributed to her pleasure in life, and by extension why art should also be important to other sensitive, intelligent people. The anecdote Hutchinson told about himself witnessing the tanagers echoed similar anecdotes that West recounted of her experiences reading and learning, analyzing and synthesizing. West also recognized the kinship between the experience of art and of science. Both were capable of eliciting deep emotional responses ranging from intense exaltation to milder everyday pleasures, and West drew an analogy between such feelings and the passions and affections that humans felt for each other. West's views about the purpose of art struck a deep chord with Hutchinson and expressed very much his own views that science was important for the same reasons that art was important. Fundamentally, he argued, there was no distinction between the values of pure science and the fine arts, although the values were expressed in different languages and the two activities had different relationships with the external world. As West concluded that "this strange necessity, art, is as essential as life," Hutchinson surely felt that this strange necessity, science, was as essential as life.

The essay in which Hutchinson unfolded his personal view of science was written in 1951 and published in 1953, the year in which Watson and Crick announced their discovery of the double helical structure of DNA. This was a critical time of change in American science. Government and industry supported science because it contributed to economic wealth and national security. Vannevar Bush's famous report *Science, the Endless Frontier* laid out a strategy for postwar research based on the logic of conquering nature: "New products, new industries, and more jobs require continuous additions to knowledge of the laws of nature, and the application of that knowledge to practical purposes. . . . Our defense against aggression demands new knowledge so that we can develop new and improved weapons. This essential, new knowledge can be obtained only through basic scientific research."[17] The two decades following the war were years of monstrous growth in science and technology. The chemical control of disease-transmitting organisms promised to solve world health problems. Industrialization of food production promised to emancipate people from the limits imposed by the amount of arable land. Air conditioning would give cities a climate independent of nature, while atomic power, it was hoped, would free societies from dependence on fossil fuels. Deciphering the genetic code, people imagined, would enable them to create plants and animals and "entire races of people" to their own liking.[18] Ecologists had to find their place in a world of rapid growth, where the dominant idea was that the main purpose of science was to control nature. This was the ideology that Rachel Carson would challenge in *Silent Spring.*[19]

Science was also important to national security. Ecology itself would be remade as a Cold War science as Americans learned to live with the bomb. Ecosystem ecology was nurtured in the bosom of the atomic age, funded largely by government agencies that evolved from the Manhattan Project.[20] The national laboratories at Oak Ridge in Tennessee and Brookhaven in New York became growth centers for ecosystem ecology and radiation ecology.[21] These developments in ecology ushered in a whole set of new ideas, conceptual schemes, and theories.

Hutchinson and his students contributed hugely to these conceptual advances, and here lies the paradox of Hutchinson's vision. One can view his sensibility as a throwback to an earlier age, the age of Humboldt. Yet Hutchinson was strikingly modern in his embrace of new methods, new theories, and new ways of conceptualizing nature. One sees Hutchinson's imprint on ecosystem science, systems ecology, the mathematical development of population ecology, niche theory, and the application of cybernetics to ecology, just to name a few areas of special significance.[22] Hutchinson was very open to new theoretical ideas, to mathematical formalizations, and to the crossing over between ecology and new subjects like cybernetics and information theory.

Can this image of modernity be reconciled with the apparently anti-modern stance of Hutchinson's warnings not to lose sight of the value of contemplation? The resolution of the paradox lies in realizing that for Hutchinson new approaches to ecology, to theory development, and to mathematical methods were all forms of contemplation that enriched science and gave insight into the beauty of the world. Hutchinson was not cautioning his readers to be wary of new theoretical developments in ecology. The point was that conceptual schemes and theories should be appreciated as contemplative activities, not just tools for manipulating and conquering nature.

Hutchinson warned that the danger in the headlong rush to develop science and technology was to lose sight of the value of contemplation and, through pride, to set ourselves on a path of destruction. One sees in his writing a critique of the utilitarian ethos of modern society, an estimation he shared with thinkers such as Simone Weil, who "cried stridently that we should take in the entire universe in an act of intellectual love extending infinitely far into the future and past and excluding nothing but our momentary sins."[23] Similar expressions of postwar angst could be found among American intellectuals as well. Lewis Mumford, the urban historian, argued that the cultivation of a broad feeling of love in all its expressions would help to rescue the world. He proposed that what we needed for our development was "not power but power directed by love into the forms of beauty and truth." As he concluded, "Only when love takes the lead will the earth, and life on earth, be safe again."[24] Science, Mumford concluded, was a kind of love affair.

Hutchinson surely had something like this in mind when he compared science to art and spoke of its power to increase the lovableness of the universe, and when he commented that ecology as it developed would ennoble the life

of man.[25] Education, he believed, should not primarily be a means of learning how to earn a living, but a quest not unlike a religious quest for spiritual enlightenment, for "everywhere in education we should aim at seeing life as a sacred dance or as a game in which the champions are those who give most beauty, truth, and love to the other players."[26]

Hutchinson explored these themes also in his contributions to *American Scientist,* in a feature called "Marginalia" that began in the summer of 1943 and continued on a regular basis for a dozen years, with occasional contributions extending to 1983. His notes covered a wide variety of topics, ranging over many fields of science. He was as likely to report on the latest symposium on quantitative biology at Cold Spring Harbor as he was to offer reflections on recent books in the social sciences, philosophy, and the arts. As Hutchinson explained, the notes reflected the attitude of the "philosophic naturalist, rather than that of the engineer, the point of view of the mind that delights in understanding nature rather than in attempting to reform her."[27] In 1943 the pressing need was to avoid destruction of the environment and to prevent the injuries that we inflicted on ourselves "with such devastating energy." These problems were no less urgent when Hutchinson penned his last set of Marginalia forty years later. He returned to the theme of human destruction and the importance of studying human behavior, writing that "it is useless to complain passively that the destructive tendencies are human nature; if we want to survive, we now have to do something about it."[28] The purpose of science, in addition to fulfilling deep aesthetic and intellectual needs, was to help us to understand ourselves—and then act on that knowledge. In that way, Hutchinson argued, we could improve our chances of evolving in a more positive direction, perhaps one that took advantage of what might be regarded as "the less aggressive, more feminine traits."

Hutchinson's ideas about the value of science were connected to his appreciation of the history of science, which he discussed in an essay about D'Arcy Wentworth Thompson, who died in 1948.[29] Thompson's classic treatise *On Growth and Form* (1942) was a study of the physical and mathematical laws that governed organic form.[30] It was prodigiously learned, with footnote references going back to Aristotle, the first biologist. Toward the end of his treatise Thompson referred to the theory of proportion of the Renaissance artist and mathematician Albrecht Dürer and used Dürer's ideas to illustrate certain problems of growth in deep-sea fishes and other animals. As Hutchinson concluded: "Though . . . we appear to have been given the beginnings of a new geometry of organic form, actually we are being told that we already had had one for four centuries or more but had not used it."[31] Thompson's ability to open up the wider vista of intellectual achievement, to present a panorama of the intellectual adventures of the human spirit for the past three thousand years, was the mark of this book's greatness in Hutchinson's eyes. It mirrored the earlier groping of the human mind far back beyond historically recorded time. The work itself dealt with contemporary problems and the latest research, but the questions it asked were of enduring interest and always important.

Hutchinson discerned in this book a statement about the nature of science, namely that science is a traditional activity, a tradition of daring and imagination, and that we are connected to the ancient world by this important intellectual thread. History and current events taught that civilization was fragile: it was hard to build, hard to preserve, and easy to destroy. While science might be blamed for some of the world's woes, the scientific mind was the light that helped to produce and preserve civilization. This idea of being connected to a long historical tradition, and the importance of realizing that there is something precious connecting us to earlier generations, was central to Hutchinson's thought and his vision of science.

This sense of connection helps to explain the very odd style of the textbook on population ecology that he published in 1978. It contains extraordinarily detailed historical footnotes that

dip back into the Middle Ages to trace the origins of ideas important in the philosophy of science and the practice of science. Stephen Jay Gould, reviewing the book, noted the plethora of adornments consisting of historical footnotes and "tangential wanderings on etymology and minutiae of natural history."[32] Gould took this exuberance as an illustration of Hutchinson's fascination with detail, diversity, and ornamentation, and he commented that this pile of information was unevaluated detail, or "love of detail for its own sake."

Hutchinson certainly took pleasure in ornamental details, but perhaps there are other motives that lie behind this extraordinary level of detail. It appears to express not just a general enthusiasm for knowledge of all kinds, but precisely this sense of being part of a grand historical tradition that Hutchinson identified in his review of Thompson's work. We should remember that scientific demography began in the seventeenth century with John Graunt's bills of mortality, because science develops gradually and requires a lot to nurture it. We should be aware that Pierre-François Verhulst discovered, or invented, the logistic curve in the 1840s but that it was neglected until rediscovered in the 1920s, after which it became a tool for the development of a new field of science, theoretical population ecology. Hutchinson was reminding the reader that scientific knowledge, once produced, could be lost, sometimes forever, and that it might take a long time for the significance of certain ideas to be grasped.

We should understand that some ideas, such as the principle of competitive exclusion, were known for several decades without having any particular significance attached to them. It was therefore an important advance when the young Russian ecologist G. F. Gause elevated a commonsense idea into a scientific principle. He did it by a marriage of theory and experiment, thereby increasing its interest and making possible the further development of the theory of competition and the niche. In each footnote, on each page, Hutchinson gave credit to those people who advanced science, who contributed to this fragile structure, who took seemingly trivial ideas and saw in them something worth developing. One needed to recognize that even small observations, if they were new or unusual, could be of great interest and importance even if one could not say what practical outcome might come from the information. It mattered to show that science had progressed, that all of this work and thought had created something rich, satisfying, and pleasurable. This was not Hutchinson reveling in detail for the sake of detail, or simply for the love of ornamentation. It was a reminder of how many people have contributed, and in how many different ways, to the concepts that enable us to experience the beauty, the lovableness, of the world.

Hutchinson's philosophy of science grew from his informal childhood explorations in natural history and from the kind of education he received in Cambridge. He remembered Cambridge fondly as an extremely beautiful city on a small river, where for many centuries there had existed a tradition of ceaseless questioning and of continuous effort to produce formalized intellectual constructs.[33] Not only was he able to benefit from the intellectual stimulus provided by his parents and their acquaintances, but the university was the sort of place where one could learn about many different fields in an informal way (a characteristic that worked very much in Darwin's favor in the 1820s). One could be self-educated and could explore general problems of theory, even through such odd routes as parapsychology and psychoanalysis, two subjects that intrigued Hutchinson as a student because they appeared relevant to the problem of vitalism versus mechanism that was being hotly debated in the 1920s. One could be a "hunter and gatherer," picking up bits of information without a strong sense of where it was leading. Only later on did Hutchinson see that this kind of accumulation was ideally suited to ecological work. This was his formative ecosystem: a physical, biological, social, and intellectual environment that stimulated his mind, imagination, and emotions.

Hutchinson also appreciated that science

could be done in many different ways and that one's approach had to match the nature of the subject. Particularly important for him was to understand that it was not necessary, or even a good thing, to reject the idea that scientific knowledge can be achieved by making logical arguments which might even appear circular or tautological. Darwin's theory of natural selection, or the idea of survival of the fittest, is an example of one such argument. Such arguments are often underappreciated. Darwin remembered that when he and Wallace first published their ideas briefly in 1858, not only did they attract little attention, but one reviewer commented that everything that was new in them was false, and what was true was old.[34] It took the longer argument in *Origin of Species* to show that there might be something important in the idea of natural selection.

Similarly, the principle of competitive exclusion was not thought to be of great interest until Gause, and then Hutchinson, developed a deeper analysis of how species did or did not manage to live together. The point was to avoid assuming that all science is based on setting up hypotheses that can be falsified. The coda to Hutchinson's textbook was his answer to those critics who were claiming that the methods of scientific investigation that he and his students had pursued were not scientific at all. This criticism, which appeared also in reviews of the book, assumed a view of science that Hutchinson thought was too narrow.[35] Drawing on the analysis of scientific inference by Harold Jeffreys, which began with a discussion of the use of logical arguments and probability in science, Hutchinson defended a style of scientific reasoning that used logical arguments to probe nature. He emphasized the idea that one can approach the truth gradually by using so-called tautologies, which are not really tautologies because one is always refining the starting premises based on new observations. If scientific method only allowed falsifiable hypotheses, whole areas of investigation would be discarded and the world would be poorer. If taken to extremes, Hutchinson pointed out, the enthusiasm for falsification of hypotheses could lead to the flawed view that the greatest development in any area of science would consist in "establishing a maximum number of negations, which perhaps may be generalized into a universal null hypothesis that everything is due to chance."[36]

Finally, the ability to construct theories required not just the right philosophy of science, but also a certain acceptance that experiencing pleasure was a good thing. Hutchinson recalled how Midwestern ecologists had been terribly suspicious of theory at the time his career was starting.[37] These suspicions had contributed to the difficulty that Raymond Lindeman had faced trying to get his now-classic paper on the trophic-dynamic approach to ecology into print.[38] Lindeman had tried to create a general theory of succession in small lakes, based on the idea that biological communities could be described as networks through which energy flowed. His argument was based on data from just one lake, and the reviewers would not accept that such a small database could justify publication of the conceptual scheme that Lindeman offered.

Hutchinson later commented upon the puritanical streak that seemed to underlie such attitudes.[39] He recalled that while working with Edward A. Birge and Chauncey Juday in Wisconsin, he had felt dissatisfied that they had never put their masses of data into some kind of informative scheme of general significance. He also noted that they viewed drinking coffee or tea for breakfast as decadent and abnormal. Later, he said, he came to see these things as connected. Appreciating the beauty of the world, a beauty revealed by the creation of scientific theories among other things, required one to let go of these puritanical inhibitions and accept the pleasurable experiences that the world offered.

Aria da Capo and Quodlibet

WITTGENSTEIN wrote, on the last page[1] of the *Tractatus Logico-philosophicus*, which, whatever else it is, may be one of the greatest poems of this century, "My propositions are elucidatory in this way: he who understands me finally recognizes them as senseless."

When one is writing science, neither Wittgenstein nor anyone else would want such a statement to be a true ending to a book. There is, however, a contemporary school[2] that is very skeptical of the validity of the kind of theoretical argument that I have used, finding it to be tautological and claiming that when tautologies are suitably mixed with empiricisms, which seem to be what most of us broadly call observations, the result of the summation is metaphysics, to be avoided like poison. Such criticism is healthy in that it continually makes us keep track of what we are doing. The criticism can, however, easily become unhealthy if it discourages certain kinds of work by too rigid an adherence to dogmas about the proper scientific procedure. Very great men[3] indeed have believed that the entire area of evolutionary biology is outside the legitimate domain of science as it cannot be approached experimentally in a direct way. Such ideas, expressed nearly half a century ago, have not prevented us from acquiring, during that half century, a vastly increased understanding of evolutionary processes.

Logico-mathematical theories are of course tautological in the sense that they are derived analytically from a set of axioms. If we are studying science, certain axioms, here called postulates, are the formal statements of something believed to be possible about the external world. Theory derived from correct logico-mathematical manipulation will produce propositions which will be true when the postulates are true statements about the external world, but may be either true or false if a postulate is a false statement. It is not possible, as is well known, to verify in an absolute sense the truth of any statement used as a postulate in this sense, but it is possible to falsify it, for all practical purposes, though even then one always might make a

1. L. Wittgenstein, *Tractatus Logico-philosophicus*. London, Routledge Kegan Paul, 1932, 189 pp. The quotation is from 6.54 on p. 189. Though this or the later English version are usually quoted, perhaps the full quality of the work is more apparent from "am Ende als unsinnig erkennt" than from "finally recognizes them as senseless."

2. References may be found in R. H. Peters, Tautology in evolution and ecology. *Amer. Natural.*, 110:1–12, 1976.

3. I am particularly thinking of my revered and much-loved former chief, Ross Granville Harrison, the greatest biologist with whom I have had a close day-by-day relationship over a number of years. I think, however, that this point of view was quite widely held in the 1920s and 1930s. Harrison, though looking at science in ways that could be very different from mine, always supported me as a junior colleague. He had a marked, if indirect, effect on ecology through his interest in allometric growth, which interest developed partly in response to Joseph Needham's work. It affected E. S. Deevey, R. Lindeman, and myself, as we all lunched with Harrison on the fourth floor, under the roof, of the Osborn Zoological Laboratory.

mistake in the process of comparison of the result deduced with the observations, and so reject an agreement that is actually adequate.[4] In a satisfactory case in which we have fairly good prior reason to think that a certain hypothesis may be true, we may find either that we are wrong, the hypothesis becoming too improbable to worry about, or that we are right, meaning that the hypothesis is a bit more probable than we previously had thought. The process is qualitatively the same whether we are increasing the probability from 0.900 to 0.999 in physics or from 0.501 to 0.510 in evolutionary biology.

Since we would not have made the deductions and comparisons if we had not initially had an intuition of being on the right track, the most practical thing to do when we have falsified a hypothesis is to modify one of the postulates used and try again. Over time, as any set of concepts develops, it will be found to have embraced a number of alternative hypotheses, some proving to be almost certainly false, others proving to be very probably true in a certain set of circumstances. The set of all these possibilities, some falsified under all, some verified under some, circumstances, would, if the subject matter were important enough, constitute what Kuhn calls a paradigm.[5] When all the possibilities of such a set that are readily conceived have been tried, before the ingenious investigator attempts to turn to something radically new, the results are filed away for future reference in papers and books. A few such results which immediately appear as important become part of the intellectual currency of science and are taught in classes and stated in textbooks; others remain nostalgically in the minds of their discoverers, on whose death they pass into the unread older literature, well summarized in the reviews of N or M.

If the work was worth doing, concepts such as the logistic may be rediscovered, after a rest of 80 years in limbo, and then may undergo a new development. This process has been amply documented in the preceding pages. More usually the best immortality that an idea can hope for is to be disinterred by a historian of science. Though historians of science probably look upon themselves as students of the development of the human mind, their value as quarrymen or miners of good forgotten ideas should not be overlooked.

All the theoretical arguments in this book are obviously based in part on postulates derived from biology. As logico-mathematical statements, they are tautological; their biological interpretation, however, is not tautological and can be falsified. Their implications can be compared with biological reality and in every case the simple initial statement is found to be wrong in at least some important circumstances. Most significantly, the logistic is wrong when the resources that determine K are not constant, as in man, or when K-selection is continually improving the efficiency with which these resources are used. The Volterra competition equations are wrong whenever the outcome of competition depends on the frequency of one competitor. This is presumably due to the competition functions being of too simple a form. As Cardinal Newman[6] put it, "In scientific researches error may be said, without a paradox, to be in some instances the way to truth, and the only way."

The concept of r- and K-selection, which is widely believed to be useful in understanding much of the behavior of populations, and the concept of frequency-dependent competition, which is likely to become a very important explanation of specific coexistence, would never have been discovered without the development of the simple theories. There are indeed also some cases where the simple theories are reasonably well confirmed by experiments.

A further criticism is that in the supposed confirmation of theory in the laboratory or

---

4. This specific point, as well as much else, I owe to Harold Jeffreys.

5. T. Kuhn, *The Structure of Scientific Revolutions.* Chicago University Press, 1962, xv, 172 pp.

6. Quoted from Henry Chadwick, presumably from Newman's *Idea of A University,* in *Times Lit. Suppl.,* August 13, 1976, p. 1003. col. 2.

in nature, the confirmation turns out merely to be entry of cases in a classification obtained by a tautological process and so covering every possible contingency. The example of frequency-dependent competition, not implied by the original Volterra equations, but recognized by the use of some of their properties, shows that such classifications may initially be deceptive, and need to be examined in terms of their own internal structure. Any classification which proves to be sound, however tautological, can be used in an important way merely by enquiring what are the empirical frequencies of occurrence of events in the different categories. The explanation and finally the prediction of such probabilities may require quite different kinds of theory from that used in making the classification. What the classification does is to tell us what sorts of phenomena we can encounter; further work may tell us which of these we will encounter, and how often we are likely to do so.

In the case of the various kinds of competition, it is reasonably clear that in a great many cases, fairly similar species cannot live together; but that when they can, it may be due to niche separation, or to frequency-dependent competition, or to the effects of predation. We have at the moment no idea which alternative is the most probable. Fifteen years ago niche specificity would have seemed the obvious answer, but now Connell[7] can write: "Predation should be regarded as being of primary importance, whether directly determining species composition or in preventing competitive exclusion, except where the effect of predation is reduced for some reason." Connell goes on to consider the two more important of such reasons, namely, refuges and defenses. They often apply, but very qualitatively the quoted sentence, though not free from tautology, implies a by no means negligible probability for the importance of predation. The volume in which the quotation occurs, presents the best

general evaluation of the state of knowledge, in 1975, of this type of problem, but it apparently contains only one mention of determination of the direction of competition by frequency dependence (p. 495–96) and part of that discussion is probably wrong. Since poppies and pomace flies are not closely related organisms and since the effect also seems to occur in daphnid Cladocera, the probability of it occurring in many other groups and having some general importance can hardly be negligible. Though in this case we cannot really get near to assessing numerical values of the probabilities, our informed guesses, such as they are, certainly have changed and hopefully have improved in a decade or so.

The present aim of the part of ecology with which this book is concerned is therefore largely to uncover possibilities, by any kind of theoretical analysis that proves helpful, and then to see how many of these possibilities are indeed realized in nature. This activity may well take the form of mapping onto any suitable classification of nature, systematic, ecological, geographic, or temporal, the frequency of occurrence of particular kinds of demographic or ecological phenomena. This is what we do when we say, for instance, that long food chains more often occur in aquatic than terrestrial communities, or make any statements about the rules that seem to govern clutch size in birds. As such maps develop and become more precise, more and more hypothetic-deductive analysis of their contents should become possible. In the case of food chains and egg numbers, this of course has largely been done, if in a rather informal way. What is most likely to impede the growth of this part of science, as of others where the approach is extensive, is not heresy in scientific method but ignorance and lack of imagination.

There is, moreover, another pressing danger. Long before we have reached even an elementary knowledge of the distribution of

---

7. J. H. Connell, Some mechanisms producing structure in natural communities: A model and evidence from field experiments. In: *Ecology and Evolution of Communities*, ed. M. L. Cody and J. M.

Diamond. Cambridge, Mass., Belknap Press of Harvard University Press, 1975, pp. 460–90; see specifically pp. 475–76.

FIGURE 142. Leguat's figure of the solitaire, *Pezophaps solitarius*.

the kinds of ecological phenomena, they may have disappeared, owing to the continual erosion of nature that is characteristic of our era.

This book has partly been written to show that there is a considerable amount of available knowledge set in an irregular way in a vast area of ignorance. We have already noted that in the case of the now extinct huia, the observations recorded suggest, but do not quite establish, the existence of a unique type of mutual feeding behavior between the sexes.

An even more tantalizing story is told of another extinct species, the solitaire of Rodriguez in the Indian Ocean. The island was colonized temporarily by a group of Huguenots in 1691. François Leguat,[8] who was the leader of the group, left a fascinating account of their adventures on the island. The book first appeared in French, but was published in London in 1707. In it Leguat gives a detailed account of the behavior of a flightless bird, a little taller than a turkey, which he called the solitaire, a name probably derived from another less well-known extinct bird, living on Réunion in the seventeenth century. The solitaire of Rodriguez was a highly territorial bird mating for life. Each territory had a radius of about 200 yards and was defended during the incubation of the single egg and during a period of several months while the young bird was not capable of independent life.

The male had a bony mass under the wing feathers that was used in defense, though Leguat says intruding females were driven off by the female, males by the male. The general form of the bird (figure 142), the moderate reduction of the wing, and the bony mass were confirmed in the nineteenth century when much skeletal material was recovered from the island. Leguat also describes the female as having a cleavage in the arrangement of the feathers on the breast, which delighted him as representing *merveilleusement un beau sein de femme*. This was doubtless an incubation patch. Except for the specific roles of the sexes in defense, part of his account is confirmed by the subfossil evidence, the rest so far seems reasonable. We now, however, reach the extraordinary part of the story. Leguat says that when the chick had left the nest for some days, a company of 30 or 40 adults brought another chick and, being joined by the parents, the two young birds were escorted to some unoccupied territory. This extraordinary ritual he describes as the marriage of the solitaire. Leguat seems so reliable on most other aspects of the bird's biology, and the marriage, if it occurred, must have been so conspicuous, that it is hard to doubt its reality. Now the bird is extinct and we shall never know the truth.

Of course, we can learn little of the ecology and almost nothing of the behavior of the several million extinct species that have existed in geological time. Nevertheless, the evidence that we have as to the limits of what evolution can do at any level of organization is precious. Moreover, it is the highly specialized, slow-breeding, K-selected species which are most vulnerable, but which give most information about such possible limits. Think what we might have learned if Gerald Durrell had been able to establish a colony of *Pezophaps solitarius* on the island of Jersey. Some well-meaning people will doubtless wonder what the preservation of odd organisms merely for their oddness does for suffering humanity. I can only reply that experience shows that the most unpromising knowledge is always proving useful and that some of us have a duty to foster and make it available. It is well to bear in mind that the niches illustrated in figure 101 are totally inadequate for man.[9] We also need[10] all the wondrous things under heaven:

> Their leaues that differed both in shape
>   and showe
> (Though all were greene) yet difference
>   such in greene
> Like to the checkered bent of Iris bowe.

---

8. See chap. 3, note 32.
9. Cf. Matth. 4:4.
10. From a poem by Christopher Marlowe, on p. 480, in *England's Parnassus; or, The choysest flowers of our moderne poets.* London, for N.L. C.B. and T.H., 1600, x, 510 pp.

have responded with a delightfully bawdy tract. Few two men could have been more unlike than Woodruff and Hill.

It always seemed to me sad that Woodruff's timidity prevented him from visiting the places that his heroes had known. He may, however, have preferred them to have lived in a space as remote and inaccessible as the time that they inhabited.

My first teaching assignment was to the beginning course, Biology 10. Woodruff was in charge; either he or G. A. Baitsell gave a lecture a week to each of three lecture divisions, two meeting on Friday and one on Saturday, while there were, I think, nine laboratory divisions to which the rest of the staff lectured twice a week. I had two such divisions. These lectures were devoted primarily to the material to be studied practically. The textbooks used were Woodruff's *Foundations of Biology* and Baitsell's *Manual of Biological Forms*. Woodruff's textbook was a very good one; it is said to have been ascribed to Aristotle, Theophrastus, Vesalius, Hales, Buffon, Lamarck, and Wordsworth by various students taking an examination in another New England college. Actually, since Woodruff was a very literate man, all these authors had no doubt contributed to it indirectly.

I was, however, greatly disturbed by two aspects of this and other courses in Yale; the disturbance to some extent still persists. The first was the use of a required textbook with assigned readings; the second was the continual succession of tests and the correlative idea that one could become educated by piling up credits for a given number of courses.

I know that at Cambridge I was exceptionally lucky in having a father who could advise me from inside so that I could easily and fairly safely take my education into my own hands. I also realize that for many students this could be quite dangerous, though not as dangerous as many responsible senior people might believe. The absolute impossibility of getting self-educated at Yale worried me; the impossibility is now less absolute but it still worries me. I once had to advise a Yale freshman student, rather as a friend than in an official connection, about planning his courses. He intended to go into biology and is now a member of the National Academy of Sciences. I told him to go on taking one or two mathematics courses each year until he got a D. I think actually he survived with a C in Advanced Differential Equations in his junior year. Today, when a computer doesn't know one D from another, one could hardly give such advice, but in the 1930s it payed off handsomely.

The Yale into which I was trying to fit myself, as far as possible by modifying my immediate surroundings rather than myself, was very different from the present institution. Intellectual excellence was far less important than social position among the undergraduates. The teaching assistants and junior faculty in fact appeared to me to be like highly intelligent Greek slaves serving young Roman patricians. On a short visit to Ithaca just before my second Christmas vacation in America I wrote to my parents: "I like . . . Cornell for its greater insistence on essentials. Good workers, good collections, and good libraries rather than endless mock gothic palazzi. We are more fortunate at Yale than any other department in having so eminent and tactful a head as Harrison. The other departments must be intolerable." This was doubtless the extreme and unfair estimate of an expatriate youth, but looking back I know exactly what I meant. Self-education, for the undergraduates that I had been teaching, would have been well-nigh impossible. Of course at Cambridge there were a number of aristocratic and rich young men, but if they were not intellectually inclined, they took a pass degree which did not interfere either with their athletics or other diversions at Newmarket and in the Hawks Club, or with the quality of the education given to the honors men.

There were, in addition to Harrison and Woodruff, two other full professors in the department, Wesley Roswell Coe and Alexander Petrunkevitch. Both were invertebrate zoologists but in every other way they differed greatly.

Coe was born in Connecticut and became a student of A. E. Verrill. He was thus the remaining link at Yale with the first great expansion of zoology in America, fundamentally due to Louis Agassiz, with whom Verrill had studied. Verrill did not remake the tradition; he was fundamentally a systematist and a museum man, and Coe started off in the same way. He became the leading student of the pelagic nemertean worms, a curious and rather rare group of invertebrates collected in small numbers by many deep-sea expeditions. He also became much concerned with sex changes in molluscs and with the feeding ecology of bivalves. None of these fields of specialization evoked much enthusiasm in his colleagues, and I think he felt apologetic about them. It is curious that in the seventh edition of *American Men of Science* experimental zoology is listed both at the beginning and the end of his areas of research. He taught histology and genetics; the former was a rigorous laboratory course, but the latter was mainly illustrated with stuffed pigeons, chickens, and rabbits. He had had very few graduate students. One, Stanley C. Ball, was the curator of zoology in the Peabody Museum when I arrived; Coe himself had earlier filled the position. Another of his students, Victor Loosanoff, later became a very important figure in the applied biology of oysters and clams. He gave Coe a panel on which were mounted shells of Connecticut oysters from the spat to a ten-year-old individual. This exhibit of growth hung on Coe's office wall and when I inherited the room I took care that it still should hang there. After Coe's retirement he moved to California and did admirable work on the role of detritus, largely derived from kelp, in the nutrition of marine invertebrates.

Some of Verrill's family went to California and were impeccable citizens of that state, where I hope and believe that they flourished, though not I think in science. One of Verrill's sons wrote well-illustrated if perhaps overly romantic popular books on natural history. There were in addition people around who claimed to be Verrills, and one of Coe's minor functions was to warn newcomers like myself against anyone trying to solicit subscriptions to an unpublished, and indeed probably unwritten, life of A. E. Verrill. There were also curious rumors of someone using the name in giving lectures, to temperance societies, on the experience of being rescued from the gutter and reformed. As in the case of Saint Hyacintha Mariscotti, whose sin was pride and whose feast is on my birthday, the fall from grace and the redemption appear to have occurred more than once. There was also a hint of a party given for the chorus of a musical show at the Shubert Theater in New Haven and paid for in an unconventional way by currency removed at Christmastime from registered letters. Coe's warnings were insistent but I was never approached nor tempted to become involved.

Of Petrunkevitch, I have written extensively elsewhere.[16] He came of an aristocratic but very liberal family and had been forced out of Russia for protesting the way in which students had been treated after disturbances in 1899. He had then gone to Germany to work

16. A biographical memoir will shortly be published by the National Academy of Sciences. See also, G. E. Hutchinson, Alexander Petrunkevitch. An Appreciation of his Scientific Works and a List of his Published Writings, *Trans. Conn. Acad. Arts Sci.* 36:9–24, 1945.

with August Weismann. In his Ph.D. thesis he gave the first statistically adequate demonstration of the truth of Dzierzon's hypothesis that drone bees are produced from unfertilized, and worker and queen bees from fertilized, eggs.

Petrunkevitch came to America in 1903, where he started his work in arachnology, at first as a private scholar, but after 1910 as a member of the Yale faculty. He was equally at home with the anatomy, physiology, and behavior of spiders, but his most significant work was no doubt contained in a long series of paleontological memoirs which dealt both with Paleozoic arachnids of all kinds and with spiders fossilized in amber. The latter specimens always fascinated him as they did the numerous students, including experimental embryologists and comparative physiologists, who attended "Pete's Tea" every Monday at 4:30 P.M. in his laboratory. In this way, he had a considerable influence in maintaining an interest in unfashionable phylogenetic and zoogeographic problems, though he often misunderstood his colleagues and they him. He never expressed any hint of an apology for the areas of science to which he was devoted.

Petrunkevitch had, early in his career in Moscow University, worked with the great geochemist V. I. Vernadsky, whose son, George Vernadsky, the eminent historian of Russia, was later to join the Yale faculty. Petrunkevitch as a student nearly lost his life with the elder Vernadsky when they were being taken down a mine, to see a mineral deposit, by an official of the mining company. An anarchist workman, thinking that this provided a fine chance to get rid of the official, cut the cable on the cage. Fortunately, the latter tilted and jammed in the shaft very close to an adit, through which the occupants were rescued.

Petrunkevitch published many of his most important works in the *Transactions of the Connecticut Academy of*

NEW ENGLAND MORAL   233

Arts and Sciences, a journal that had earlier been im-
mortalized by J. Willard Gibbs. During the 1930s he was
president of the Connecticut Academy, and the meetings,
with the rere-suppers that followed them, became a great
gathering place for intellectual émigré Russians. They
were also one of the few occasions where the new and
unimportant could meet interesting people outside their
own departments.

Learning about V. I. Vernadsky from Petrunkevitch
and from George Vernadsky, I became much interested
in some of his ideas, notably his conclusion that each
species of animal or plant had an elementary chemical
composition, varying around a mean characteristic of the
species. Later, I was able to study this possibility rather
extensively among the club mosses or Lycopodiaceae,
where such a situation is more or less true of the aluminum
content; *L. flabelliforme*, from which Christmas wreaths
are often made, contains 0.6 to 1.25 percent of the metal
in the dry leaf. These fascinating plants also show great
variation from species to species in their alkaloids, and at
least three classifications, one based on morphology, one
on aluminum content, and one on alkaloids, should be
possible. Though the chemistry is still not fully known,
the three approaches clearly suggest the same general
kind of taxonomy.

Vernadsky had a strong influence on other aspects
of my research, and I did my best to help Petrunkevitch
and George Vernadsky make his ideas about the biosphere
better known in English-speaking countries. Though I
came to biogeochemistry through Vernadsky, I soon
realized the great importance to biology of the concepts
introduced by my father's friend Viktor M. Goldschmidt.
Putting these two together in an ecological context I
think did something to further the more chemical aspects
of ecology.

The associate professors in the zoology department

in 1928 were G. A. Baitsell and J. S. Nicholas. The latter had worked with Harrison, on salamander eggs, but had also developed operative procedures that permitted the methods of experimental embryology to be applied to mammals. However, he never utilized his methods to their full capacity and later in life became too much engrossed with worldly power to continue as an effective scientist, which distressed all who knew him at all well.

Baitsell is still an enigma to me, though I owe him a great deal. He had been a student of Woodruff, but later he became interested in Harrison's approach and did work on the genesis of connective tissue fibers in tissue culture. He was, however, temperamentally much closer to Woodruff than to Harrison, a Republican Masonic Baptist and an ardent prohibitionist, but with a strong sympathy for the sons of wealthy alumni who might make great gifts to Yale. Later when I came to understand the rigors of his early life and of that of the pioneer Iowa family from which he sprang, the differences in our outlooks seemed unimportant. In the 1930s he became mainly a teacher and administrator. But he was clearly mulling over his earlier scientific work, for when in 1938 he came to give an address to the symposium arranged by the American Association for the Advancement of Science, marking the centennial of the cell theory, he seems to have been the first person to use the expression "molecular biology."[17] Although he considered the cell, as the smallest complete functional unit in biology, to be in a sense a single molecule, his ideas were in many ways strikingly modern. During his later career he became the secretary of the Society of Sigma Xi and played an enormous part in revivifying the society and in estab-

17. G. A. Baitsell, The Cell as a Structural Unit, *Amer. Natural.* 74:5–24, 1940; reprinted *Amer. Scient.* 43:133–41, 147, 1955.

NEW ENGLAND MORAL   235

lishing the *American Scientist* as a unique periodical. Although initially I must have represented much of what someone of his background would dislike in an English intellectual, he realized that there might be something that I could bring to his journalistic venture. The result was a partnership that gave me a unique opportunity to try out ideas under the heading of "Marginalia."

Though sometimes greatly disturbed by the attire of the young women who worked in the laboratory, he was a strong believer in the freedom of the press, even if it involved, as it once did in one of my pieces, a naked bride with a garland of roses round her waist trespassing from the inside of a fifteenth-century *cassone* lid on to one of his pages. I think he was badly shocked, but I felt that the lady made my point perfectly and there she has remained.

The Peabody Museum began to play a great part in my life as soon as I stepped into it. In those days, there was a Foucault pendulum hanging suspended from high in the tower over the entrance. Every morning it was set swinging in a plane defined by a protractor lying below the bob. As no other significant force acted on it, it swang backwards and forwards in the same direction in space. The earth, however, was rotating, so the path of the pendulum appeared to move. Though I had been brought up from infancy to accept the rotation of the earth, seeing it demonstrated in this way for the first time was curiously exciting, something intellectually so simple and obvious, yet so unexpected to the purely practical part of a human being.

The museum had been founded by George Peabody of London and its first building was opened in 1876. It acquired enormous collections in vertebrate paleontology from Professor Othniel C. Marsh, the founder's nephew, and of marine invertebrates collected by A. E. Verrill for the United States Fish Commission. A very rich anthropological and archaeological collection was also built up, initially also largely by Marsh. The museum originally stood at Elm and High Streets, near the center of New Haven. This building was demolished in 1917 and the collections were largely inaccessible until 1925, when the present building was opened. In 1928, though the new building was complete, apart from the caulking of numerous leaks, the largest specimen that it contained, the reconstructed skeleton of *Brontosaurus*, was only partly mounted. The neck vertebrae were, I think, being put in place when I first saw it; illuminated by a powerful floodlight, the rigidly geometrical curvature of the back and uncompleted neck gave the whole scene a constructivist theatrical quality. Many years later, this theatrical quality momentarily reappeared at a dinner celebrating the opening of the first, though very small, American exhibition of King Tutankhamun's treasure. There was a belly dancer to entertain us, and she danced along the entire length of the skeleton between performances at either end of the Hall of Reptiles, where the dinner was held. Seen below the enormous, geometrical, static, reptilian skeleton of uncertain sex, the same basic tetrapod structure repeated in the small, living, and moving feminine human body of the dancer provided one of the most beautiful contrasts I have ever seen.

Near the *Brontosaurus* is a case containing the skeletal remains of the remarkable toothed, but apparently secondarily flightless, bird *Hesperornis regalis*, described by Marsh, the Cretaceous ecological equivalent of our penguins. When the great German ornithologist Erwin Stresemann came to Yale and went to see these fossils, he was apparently so surprised by their perfection that his eyebrows raised automatically and his monocle crashed on the cement floor.

When I first knew the museum, the great murals of the Age of Reptiles and the Age of Mammals, painted by Rudolf Zallinger, had not been begun, nor had the lovely habitat groups by the great museum artist J. Perry Wilson, illustrating the ecology of southern New England and some of the larger North American mammals in their natural habitats. These works, though in a sense applied, and certainly outside the mainstream of twentieth-century art as ordinarily understood, ultimately may prove to have far greater aesthetic significance than most critics would concede to them today.

In 1928 the intellectual significance of the museum was considered primarily to reside in the department of geology. Richard Swann Lull, a superb mammalian paleontologist, and a very popular undergraduate teacher, was director, and the most powerful mind associated with the collections was that of Charles Schuchert, a largely self-taught invertebrate paleontologist and stratigrapher. Schuchert's work led to important conclusions about Central American paleogeography. He was regarded as authoritative and was intellectually somewhat unbending, perhaps on account of his youthful struggle to become a scientist. He greatly disapproved of Alfred Wegener's theory of continental drift and imposed his disapproval on others so effectively that mention of the moving continents in geological circles in Yale induced a deeply silent response comparable to what might be elicited by a grossly obscene remark at a church supper. Coming from South Africa, where A. L. du Toit had expounded the theory so ably, this seemed to me extremely odd.

An excessive authoritarianism was perhaps a little too prevalent among the geologists and paleontologists of the time. In 1932 G. E. Lewis, then a young paleontologist on the Yale North India Expedition, collected from Tertiary beds in the Siwalik Hills, now in Pakistan, a fragment of the upper jawbone of a primate which he described as *Ramapithecus brevirostris*.[18] He pointed out that though there was little enough of the fossil, the animal that it represented must have had a smaller canine, a less parallel set of premolar and molar teeth, and a shorter muzzle than any Tertiary primate then known. All three of these features are man-like. He accordingly, and to me quite convincingly, concluded that *Ramapithecus* was allied to the ape-men then being described by Dart and by Broom as *Australopithecus*. Its age made it the oldest member of our own family, Hominidae, to which the African fossils were generally ascribed. Authorities in America rather pointedly ignored Lewis's claim and the fossil was largely unappreciated for a quarter of a century. Although it is now known that teeth of the same species had earlier been described as *Dryopithecus punjabicus*, the validity of Lewis's claim is now universally admitted and, as *Ramapithecus punjabicus*, it has an honorable place in our family tree. I am proud of the fact that in many lectures to elementary biology classes on the evolution of man I always showed a slide of this little Yale specimen as a relic of our oldest known ancestor, long before Elwyn Simons fully rehabilitated it.

In 1928 the status of zoology at the Peabody was at a low ebb. The curator, S. C. Ball, was an excellent field naturalist, with wide experience in the Pacific. He was at his best in the planning of the hall of southern New England natural history that I have mentioned. The design and placing of organisms in the diorama was so good that I had no difficulty in writing a pamphlet on the

18. G. E. Lewis, Preliminary Notice of New Man-Like Apes from India, *Amer. J. Sci.* (ser. 5) 27:161–79, 1934; E. L. Simons, The Phyletic Position of *Ramapithecus*. *Postilla* 57, 1961; E. L. Simons, On the Mandible of *Ramapithecus*. *Proc. Nat. Acad. Sci., Wash.* 51:528–36, 1964.

ecological principles easily seen in the exhibits, though I do not think Ball had any idea what I was writing about. He was admirable as a popularizer but had no capacity to appreciate the interests of advanced students engaged in research. He was, moreover, better at refusing than accepting collections, to our great loss.

The process of rebuilding zoology in the museum, where in the nineteenth century Verrill had done so much, bringing it up to the level of paleontology, was in a sense begun by Albert Eide Parr in the Bingham Oceanographic Laboratory, affiliated to the Peabody, in the late 1920s; but its real flowering was due to S. Dillon Ripley and in a quieter way to Willard Hartman. As all are happily alive and working, they only qualify for brief though very affectionate mention in this book.

The historical problem of the decline and restoration of museum zoology during this century is probably a general one and not merely something related to a single institution and resulting from local circumstances. The causes actually are not far to seek. The production of mathematical models of natural selection in the late 1920s and early 1930s gave grounds for believing that evolution could be studied scientifically, if not experimentally, and evoked, in the "New Systematics," a study of the nature of the barriers that developed between species and of the nature of the variation within them. Meanwhile, the rise of population genetics and of ecology, with the realization of the species, in the sense of an interbreeding population, as the real unit, put a premium on systematic work that it had not had since the early exploration of the biota of the earth in the eighteenth and beginning of the nineteenth centuries. The astonishing growth of zoology in the Peabody Museum was certainly part of this historic trend, which could probably be noted in many parts of the world. One aspect of the trend is of some interest, though it does not seem to have been much discussed. In the nineteenth century, when George Peabody was founding our museum, right on through the early part of this century, the development of natural history museums was a favorite aim of private charity. Now it depends very largely, though until recently not entirely, as the Coe and Bingham addition to the Peabody testify, on the National Science Foundation. It is interesting to inquire, and the answer might be of great practical import, if donations to natural history museums have decreased proportionately to other charities and why this should be the case. In this connection it is worth bearing in mind not only the purely scientific work of the museums but the extraordinary number of school children, many from the inner city, who may, as a friend teaching them once said to me, have their first truly intellectual experience in such an environment.

The second year I was at Yale I was asked to give a course on freshwater biology in the second term, Albert Parr having initiated the teaching of oceanography in the first term of 1929–30. In preparing my lectures I distinctly remember having the outline of a book on limnology in mind.

During my first year I had given a departmental seminar—always called a journal club in those days—on my research. I talked about the pans of the Transvaal. The whole approach, based primarily on Thienemann's point of view, was, I suspect, utterly unlike anything that had hitherto been presented to the department, though I think its respectable German origin probably appealed to Harrison. To the limnological community of North America, had any of its members been present, it would have given the same impression as a Dominican or Jesuit sermon would have done to a seventeenth-century Puritan congregation. The only professed ecologist in Yale at that time was G. E. Nichols, chairman of the department of botany, but though botany and zoology shared a common library and both contributed to the elementary course in biology, there was a great gulf artificially fixed between the two disciplines; this could not be crossed officially during working hours.

My own research was largely still involved with Africa as there was much to be done in writing up the work on the pans. I did try out unsuccessfully an idea that had been rattling around in my mind for a year or two, attempting to establish a conditioned reflex with a time lag in a poikilothermous or cold-blooded animal, such as a salamander, in order to see what its temperature characteristic might be. During the period of which I write, following the work of Crozier, such a determination would have seemed much more interesting than it would now appear to be.

Failing in this, I started an investigation of the supposed effects of magnesium salts on Cladocera, effects which were supposed to keep this group of crustacea out of Lake Tanganyika. As I wrote on 18 November 1929 to my parents: "I have several families of the most sensitive species in water 6 times more concentrated than the lake swimming about here in this room. Hydrobiology satisfies both my physiological interests and naturalist's instincts more than any other branch of zoology; I can get more work done in this field than any other for that reason. So far it has been the dumping ground for inferior off-scourings of the profession of zoologists. . . . All the best things are missed save by a few men like old Birge." I am pretty sure that my arrogant remarks about hydrobiologists were meant to apply primarily to those of the English-speaking world; even for such a subset they were grossly unfair. It took me some time to realize where the difficulties that I seemed to encounter, and which I was later to share with Ray Lindeman, actually lay.

Embryology had been taught for several years by J. W. Buchanan, a general physiologist who had worked with C. M. Child at Chicago and was, therefore, regarded as slightly heretical, though tolerated in the interests of free speech. Buchanan got a good offer and left, and I was asked to take the undergraduate embryology course that he gave. On paper this was ridiculous for someone whose only formal instruction in the matter had been Shearer's fragmentary and inaccurate course in the Cambridge department of anatomy. The fact that I was asked to give the course in a department famous throughout the world for its embryological research really made my situation quite ludicrous. Fortunately, I was pretty well self-educated in mammalian development and had followed Needham's early work in chemical embryology ever since I had met him in Sydney Cole's laboratory. I soon found that I was looking at embryos in what Needham called cleiodic, or shut-up, eggs as self-contained systems like so many South African lakes. At times I was terrified, but with Miss Pelham's hand metaphorically on my shoulder, I think the course soon went well; I hope at least some of the students learned almost as much from it as I did.

Though in those days it was regarded as very dangerous to let women take undergraduate courses, an exception was made to allow a lady called Dorothy Benton to take embryology. She had worked on *Biological Abstracts* in the early days and had many friends in biological circles in Philadelphia. Her husband had, for a few years, an instructorship in mathematics at Yale and was also a keen naturalist. I got to know them both quite well, and we had extensive conversations about population biology and mathematics.

About this time Volterra's work on competition and on prey-predator relations was becoming known. Volterra,

who was one of the great mathematicians of the early
twentieth century, had become interested in population
mechanics, as a result of his daughter Luisa, who was a
limnological zoologist, marrying Umberto d'Ancona, a
fisheries biologist of great insight. D'Ancona had noticed
that the limitation of fishing in parts of the Mediterranean during the First World War had apparently
produced a change in the proportions of the various
species of fish caught commercially. He asked his father-
in-law about the possibility of constructing a theory to
explain this kind of phenomenon. Though Volterra's
initially best known results concerned cyclical oscillations
due to predation of one species on another, he showed
at the beginning of his most important paper, in 1926,
that two species feeding in unrestricted competition on
precisely the same food under the same conditions could
not coexist indefinitely as equilibrium populations, unless
their relevant properties were identical, which is most
unlikely in validly separable species. J. B. S. Haldane
had actually published the same result in 1924, but none
of us realized its significance. The result now sounds a
little obvious, but it is actually the theoretical basis of a
very large part of modern ecology. In 1934, G. F. Gause
stated it in the form that "as a result of competition two
similar species scarcely ever occupy similar niches, but
displace each other in such a manner that each takes
possession of certain peculiar kinds of food and modes of
life in which it has an advantage over its competitors."
This conclusion is now generally denoted by G. Hardin's
term, the "principle of competitive exclusion."

The whole history of this simple, powerful, but
sometimes incorrect generalization is most curious and
bears greatly on the state of biology at the time of which
I am writing. It is now known that statements equivalent
to the principle had been made, sometimes very casually,

sometimes quite formally, by at least six zoologists and one botanist between 1857 and 1924.[19] Of these, Joseph Grinnell of the University of California at Berkeley was the most important. He had developed the idea quite fully in 1904 and had used the word *niche* since 1914 to express that space, in an unconsciously abstract sense, in which two closely allied species did not co-occur. Grinnell was an extremely active and well-known student of the mammals and birds of western North America, but his interests lay so far from those of the greater number of intellectually significant biologists of the time, that no one in the biological establishment realized that he had something to say of great theoretical importance. Most of his colleagues in vertebrate taxonomy and ecology were unprepared for any deep theory, and, in fact, at least the ecologists were deeply suspicious of any generalizations other than those implied in classificatory schemes. Although Grinnell was recognized as an outstanding student of birds by the British Ornithological Union, he received no honor in his own country for having made major contributions to the general understanding of the ways in which populations of more than one species might interact. It was not until the publication of Volterra's mathematics, with the comparable contributions of A. J. Lotka, and of Gause's experimental confirmation of their results, that what had seemed obvious to Grinnell and his students came to be regarded as a basic part of biological science.

Raymond Pearl of Johns Hopkins undoubtedly played a major part in making these advances possible, though I think that in his own experimental researches he kept

19. Most of the available material on the history of the concepts of competitive exclusion and the ecological niche is summarized in G. E. Hutchinson, *An Introduction to Population Ecology* (New Haven and London: Yale University Press, 1978).

to single-species populations. W. W. Alpatov, who taught Gause in Moscow, had been a student of Pearl; Pearl himself had studied with Karl Pearson, one of the founding fathers of modern statistics, in London. Gause's famous little book *The Struggle for Existence* had a foreword by Pearl and was published by Williams and Wilkins, who did the *Quarterly Review of Biology*, which Pearl edited.

I only met Pearl once, but he had greatly influenced O. W. Richards, a student of population growth in yeast, who was a colleague of mine at Yale during the thirties. We eagerly discussed Gause's work together and I then came to feel the beneficial influence of Pearl. He was well known for his wit, and as editor of the *Quarterly Review of Biology* invented a character called Reginald the Office Boy, who occasionally reviewed second-rate books on sex. He also, on at least one occasion, used the expression "There is no index" as the whole of one of his own reviews. Pearl, as a very good statistician, was highly acceptable to the biological establishment and probably played a greater part in bringing together disparate areas of biology than anyone else at that time.

Quite apart from the concept of competitive exclusion, the idea of the ecological niche has had a queer history.[20] The expression was apparently first used by Roswell H. Johnson (1877–1967), who had a most curious career, starting with evolutionary zoology, then turning to

20. Roswell H. Johnson's part in the establishment of the concept of the ecological niche was rediscovered accidentally by R. M. Gaffney; see Roots of the Niche Concept, *Amer. Natural.* 109:490, 1973. His role is discussed briefly in *An Introduction to Population Ecology* (p. 155), but it has as yet been impossible to form any idea as to his influence or lack of influence in the period from 1910 to 1914. My friend David Cox tells me that there may be a body of Johnson papers in existence, though he has as yet been unable to consult them.

246   THE KINDLY FRUITS OF THE EARTH

petroleum geology, and finally becoming director of counseling in the American Institute of Family Relations in Los Angeles. In 1910, he introduced the term *niche* in an essentially modern sense, but did so only to point out the general invalidity of the concept.

There is no evidence that Grinnell knew of this discussion, though it appeared in a publication of the Carnegie Institute of Washington on evolution in ladybug beetles. Johnson clearly disliked competition and the "strong Malthusian leanings of Darwin." He evidently only became happy in his career when he felt he was an intentional promoter of harmony. He retired from that profession in 1960 in his eighty-third year.

Elton developed the idea of the niche rather differently in his *Animal Ecology*, with no thought of competitive exclusion. To him, it meant the role that an animal played in a community. It is curious that he certainly knew of at least one book in which Grinnell had used the word, without noticing that latter worker's meaning.

A deep suspicion of theoretical formulations was probably most marked among the biologists of the middle western states, where plant ecology was rapidly growing. It came very strikingly to my attention later in my career, when I was attempting to get Raymond Lindeman's famous paper "The Trophic-Dynamic Aspect of Ecology" published.[21] This paper was the first one to indicate how

21. Through the kindness of Frank Ruddle, the present chairman of the department of biology at Yale, I have been able to see the letters written when I was being considered for promotion to associate professor, which are still in the files of the department. Two of these, from Paul S. Welch and A. C. Redfield, were quite favorable, though probably not favorable enough to satisfy the standards of today. One was from Chancey Juday, who wrote, "Professor Hutchinson has some very good ideas and, if his mathematical treatises were based on

biological communities could be expressed as networks or channels through which energy is flowing and being dissipated, just as would be the case with electricity flowing through a network of conductors. Though the concept is now regarded as both basic and obvious, like the principle of competitive exclusion, it roused extraordinary opposition. The resistance to publication was the more poignant in that the young author was dying

---

much larger amounts of observational data, his contributions to limnology would be much more valuable in our opinion." This is in the same spirit as the criticism from the same source that delayed the publication of Lindeman's trophic-dynamic paper (see n. 13).

It is ironical that the most important conclusion in those of my contributions submitted, in part in manuscript, to Juday, was largely based on a table giving the mean temperatures at each meter depth in Lake Mendota throughout the summer, based on a very large number of individual determinations over many years. The table had been most kindly sent me by professors Birge and Juday, to whom it had not revealed its secrets. The mechanisms of heating that it suggested were also apparent in Connecticut lakes and are probably general in small lakes in temperate regions.

The fourth letter, from an unidentified correspondent called "Bill," possibly Dr. William Balamuth, writing to George A. Baitsell from Evanston, Illinois, indicated that two distinguished ecologists had never heard of me, that a third had no very high opinion, though one or two of my reprints "contained significant contributions," while a fourth had remarked that I had confined my work to limnology and "did not stand out in that." The last communication was clearly from someone who was not a regular reader of the *International Journal of Psychoanalysis*. The writer of the letter himself, after quoting these opinions, felt that Baitsell had on his "hands a problem in which the only accusations that can be levelled are a general tepidity." I think that this letter, even though accompanied by three fairly good ones, would today be enough to damn any candidate before some committee or other engaged in evaluating promotions. I therefore hope that everyone who does get promoted this year will have a proportionately more brilliant career than I have had. My own promotion clearly must have depended fundamentally on Harrison's judgment.

of an obscure hepatitis as the paper was finally accepted and went to press. The whole history has been recently recounted by Robert Cook. Thinking about the matter, and about a similar difficulty that befell my first graduate student, Gordon Riley, when he submitted a paper on plankton productivity, containing a good deal of statistical theory, to *Ecological Monographs*, I began to wonder whether he and Ray and I had not been suffering from a sort of commonsense backlash generated at the Reformation by the ultra-intellectual and antiempirical aspects of medieval scholasticism, which backlash had flourished in America wherever a Puritan attitude was still strong. I then remembered how when Professor E. A. Birge and Professor Chancey Juday were kind enough to let me spend a week at the Trout Lake Laboratory in Vilas County, in northeastern Wisconsin, I had learned a fabulous amount about limnological technique but had come away with two feelings of dissatisfaction. One was that it would be nice to know how to put all their mass of data into some sort of informative scheme of general significance; the other was that it would be nice to have either tea or coffee, without seeming decadent and abnormal, for breakfast. I now suspect a connection.

My unsuitable assignment to teach embryology lasted until 1936; after that I took over from Petrunkevitch an undergraduate course on the natural history of animals. I tried to make it into an ecology course organized entirely in terms of interaction between individuals and species, as it seemed that at an elementary level it was paradoxically easier to understand what happened when two fantastically elaborate systems met than what happened when one such system encountered its much simpler physiochemical environment. This is a sort of inversion of a theme by that wonderfully philosophical zoologist Carl F. A. Pantin, who said that a physicist was a person who only

tackled the easy questions on the examination paper set
by nature.[22] My graduate course on ecological principles
began in 1939 and was followed by biogeochemistry in
1946. The latter course, which toward the end of my
career I happily shared with Catherine Skinner, was, I
imagine, the first one taught on the subject outside Russia.
All these activities, as well as the informal ecology seminar
which reached its apogee in 1957 and the underground
seminar in science, philosophy, and art when I had
retired, developed long after the date intended for the
formal end of this book. A few further thoughts, however,
clamor to find expression before that date is reached.

The period of which I have written saw a transformation of the zoology department, which finally fused with
botany to form a biology department. Each side felt
threatened by the other, but the threats were illusionary.
All that happened was a recognition of the end of the
usefulness of the nineteenth-century descriptive paradigm.
What could have been a threat was not the realization that
plants and animals have much in common, which is true,
but the belief that only *Escherichia coli* and the T_4 bacteri-
ophage are illuminating objects of study for biologists. Yale
escaped being totally inundated by the culture media
of such minute beings, partly I suspect because the
administration could not persuade their most distinguished
proponents to leave their lush fields; our home-grown
molecular biologists have subsequently done remarkably
well.

In spite of the hopes that all universities have of

22. I had hoped to find this remark in Carl Pantin's admirable
posthumous *The Relation Between the Sciences* (Cambridge: Cambridge
University Press, 1968), but I can discover it nowhere in the book, nor
in his other general writings, though the thought is continually expressed in other ways. It is so characteristic that I am sure he made it;
it is probably somewhere in print.

cultivating Nobel Prize winners rather than learning, biology at Yale has come to be more diverse than ever before. In this the overall field is far closer to that in which I grew up than was the case in America in the 1930s. Basically, I think the reason for this has been that as the harder questions on nature's examination paper came to be tackled, the number of interpenetrating aspects of organisms that needed to be known increased. This, as I have noted, was very true in the resurgence of systematics, when deep and difficult genetical and ecological questions arose which depended on taxonomic distinctions. I suspect, moreover, from what my more sensitive friends in the humanities and social sciences tell me, that the whole of learning is evolving in this way.

The need for variety in knowledge increases if we are to continue learning about ourselves and the world in which we live, but a new problem is also arising. Most new basic knowledge is produced in institutions that were founded ostensibly for teaching. This is particularly true of really new ideas, which are unlikely to be produced in any sort of institution set up purely for investigation and thought, for all such institutions, except perhaps All Souls College, Oxford, have their activities defined by certain limits. In universities the concept of academic freedom has arisen as a sort of by-product of teaching activity. Good advanced teaching requires good scholars, who can only be induced to come to teach if they have time for other learned activities, to which intellectual limits are not supposed to be set. As populations stabilize, the number of such positions is declining, and is likely to reach an equilibrium value well below the number of people who by ability, education, and temperament can fill them at a very high level. Meanwhile, general unemployment seems chronic and, in attempting to reduce it, more and more people are likely to be forced into retirement while they are still highly active human beings. It would seem to me that what we need is the return of the really able amateur, because that is what, in the true sense of the word, we should all be. Perhaps ultimately all scholarship that does not involve expensive physical installations should become the province of the amateur virtuoso. The standard of publication could remain the same; many amateurs would, of course, still be teachers. With a slightly reduced working week everyone would be expected to take part in various sorts of volunteer activity, of which scholarship would, for the very gifted, be a particularly attractive one. There are obvious difficulties which have to be expressed so that they can be overcome. The idea at first would probably be academically very unpopular, as the professional status of learning has often been held to be far more important than its beauty and interest. Not long ago the *Yale Alumni Magazine* asked a number of the faculty to write briefly their thoughts on the meaning of education. Two of us quite independently included a plea for the amateur in our replies. Many letters were later published about the series, but no one commented on this matter. I suspect that last year it was unworthy of notice; perhaps it will do better on a second presentation.

I like to think that as a fellow of the Linnaean Society of London, I am a member of one of the few local natural history societies whose province is the whole world.

In education, as in the design of all sorts of social action, I believe that those in charge aim at far too low a standard of achievement. It is obviously desirable to learn how to earn one's living. Sickness, hunger, and cold prevent many people from living satisfying lives and must be alleviated. If it is not realized that this is only a beginning, we are liable to get ant-like societies where painless persistence is the highest value. Everywhere in education we should aim at seeing life as a sacred dance or as a game in which the champions are those who give most beauty, truth, and love to the other players.

Since Sir Thomas Browne accidentally reversed the rotation of the earth when the Quincunx of Heaven ran low and it was time to close the five ports of knowledge,[23] perhaps I may end by looking to the rising sun. In 1931, Hellmut de Terra, a German geologist of great geomorphological insight and imaginative power, persuaded Yale to let him organize an expedition to the western end of the Tibetan Plateau. I was asked to be its biologist. I have already given a detailed account of this extraordinary journey in a book called *The Clear Mirror*.[24] In it, I tried to write of travel solely in terms of the impressions made on the traveler, who is otherwise unidentified. Only what he sees or hears exists, not himself. Looking back on this journey, an incident that cannot be reported in this way, without living human beings, insistently comes into my memory.

We had started from Kashmir and crossed a snow-covered pass called the Zoji-la before dawn. In doing so, we had passed into the great Himalayan rain shadow; the forest of Kashmir was giving way to the semidesert of Central Asia. We traveled through small towns or villages, Kargil and Dras, where the inhabitants were Mohammedan. Then we reached, at Lamayuru, the beginning of Tibetan Buddhist country, though politically we were

23. This is in the last chapter of *The Garden of Cyrus*. As I pointed out in Tuba Mirum Spargens Sonum, Per Sepulchrum Regionum, *Amer. Scient.* 39:145–50, the problems raised by Browne in chap. 5 about the existence of five-rayed echinoderms and flowers still have no real solution in developmental terms.

24. G. E. Hutchinson, *The Clear Mirror: a Pattern of Life in Goa and in Indian Tibet* (Cambridge: Cambridge University Press, 1937; reprint ed., New Haven: Leete's Island Books, 1978).

NEW ENGLAND MORAL   253

still in Kashmir. After establishing our camp, we started
walking about, to find our way to a *gon-pa*, or monastery,
on a hill overlooking the village. A small boy of about
six took my hand. Led by this little child, who seemed
fully to understand that we could not know each other's
languages, I was taken into one of the few parts of the
world which for better and for worse still remained in the
high Middle Ages. He stayed with me, still holding my
hand, till dusk, when he returned home and we went to
the tents to dine. It was the most moving human relationship that I experienced on the journey until the
voyage home.

2 *Biography*

From English Schoolboy to America's Foremost Ecologist

Nancy G. Slack

In G. Evelyn Hutchinson's life span, 1903–1991, remarkable changes took place in the study of biology as a whole, and during Hutchinson's most productive years, a whole new science of ecology, particularly theoretical ecology, developed. Together with his students, from the early 1930s to nearly the end of his life, he infused new ideas and methods into limnology, biogeochemistry, systems ecology, and population biology. He is often referred to as a polymath; his broader interests and writings encompassed history of science, art history, psychoanalysis, and numerous other fields. But Hutchinson's greatest impact was on ecology in nearly all its aspects.

How did this come about? Why Hutchinson? In a review of biographies of Charles Darwin and of Alfred Russel Wallace, Frank Sulloway has noted both the similarities and considerable differences in the upbringing of these two scientists, and how they relate to scientific creativity, a subject of special interest

both to Darwin and to Sulloway.[1] Hutchinson's upbringing was closer to Darwin's than to Wallace's: a relatively wealthy family, a father with position, a Cambridge education. All three men, Darwin, Wallace, and Hutchinson, had an intense early interest in beetles (Hutchinson's in aquatic beetles and bugs.) In addition, they all had a wide range of interests. All traveled at relatively early ages to remote, little-studied parts of the world.

All three were interested in the "why" questions, leading them to theorize about what they saw. Hutchinson worked and traveled a century later than Darwin and Wallace, but in the mid-twentieth century the "why" questions in ecology and biogeography still needed asking and answering. Wallace's biographer Michael Shermer called this 1861 comment by Darwin on how observation and theory should interact "Darwin's dictum": "There was much talk that geologists ought only to observe and not to theorise: and I well remember someone saying that at this rate a man might as well go into a gravel-pit and count the pebbles and describe the colours. How odd that anyone should not see that all observation must be for or against some view if it is to be of any service."[2] Hutchinson arrived at this view early in his career. Substitute "limnologists" for "geologists" and it is essentially his defense, to the editor of *Ecology,* of Raymond Lindeman's pioneer work in ecosystem ecology. Lindeman's paper, influenced by Hutchinson's views, had been turned down by two prominent limnologist reviewers who wanted more "pebbles" and no theorizing.

EARLY LIFE AND CAMBRIDGE UNIVERSITY

Hutchinson's earliest recorded ecological activities included collecting pond creatures and discovering their preferred habitats at age five. As a schoolboy, he turned from collecting butterflies to the less prosaic aquatic bugs and beetles.[3] He published his first paper, about a swimming grasshopper, at age fifteen, beginning 70 years of publishing.[4] His was a privileged youth, not in terms of family money, but in terms of special opportunities provided by his family and by Cambridge itself. His father was a well-known mineralogist who took Evelyn on local geology expeditions and introduced him to many Cambridge biologists; some were very helpful to the inquisitive boy. Throughout Hutchinson's childhood and early life Cambridge was inhabited by genius, possibly more than anywhere else before or since.[5] Hutchinson described Cambridge University in his day as the best in the world: "Witness the discovery of electrons . . . sex-linked inheritance, glutathione, cytochromes." And not only in science—he included many other fields: "the writing of Frazer's *Golden Bough,* Jane Harrison's Greek Religion . . . Keynes on Economics."

Evelyn spent his secondary school years at Gresham's School, a public school (in the English sense) in Norfolk. Gresham's was unusual in emphasizing the physical sciences and math, as well as modern languages, in contrast to the usual concentration on classics. Hutchinson realized at Gresham's that chemistry was useful.

He needed to be able to do chemical tests since organisms differed in their chemical environments. He made himself practice titrations over and over before he let himself collect in his favorite ponds and woods. His training in physics, chemistry, and math later served him well.

From Hutchinson's very earliest schooling through his university years, he was always a member of natural history clubs. He always had like-minded peers with whom to explore and make collections. Gresham's had an extensive and research-oriented natural history club, in which Hutchinson was a leading participant. He wrote reports and gave talks on current questions. One of these talks, he admitted later, was quite Lamarckian. The "why" questions were already apparent in his thinking at age seventeen. "My mind was full of problems of distribution and variation [of insects]. In a great state of excitement, I left school to start a real scientific career at Cambridge."[6] Hutchinson was writing in 1979 about his state of mind in 1921, but letters written about him by his Cambridge University zoology professors corroborate his zeal. He was about to "read" zoology at Cambridge, but biology was not a regular secondary school subject as yet, even at Gresham's; Hutchinson took his Cambridge University entrance exams in math and physical science.

At Cambridge there were two natural history societies as well as the Biological Tea Club, of which he and a fellow zoology major, Grace Pickford (later his wife), were founding members. These societies all served both intellectual and social ends. Hutchinson gave his share of talks and consumed his share of food and drink. He and Grace made excursions to nearby Wickham Fen and sometimes, with other student friends, to islands in the English Channel, all in pursuit of answers to biological questions.

Hutchinson got "firsts," the highest grades, in both part I and part II of the zoology finals (tripos) and surely could have stayed on for graduate work. Although he was later awarded many honorary PhD degrees, he never went back to do graduate work at Cambridge or any-

Hutchinson at age 15. Yale University Library.

where else. He received a Rockefeller fellowship when he graduated in 1925 and left Cambridge to do research at the Stazione Zoologica in Naples. There he did not succeed in learning anything important about the endocrine system of the octopus, as he had proposed, but he did learn a great deal about Italian culture.[7] He loved Italy; later in life he returned to study Italian lakes.

SOUTH AFRICA; BECOMING A LIMNOLOGIST

After Naples Hutchinson went on to South Africa, to his first real job at the University of Witwatersrand in Johannesburg. He went despite a warning from his parents about the despotic zoology Professor H. B. Fantham there, who did indeed fire Hutchinson from his teaching post for supposed incompetence. Fantham had to pay his salary altogether, however, until his contract was up. Grace Pickford had come to South Africa on her own research fellowship and they married there. Together they studied the dry lakes, or "pans," near Cape Town. They carried out both physical and biological research on these lakes and asked ecological questions. Evelyn had discovered his new research field: limnology.

It was Lancelot Hogben, then department

chair at Cape Town University, who sponsored the limnological work, and it was also Hogben who told Hutchinson about a fellowship at Yale, for which he applied, but too late. All the fellowships had already been given out. However, there was a newly vacated instructorship, and Hutchinson applied for that by cable. Hogben and two of his Cambridge professors wrote excellent recommendations and he received the position—also by cable.[8]

He went to Yale in 1928 via Cambridge, where he saw his limnologist friend and fellow Cambridge zoology student Penelope Jenkin. She gave him August Thienemann's newly published *Die Binnengewasser Mitteleuropas,* an important work on European limnology.[9] He also read Charles Elton's recently published *Animal Ecology.*[10] Elton's work on food chains truly impressed Hutchinson. He, together with Raymond Lindeman and other students, was later to describe energy relationships both quantitative and theoretical of food chains in lakes. Even before arriving at Yale, but under the influence of his South African research and these two seminal books, Hutchinson, in his own words, "had in fact, become a limnologist."[11] Late in Hutchinson's life, when newspaper articles touted him as the "Father of Ecology," he himself said no and nominated Charles Darwin or Charles Elton for that honor.

YALE AND THE NORTH INDIA EXPEDITION

Hutchinson, with Grace Pickford (who kept her own name), arrived at Yale in 1928. The young zoology instructor, hired almost accidentally, was soon involved in teaching a variety of undergraduate courses, including embryology, the field of the renowned department chair, Ross Granville Harrison, with whom, at Hogben's suggestion, Hutchinson had originally hoped to study.

Hutchinson later credited his own highly successful research career to Harrison, whom he called "the greatest scientist whom I have known continuously, week by week over a long period of time."[12] Harrison provided much-needed support in Hutchinson's early years, even though limnology and ecology were far from his own area of interest and expertise. Hutchinson credited Harrison in a 1990 interview with allowing him much freedom of exploration, encouraging him in his early days at Yale to explore various fields instead of staying with just one research topic.[13] Hutchinson certainly took this advice, publishing papers in limnology, entomology, and chemical ecology and on the Burgess shale fossils in his first four years at Yale.

Grace Pickford was able to pursue her graduate studies at Yale and completed a PhD in 1931. She subsequently had an important research career in a number of related fields, but particularly in the endocrinology of fish. Many researchers from Europe and elsewhere came to work in her Yale laboratory. Late in her career she was the first woman to become a full professor of biology at Yale.[14]

In 1932 Evelyn Hutchinson followed another important aspect of the Darwin-Wallace life path and went on an expedition to little-known parts. He was the head biologist on the Yale North India Expedition, which traveled first to Kashmir and then to Ladakh. Helmut de Terra, a German geologist, was the leader of this expedition. While they were still in Kashmir, de Terra wrote back to President James R. Angell of Yale and to Ross Harrison that he and Hutchinson had worked two weeks in the Salt

Hutchinson (at right) during the Yale North India Expedition, 1932. Yale University Library.

Range and that "Mr. Hutchinson's glass tubes and tins begin to get filled with queer water animals and I found the first data on the youngest unfolding of the Himalayas." Hutchinson had rented a houseboat and installed a biological laboratory in it, "a floating offspring of the Osborn Laboratory at Yale."[15]

On this expedition Hutchinson collected insects, aquatic life, and other animals and plants, as well as fossils. He was even called upon to skin large animals, a skill he had acquired while a student in England. These were for the Peabody Museum, "for the children of New Haven," he later said. His own particular interest was in the ecology and biota, including people, of the very high areas of Ladakh. He studied lakes at over 17,000 feet and took many chemical and other measurements. He sent a great many letters to Grace Pickford describing his work and the people of Ladakh. After the expedition he recruited specialists from all over the world to identify the specimens collected, while making sure their work as well as his own was published. His research provided insights into biogeography and paleolimnology, and especially much new data on high-elevation lake ecology. Many of these lakes did not support fish; small crustaceans were the top predators.

At one such lake he wrote that he was doing ultraviolet light determinations, on which he was anxious to get more data, in Darwinian fashion, in support of a hypothesis. He thought there might be a correlation between high UV light intensity and the black color of the water fleas (*Daphnia*). At this high-elevation lake the "black Daphniid and a copepod, scarlet red, are the only plankton animals one sees coming within 1 or 2 cms of the surface," he wrote to Grace Pickford.[16] He also wrote his general conclusions about the water chemistry of the high-altitude Ladakh lakes, which were so different from the South African pans on which they had both worked. Hutchinson was not yet thirty but he had already studied lakes on four continents, both their physical chemistry and their biota. From Ladakh he had "plenty of data which will show the qualitative variation of plankton with chemical composition." By September, toward the end of the expedition, he had described a bay where river and lake water mixed. It was "incredibly full of plankton," with different planktonic crustaceans occupying different lake depths, which he had correlated with UV light levels. All of this was long before the Hutchinson niche model, but he was already searching for niche dimensions in the lives of aquatic organisms in a difficult environment.

On this expedition Hutchinson proved himself very adept at making and repairing equipment. Much of the equipment he brought with him had been delayed, lost, or damaged. The transport methods were primitive, mostly four-legged. He made much of his own equipment en route, quite ingeniously, from whatever he could scrounge for a limnological tool kit, including a collection of different sorts of wire. Yet with this he was able to take much environmental data and to conduct many experiments. In later years stories abounded about broken glassware and Hutchinson's general clumsiness in the lab. Did he lose this adeptness, or was it no longer necessary to do all the careful technical work himself?

An end to the limnological research in Ladakh came in September 1932. "After tomorrow no more lakes. I'm going to take a holiday, collect little but bugs and beetles like an English gentleman . . . and visit all the monasteries on the way back." Buddhist art and religion fascinated him, as did Ladakhi women and their polyandrous way of life in high-elevation villages. This letter, however, also contains a graph of temperatures and oxygen levels at various depths in one of the Ladakh lakes.[17]

After five months, the expedition returned safely to Srinagar in Kashmir. They had moved all their scientific apparatus and provisions on ponies and yaks over snow-covered passes. The expedition had traveled 1,300 miles and mapped 4,600 square miles. "Contrary to rumours at times circulating in North Indian bazaars that all our transport animals had perished on the high storm-swept plateaus," de Terra wrote that they had covered "a long and difficult trail across

6 passes above 18000 ft. . . . and sometimes over a region which has hardly seen a white face. Our scientific results are most successful in every way. . . . I hope that both the Peabody Museum and the Osborn Zoological laboratory will be pleased with the things which we will add to their treasures."[18] Hutchinson, however, was frustrated that de Terra did not want him to do anything with the collections until de Terra's return at the end of the following April. Hutchinson was to return to Yale at the end of January. He wrote to Grace, "I foresee that my now polychrome beard will be white, India independent and five Russian expeditions achieved and published on Ladakh lakes before my stuff is finished." In addition, Hutchinson wrote that he was "pushing on with experimental ecology . . . because that is where the results will come from that are really exciting now."[19]

One does not typically think of Hutchinson as an experimentalist, but he had been doing ecological experiments with zooplankton from his earliest days at Yale. On this expedition we see his true excitement about the "why" questions in freshwater ecology. He could have simply collected specimens in a part of the world in which almost everything was new to science, as many others had done. Hutchinson, like Darwin, always seemed to need a hypothesis to test. (Darwin is still not well known as the excellent experimentalist he was.) In one case his hypothesis was that the pond fauna in Kashmir was temperature determined. He tested three species of *Anisops* (water bugs) to determine their behavior at 40 degrees Centigrade and their thermal death point.[20] The data chart, sent in a letter to Grace, fit his hypothesis. He hoped to find another *Anisops* to test in Goa, then a Portuguese colony, his final destination. There, however, he instead conducted experiments on *Halobates,* a water strider, to see the effects on its behavior of salt versus fresh water, which differ in surface tension. In this case his hypothesis proved wrong. But he continued his experiments until the very end of his travels in India and Goa.

By the end of that year Hutchinson had al-ready published a paper in *Nature,* "Limnological Studies at High Altitudes in Ladak."[21] In it he compared the Ladakh lakes to central European alpine lakes. The lake he studied most extensively in Ladakh, Yaye Tso, at over 15,000 feet, was very different from the European alpine lakes. The latter were oligotrophic with high oxygen in the hypolimnion (lower layer) and a poorly developed bottom fauna. Yaye Tso was a eutrophic lake with an oxygen deficit in the hypolimnion and a rich bottom flora, also true of several other Ladakh lakes he studied. In Yaye Tso there were approximately 4,800 organisms per square meter in the bottom fauna, largely chironomid (midge) larvae and tubifex worms. The lakes also differed markedly from lakes in another semiarid region (Ladakh is semi-desert), the "pans" in South Africa, which he had studied and published on with Grace Pickford and a South African graduate student, Johanna Schuurman.[22] Further limnological publication awaited the taxonomic studies of the organisms, made by Hutchinson himself and many other experts worldwide. Even two of the species of *Anisops* he studied in Kashmir were previously unknown.

John Nicholas, longtime Yale zoology department chairman, who had been Hutchinson's severest critic in his tenure and promotion proceedings, recalled in a letter: "When his advancement came up I opposed it strenuously, Pete [Petrunkevitch] said 'viciously.' . . . Then the North India expedition came along and instead of retiring or folding up, Hutchinson became the backbone of the whole expedition. When he returned he . . . had matured and found himself. . . . Since then his gifts of intellect, wit and breadth of knowledge have become steadily more apparent. . . . The North India reports [were] due in no small way to Hutchinson's editorial insistence that laggard systematists and others get their work finished."[23] Nicholas's observations were not always reliable, and Hutchinson had surely "found himself" as a limnologist earlier. But Nicholas's view of Hutchinson's role in the publication of the scientific results of the expedition appears correct, validated by voluminous letters to and from scientists all over the world that testify to Hutchinson's encouragement, persistence, and ability to find publication funds.

This North India expedition was also important in another respect: it marked the beginning of Hutchinson's literary career. Few twentieth-century biologists can claim literary achievement; very few wrote as well as Hutchinson about biology. Many of his essays for the *American Scientist,* starting in the 1940s, were read by biologists and non-biologists alike, and they were later collected into several books. But it was his book about the North Indian expedition and his subsequent time in Goa, *The Clear Mirror: A Pattern of Life in Goa and in Indian Tibet,* published in 1936, that was his first truly literary venture. The back cover of the 1978 reprint edition reads: "In 1932 a young biologist traveled to Goa and Tibet as a member of the Yale North India Expedition. He recorded his observations with the eye of a trained scientist and the style and sensitivity of a novelist."[24] This book was also the first to reveal the immense breadth of Hutchinson's interests and knowledge. It is about not only biology, but also archeology, religion and ceremony, painting and sculpture. But there is also much ecology, including the ecology of landscapes and human ecology in the difficult environments of Ladakh. It is in the description of the high-elevation lakes and their fauna that Hutchinson's writing skills are often most evident. From water striders to the red, black, and gray crustaceans to the colors of the lakes, he seems to be painting word pictures. To an ecologist, the "Lakes in the Desert" chapter of the book is perhaps most memorable: "Where the valley of the Indus narrows, the country is sparsely settled, but north and south of the gorge, lying in a dusty land like a few remaining stones on a worn-out headdress, many lakes are scattered. They vary in colour within a restricted range of blues and greens, just as the turquoises that they resemble [the turquoise jewelry that Ladakhi women wore] differ one from another. Each shade of blue or green sums up in itself a structure and a history,

for each lake is a small world, making its nature known to the larger world of the desert most clearly in its colour."[25] Later in this chapter we meet the inhabitants of these high-altitude lakes after the last snows of June. There are no resident fish but three dominant crustaceans, one black, one red, one pale gray. "In the blue water small black spots move leisurely here and there, and still tinier bright red points can be seen skipping among them. . . . In the deeper water and along the shore, small pale shrimps scurry about." Even higher up were lakes that were ice-bound even in summer. The ice melted from below on the lake edge in July: "Underneath it is sculptured into thousands of spikes and icicles, hanging into the green water of the lake like stalactites. . . . Below lies the greenish water, in which tiny red crustacea swim, more numerous than in the turquoise Pangog Tso, for here, in this shallow lake [Ororotse Tso] the greenness of the water indicates that there is an abundant supply of food."[26] These painterly descriptions are not simply literary; they contain a great deal of ecology. He explains why, for example, those three crustaceans can thrive in the icy waters of Ororotse Tso in Ladakh, even at forty feet below the surface, whereas in the icy lakes of the Alps few such aquatic creatures are able to exist.

When describing his experiments on water striders in Goa in relation to the surface tension of the water, Hutchinson wrote: "In the heat of the day, as if propelled by the sun itself, water striders shoot about like balls of quicksilver on the surface of the water. Later, in the cool evening, when the human inhabitants of the town come out again, the precarious progress of these insects on the boundary of sea and air will be so slow that they will seem like clockwork toys, whose machinery, wound up by the sun, has now almost run down."[27] Hutchinson proved to be one of a small number of biologists able to write extremely well both for a wide audience, as above and in several later books, and for scientists. In addition to his own doctoral students whom he supervised over a forty-year period at Yale, he also influenced, or one could say taught, many younger ecologists, including this author, who did not study with him personally but learned so much from his lucid scientific publications.

THE FIRST CROP; HUTCHINSON AND HIS GRADUATE STUDENTS

In 1948, when Hutchinson was better known for his pioneer work in the biogeochemistry of lakes than for theoretical ecology, he received a tempting offer from the United States Geological Survey. He was asked to become a full-time member of their research staff, as the letter stated, "to spend 12 months of the year, year after year, in laboratory or field research along whatever lines you think need to be pursued. Our responsibility would be to provide you with what facilities you would need and wish, and to shield you from red tape and chores."[28] The salary offered was high, surely more than he received at Yale. Hutchinson replied that of the various offers he had recently received to persuade him to leave Yale for other institutions, this was the only one that really tempted him. His objection to their very generous offer was his commitment to his present and future Yale graduate students: "The chief thing that I am doubtful about is that I have had very considerable success in turning out young investigators of quite outstanding ability. These men . . . are rapidly becoming leaders in their field. The prospect of not being able to train such people is the most cogent . . . argument against your otherwise very attractive offer."[29] Hutchinson always thought of himself not only as teacher but as a collaborator with his graduate and postdoctoral students, later young women as well as young men. He never ran a "research school" in the official sense of that term, although Ross Harrison had done so and was still doing so when Hutchinson arrived at Yale. Harrison's students all worked on one set of problems, usually using the techniques that Harrison had himself invented or perfected. Harrison was referred to as "the boss." The vast majority of the zoology graduate students when Hutchinson came to Yale were Harrison's.[30]

Hutchinson's own research group, right from the start, worked on a great variety of problems. Many, though decidedly not all, were related to lakes, using a great variety of techniques, many of which were invented by the graduate students themselves. Most of Hutchinson's graduate students called him "Hutch" or even "Evelyn." Although he certainly influenced their work, Hutchinson rarely put his name on their papers, not even the seminal paper in systems ecology by his postdoctoral student Raymond Lindeman. Hutchinson's ideas were important in the development of that paper. Hutchinson defended the paper to the editor of *Ecology* after negative reviews by senior limnologists and was instrumental in getting it published, but Lindeman remained the sole author.[31]

The diversity of doctoral research performed under Hutchinson's guidance is evident even among his first students, Gordon Riley, Edward S. Deevey, W. Thomas Edmondson, and Max Dunbar, all of whom had outstanding subsequent scientific careers. They all arrived in the 1930s. The next generation, arriving in the 1940s, included two future Yale professors, John Brooks and Willard Hartman, and other well-known ecologists and limnologists including Fred Smith, Larry Slobodkin, H. T. (Tom) Odum, and Vaughan Bowen. There was a particularly outstanding group of graduate students in the late 1950s, including Robert MacArthur, Peter Klopfer, Alan Kohn, Joseph Shapiro, and Peter Wangersky. Hutchinson credited this group in print, together with Jane and Lincoln Brower, with contributing to one of his own most important papers.

In the 1960s outstanding women graduate students, including Donna Haraway, Karen Glaus (Porter), Maxine Watson, and Allison Jolly, became part of Hutchinson's research group, although several did not receive their degrees until after he had officially retired in 1971. In that year an entire issue of *Limnology and Oceanography* was "dedicated to G. Evelyn Hutchinson," and a "phylogenetic tree" of his intellectual descendants was included. By that time many of his original graduate students had research groups of their own, shown on the tree, in the United States and Canada.

The postdoctoral students are there, too. One former graduate student, Alan Kohn, presented a quantitative analysis of the "intellectual radiation of G. Evelyn Hutchinson's postgraduate descendants into new adaptive zones" with graphs to show the trends and rates of evolution of this interesting and diverse clade.[32] Many later graduate students came to Yale specifically to work with Evelyn Hutchinson, but Gordon Riley, who arrived in 1935, did not. He had been given the same advice that Hogben had given the young Hutchinson in South Africa: Go to Yale and study experimental embryology with the great Ross Harrison. Riley came to Yale, but Harrison already had about fifteen graduate students in experimental embryology. Riley could not find an untapped research problem. He had not even heard of Evelyn Hutchinson before. "He was a young instructor—a nobody—but a great nobody," Riley later wrote. He took Hutchinson's limnology course: "One lecture was enough to make me begin to sit up straight and bright-eyed and to struggle to assimilate every word of his thick British accent. He was dynamic and obviously very bright, full of new ideas, and was dissecting the literature with keen and frequently witty comments. Within a week, I knew that limnology was where I wanted to be. Unsolved problems stuck out all over the place."[33] Characteristically, Hutchinson was excited, too. "Gordon was my first graduate student. The excitement I felt when he asked, I think after a lecture in my limnology course, if he might work with me, is still a vivid memory."[34]

Uncharacteristically, Hutchinson did choose a thesis topic for Riley, who had come straight from a master's in embryology into a completely new field. Hutchinson questioned him about his chemistry background, which was excellent, and then suggested he study the copper cycle in Connecticut lakes, something that had never been attempted before. Hutchinson thought this project important because copper is essential to crustaceans, whose blood contains hemo-

cyanin, not hemoglobin. Riley picked out Linsley Pond, the more or less natural lake closest to Yale, in which many innovative studies were subsequently carried out by Hutchinson and his students and research associates.

Hutchinson worked closely in the field with his early graduate students. He was busy teaching undergraduates during the week, but on Saturdays he and Gordon did their research together, Riley analyzing the various types of copper found in Linsley Pond, Hutchinson determining levels of phosphorus and nitrogen. They returned to the laboratory in the afternoon to do their chemical analyses. Gordon Riley described their adventures memorably. As soon as Linsley Pond was ice-free the two of them took to the lake in the same inflatable rubber boat Hutchinson had used in Ladakh. It leaked. They both got cold and wet, but they managed to secure a few samples. Then the boat collapsed. Hutchinson bought a better boat, but there were still perils, including thin ice, which Hutchinson insisted on walking out on to secure samples, while Riley watched in terror. There were good memories, too: "The best part of course was that Evelyn was collecting samples, too, so that most of the time we went together, and this was a rare opportunity for me to further my education . . . with a man whom I have always regarded as having the keenest and best informed mind of any scientist I have known. We talked at length about new papers that were coming out and about limnological problems. . . . Ed Deevey and I, his first students, had the best of it in the old days when he could afford to be generous with his time."[35]

W. Thomas (Tommy) Edmondson, future graduate student but then an undergraduate at Yale, helped Riley with his field collecting one summer. Hutchinson was in Europe and had loaned Gordon and Tommy his old truck. It had dysfunctional brakes, which at this time was typical of Hutchinson's life, in Riley's view: "Evelyn lived in that magnificent, well-ordered mind, which was a good place for him to be, for he was surrounded by chaos. His clothes were shabby, his car decrepit. Every surface in his office was piled high with books and papers, although he could instantly locate anything he wanted. . . . One day he came to my lab to borrow some Nessler tubes . . . used in visual colorimetry. He stuck a half dozen in his jacket pocket. They went through a hole in his pocket and smashed on the floor."[36]

Before Riley had completed his PhD, Albert Aide Parr, director of the Bingham Oceanographic Laboratory in New Haven, invited him to join an oceanographic cruise on the ship *Atlantis* off the mouth of the Mississippi River. Parr was a Norwegian who had trained in Bergen as a fisheries biologist. When he came to New York he worked at first at the New York Aquarium feeding the fish. There, however, he soon met Henry Payne Bingham and persuaded this rich yachtsman to set up a program collecting deep sea fish, using Bingham's yacht. Parr later got him to start the Bingham Oceanographic Foundation. It had a small laboratory in New Haven next to the Peabody Museum, and later it became an important site for Yale research. After his research cruise Riley came back to Hutchinson's laboratory and to limnology as a postdoctoral student.[37]

With Hutchinson's encouragement Riley introduced a number of new methods to American limnology. He was the first to introduce light and dark bottles in the United States to measure photosynthesis and respiration, at Linsley Pond. They were initially used in Norway and then extensively in England at the Plymouth laboratory by Penelope Jenkin, Hutchinson's Cambridge classmate and correspondent. Riley also measured plant pigments, initiating carotenoid separation techniques. His innovative application of statistics, such as multiple regression analysis, to these limnological studies was too much for the editor of *Ecological Monographs,* a senior limnologist. Parr, however, saw the value of Riley's work and published the statistical results in his own *Journal of Marine Research*—even though it was on freshwater research!

When Riley completed his postdoctoral fellowship in limnology and needed to find a job,

Parr sought him out for a position at the Bingham Laboratory at Yale. He accepted. Thus Gordon Riley had now become an oceanographer. Riley remained Hutchinson's colleague at Yale until he left in the 1960s for Dalhousie University in Halifax, Nova Scotia. There he developed a research group of his own in oceanography. Several former Hutchinson students eventually joined him. Indeed, a number of Hutchinson's freshwater graduate students turned to salt water, from limnology to oceanography, in their later careers. Riley wrote of Hutchinson's influence on him: "I am deeply indebted to Evelyn for introducing me to his own scientific philosophy about ecology, which was ahead of its time and has permeated all my plankton work. He maintained that populations needed to be studied in terms of dynamic processes—rates of production and consumption and the way these are affected by ecological factors."[38]

After Riley came Edward (Ed) Deevey. He and Riley actually started graduate school at Yale at the same time, but Deevey started in botany, which was at that time a separate department from zoology. He had his own ideas for his graduate studies; he wanted to work on the history of human impacts on lake environments and thought he could do this using lake cores. Unfortunately, no one in the Yale botany department seemed much interested in his cores from Linsley Pond.

Deevey wanted to combine his own interests in botany and zoology, paleoanthropology and paleolimnology. Pioneering work in paleolimnology had been carried out by Gosta Lundquist in Sweden in the 1920s. Later, pollen cores from bogs in North America had been used by Paul Sears and others to determine post-glacial vegetation. (Pollen grains have very distinctive and resistant outer or exine layers that, even after thousands of years, enable the pollen to be identified, usually down to species.)

Deevey was at that time giving Hutchinson rides to Linsley Pond for his own work. No comprehensive study of lake cores, such as Deevey suggested, had ever been done. Hutchinson was not only interested in Deevey's cores but excited about his project, and he offered to take him on as a zoology graduate student; Deevey agreed.[39] Hutchinson later recalled that although the botanists and zoologists at Yale shared the freshman year courses and the library, "there was a great gulf artificially fixed between the two disciplines; this could not be crossed during working hours."[40] Nevertheless, Deevey did get help from the botanist G. E. Nichols, who loaned him a Davis peat sampler to take his cores and helped him with his pollen analysis.

Apart from pollen analysis, Deevey's voluminous data covered two other related subjects. One was the contemporary limnology of Linsley Pond, including its living fauna, especially the bottom fauna. The other was the paleolimnology of the lake sediments; that is, the postglacial lake history as recorded in fossil fauna. Several other Connecticut lakes were included in Deevey's study. Although these data were used primarily to study the developmental history of Connecticut lakes, the presence of human culture and its effects on the lakes during the post-glacial period was a continuing interest of Deevey's. His work combined original data from fields that were earlier considered separate. It raised as many questions as it answered; he carefully pointed out in his dissertation, surely in consultation with Hutchinson, the areas needing more research, and what techniques might be used to find definitive answers.

What was Hutchinson's view of all this work? He was clearly excited about paleolimnology and Deevey's comparison of the fossil and present microflora. In the 1930s Linsley Pond was a eutrophic lake with poor visibility, high alkalinity, and more phytoplankton than zooplankton. Deevey found the latter dominated by a cladoceran, *Bosmina,* and by rotifers; the bottom fauna was dominated by two types of midge larvae. Although Hutchinson and Riley had already made many measurements in Linsley Pond, Deevey made his own: temperature readings, depth soundings, and determinations of dissolved oxygen and alkalinity. Hutchinson was

knowledgeable about zooplankton and no doubt gave advice on methods. Riley cited the help both he and Deevey got from Hutchinson in the era when he was often a field companion to his graduate students.

Much of the material in Deevey's dissertation cried out (in hindsight at least) for an independent method for determining chronology. Such a method, carbon 14 dating, became available only after World War II. Deevey eventually obtained a Rockefeller Foundation grant to set up a geochronometric laboratory at Yale in 1951. Thus the ages of his cores, previously only relative, became actual, and it became possible to use lake sediment pollen cores to date climate change and to understand the limnological history of lakes. Willard Libby, at Columbia University, had first used carbon 14 for dating archeological samples. Hutchinson wrote that he, Deevey, and Yale geologist R. F. Flint were all close to Libby and worked together in "getting carbon 14 on its feet. Very largely, the extent and speed of the spread of its use was due to Ed Deevey."[41] Hutchinson was probably the first after the war to obtain radioisotopes for use in ecological studies. Together with a 1940s graduate student, Vaughan Bowen, Hutchinson studied phosphorus cycling in lakes using phosphorus 32, the first such radioisotope study, leading to the new field of radioecology. Hutchinson disliked "big science"; he preferred to find research funds for himself and his students, at which he was very successful. He left this field to others, including his former graduate student H. T. Odum. Radiation ecology was amply supported by the Atomic Energy Commission and became a major field of research.

Deevey and Hutchinson did some research together. Deevey analyzed pollen from interglacial beds in the Pang-gong Valley of northern India, in cores that Hutchinson had collected on the North India Expedition, resulting in the first such paper from this geographical region.[42] Deevey later studied paleolimnology in Guatemala and San Salvador. He kept in touch with Hutchinson throughout this period, a fascinating correspondence. Deevey taught at Rice University from 1939 to 1943 and spent the rest of World War II at Woods Hole, one of a notable group of Hutchinson's former students working on war-related marine projects. After the war, Deevey took a position at Yale in zoology with Hutchinson, although he had hoped for the formation of a separate department at Yale, perhaps in biogeography. Eventually he, too, left Yale for Dalhousie University and later went on to form his own group at the University of Florida. He and Hutchinson were both among the Ecological Society of America's "distinguished ecologists" as well as members of the United States National Academy of Sciences.

The third early graduate student was Tommy Edmondson. He was actually the first to arrive in Hutchinson's laboratory, as a New Haven high school student with a burning interest in rotifers. With his high school biology teacher's microscope and a copy of Ward and Whipple's *Freshwater Biology* that his brother had found for him, he proceeded to collect and identify the local rotifers. The teacher, Ruth Ross, introduced him to Yale professor L. L. Woodruff, a protozoa man, who heard the word "rotifers" and took him down the hall to meet Hutchinson. Hutchinson, who had just been out collecting, poured out some water from a thermos into a finger bowl, and Tommy beheld a colonial rotifer that he had read about in a book by Hudson and Gosse but never seen before. Hutchinson gave him a corner of his lab and loaned him a microscope. He spent every minute he could there throughout his high school years. He also took Hutchinson's invertebrate zoology course, examinations and all.[43]

Edmondson quickly became proficient at identifying all kinds of rotifers. He was not the first. There was a nineteenth-century English tradition of serious amateurs, particularly tradesmen, studying some specialized aspect of the natural world. Hutchinson himself, always interested in the history of his sciences, wrote about this subject.[44] One such amateur, an English wine merchant, had composed long rotifer monographs. Edmondson recalled Hutchinson in the laboratory in the early 1930s: "As I was

sitting at one end of the room looking at rotifers, Hutchinson would be running chemical analyses at the other end, calculating the data, and expressing surprise and pleasure at some new finding."[45]

When Hutchinson came back from North India it was Tommy, the high school student, who was selected to work on the rotifers he brought back. That collaboration eventually resulted in a paper by "Edmondson and Hutchinson."[46] Edmondson was amazed that his name was first. Hutchinson remarked that their names were simply in alphabetical order. Edmondson told the author later that Hutchinson was "very sharing. He didn't hog the spotlight. He acknowledged the people who helped him."[47]

Ross Harrison was still head of the zoology department at this time. One day he called Tommy Edmondson into his office. "I understand that you do a lot of work here [in Hutchinson's lab] at night," he said, and he handed Tommy his own key.[48] Edmondson graduated from high school and became an undergraduate zoology major at Yale. He took the embryology course not from Harrison, but from Hutchinson. "This course that [at other universities] normally would just be looking at chicken slides, had a lot of experiments. Hutchinson taught it as a functional thing."[49] Ray Rappaport, now a well-known developmental biologist, took a seminar course with Hutchinson. He commented that Hutchinson was teaching students to be scientists early in their careers, even in his undergraduate courses.[50] Although Edmonton's interest in rotifers was originally in identification and taxonomy, he started collecting in Linsley Pond and "every other body of water" near New Haven and soon became interested in their ecology and population biology, interests he pursued in his graduate work and thereafter. He remembered discussing these subjects with Hutchinson in the lab they shared. Hutchinson worked on water bugs from an early age, becoming an excellent taxonomist of this group of insects while still a high school student. But in tracing the development of Hutchinson's scientific ideas, it is clear that almost from the beginning Hutchinson was asking ecological questions about his water bugs. This was not true of many of his contemporaries, who continued doing pure taxonomy of whatever group of organisms they worked on.

Edmondson suggested in an interview that one might trace the development of Hutchinson's thought from his work with Pickford and Schuurman on the South African pans, where he was identifying "not just the bugs but the whole biota . . . and trying to relate it to the chemical characteristics of the situation."[51] Edmondson thought the ecological aspects of that South African paper were original, although Hutchinson might have been influenced by Charles Elton's book *Animal Ecology.* The South African work had been done before Hutchinson read that book, but he read it and probably incorporated it into his thinking before publishing the paper. Edmondson also mentioned S. A. Forbes's paper "The Lake as a Microcosm," much cited as a classic, but largely in hindsight. It was published obscurely in 1887 and reprinted in 1925, but again in a local journal. It is unlikely that Hutchinson would have read it at an early date, although he quoted it more than thirty years later on the second page of volume 1 of his *Treatise on Limnology.* Hutchinson was well known to his students as a voracious reader who had a "cross-indexed brain," a comment of postdoctoral student Clyde Goulden.[52] Everything he learned he related to his previous knowledge—and it all no doubt influenced his own and his students' work.

Edmondson returned to Yale as a graduate student with Hutchinson in 1939 after a research assistantship with Chauncey Juday at the University of Wisconsin's Trout Lake. During his doctoral work Edmondson found a method for marking rotifers with carmine dye to measure their growth rates. More work on rotifer population ecology followed. Edmondson also worked at Woods Hole during World War II, and afterward he became the first Hutchinson student to bring limnology to the West Coast, developing a large group of graduate students at the University of Washington. He is best known

for his and his students' research that led to the understanding and subsequent reversal of pollution in Lake Washington in Seattle. Edmondson was also elected to the U.S. National Academy of Sciences.

One additional student, Max Dunbar, worked with Hutchinson in the 1930s.[53] He was not officially a graduate student but an advanced undergraduate who came to Yale from Oxford on a Henry Fellowship from 1937 to 1938. Although he had studied zoology at Oxford, he had come to a different Yale department. But he walked into Hutchinson's office in Osborne Memorial Laboratory and told him about a project he wanted to do on osmotic effects in marine or brackish water animals. Hutchinson said, "Come and work here." He did, with *Vorticella,* the one-celled ciliate protozoan that attaches to algae or other substrates with a stalk that expands and contracts like a Slinky toy. Dunbar was able to watch its reactions to metallic ions in seawater.

Dunbar completed his Oxford degree and then went to McGill for a PhD, working in the Canadian Arctic. He was named Canadian consul in Greenland, but he was able to continue his research. He kept in touch with Hutchinson and wrote him a long letter from Greenland in 1943 about plankton production, about size differences in two coexisting subspecies of zooplankton, about his methods of measuring salinity, and even about a young sea-eagle he had weaned who could recognize humans by the color of their pants. Dunbar obviously knew that Hutchinson was interested in everything. Charles Elton, whose book had so influenced Hutchinson at age twenty-five, was at Oxford during Dunbar's undergraduate days. Dunbar related that someone had seen Elton and Hutchinson "plotting together on the Broad" the first time they had met, "two people responsible for most of ecology."[54]

Elton pioneered new ways of looking at the interrelationships of animals. Hutchinson and his students took this much further and made it more quantitative. Dunbar thought Hutchinson was the first to recognize the significance of size difference within different groups, including copepods. "He wrote a magnificent paper," said Dunbar, "'Copepodology for the Ornithologist.' . . . Elton was on to the same idea but he didn't write about it as Hutchinson did."

Max Dunbar and Ed Deevey were among a small group of students that Hutchinson took on local field trips in the 1930s, "to chat about what he was seeing, like birds and spiders." When asked whether Hutchinson knew about birds, Dunbar replied, "Hutch knew about everything." Such was the folklore. Actually Hutchinson continually added new areas of knowledge. The year Dunbar arrived he was studying music before Palestrina. But his old understanding of ecology could always be altered. Dunbar told the story of a graduate student who had come in and asked about a paper Hutchinson had written: "Hutch talked about it for quite a while and the graduate student said, 'That's not what you wrote in this paper.' Hutch said, 'No, that's what I was thinking six years ago.'"[55] Dunbar himself went on to become a professor at McGill and to a distinguished career of Arctic research. In rather Hutchinsonian fashion, however, he also made recordings of Scottish folk songs that are now at the Smithsonian Institution.

PERSONAL LIFE, THE THIRTIES AND BEYOND

The 1930s and 1940s saw many changes in Hutchinson's personal life. In 1933 Grace Pickford and Evelyn Hutchinson were divorced. Hutchinson went to Nevada, where only six weeks' residence was required for a divorce. While in Nevada he did an important piece of research on arid lakes, related to the work that he and Grace had carried out together on the South Africa pans. Subsequently he married another English woman, Margaret Seal. This was to be a marriage of fifty years. Margaret did not share Evelyn's scientific interests, but they shared deep interests in music and art. Evelyn had written to Grace from North India that he was looking forward to being reproductive in a future marriage, but this did not happen. The only child he and Margaret shared was Yemaiel,

a two-year-old refugee sent from England at the beginning of World War II and returned to her own parents near the end the war. Evelyn and Yemaiel and her own family became close friends again only in his old age.

In the 1940s Hutchinson attended, together with his close friends Margaret Mead and Gregory Bateson, the Josiah Macy "feedback" conferences in New York City. Bateson had been a classmate of Hutchinson's at Cambridge University. He and Mead were both on the staff of the American Museum of Natural History, Hutchinson as a consultant in biogeochemistry. Cybernetics was importantly involved in the conferences. Norbert Wiener was a leading member, as were mathematician John von Neumann, who wrote *The Computer and the Brain,* and psychiatrist Lawrence S. Kubie. Hutchinson wrote that he was "trying to learn quietly and unobtrusively from this extraordinary group of genii."[56] However, he himself gave an ecological paper that was well received by the "genii," later published as "Circular Causal Systems in Ecology."[57]

Hutchinson was elected early in his career to the National Academy of Sciences, as were several of his students and coworkers, including one of his earliest collaborators, Ruth Patrick. He became interested in her early work, and she subsequently did research with him on the diatoms of Linsley Pond. She went on to do innovative work on diatom ecology. They were lifelong friends and colleagues. Hutchinson received a great many honorary doctorates, but never applied to Cambridge University for an earned doctorate based on his research papers; he most certainly would have been awarded one. In 1981 he received an honorary Cambridge University doctorate with much pomp and ceremony, together with Max Perutz and other notables.

Hutchinson lost his beloved wife Margaret to Alzheimer's disease in 1983. His close friend and constant correspondent, the English novelist Rebecca West, died the same year. In his eighties he married a much younger woman, Anne Twitty Goldsby, a biologist of Haitian origin. She had formerly run the freshman biology laboratory at Yale. Anne was able to accompany him to Japan, where, in 1986, at eighty-three, he received his last major honor, the Kyoto Award, called the Japanese Nobel Prize, for his work in ecology.

RECOGNITION, LATER STUDENTS, AND CONTRIBUTIONS TO MODERN ECOLOGY

Hutchinson received many awards, but he himself was most excited (but also humble) when he received the Benjamin Franklin Medal in 1979. He wondered what he had invented to find himself in the company of previous winners like Thomas Edison, Max Planck, and Albert Einstein (and of Ben Franklin himself). He had of course invented much of modern ecology, with seminal contributions in systems ecology, radiation ecology, population biology, limnology, and biogeochemistry. The Franklin award, did, in fact, cite him for developing the scientific basis of ecology.

He was also happy to receive two awards for his environmental work, the Tyler Prize for Environmental Achievement and the National Academy of Sciences Award for Environmental Quality, particularly for his basic research. This kind of research was the most important to him, and the most satisfying. Whether or not it was "useful," it could illuminate and make more understandable the processes and beauty of the natural world. However, he also did important work behind the scenes on practical ecological problems, from the deleterious effects of Agent Orange in Vietnam to a major role in preventing the destruction of the island of Aldabra in the Indian Ocean, home of giant tortoises. It had been slated to become a military base.[58]

Hutchinson's environmental philosophy was evident in his very first "Marginalia" column for the *American Scientist* in 1943: "These notes will reflect the attitude of the philosophic naturalist, rather than that of the engineer, the point of view of the mind that delights in understanding nature rather than in attempting to reform her. . . . The writer believes that the most practical lasting benefit science can now offer is to

teach man how to avoid destruction of his own environment, and how, by understanding himself with true humility and pride, to find ways to avoid injuries that at present he inflicts on himself with such devastating energy."[59]

Hutchinson continued to publish nearly until his death in 1991. Major contributions in limnology were published in the 1950s and thereafter. The first volume of *A Treatise on Limnology* was published in 1957; it is still in use and has enhanced his reputation as a leading limnologist worldwide.[60] Three other volumes of the *Treatise* followed, the fourth published posthumously. He had finished the writing before he died, but former students and others collected the illustrations and other missing pieces under the editorship of Yvette Edmondson.[61]

Hutchinson's best-known papers in population and theoretical ecology were published in the 1950s and 1960s. The 1950s crop of graduate students was outstanding. It was with this group, both men and women, that Hutchinson worked out some of his most important ideas, ideas and models that spawned a great deal of research by young ecologists. These papers included "Concluding Remarks."[62] Others were "Homage to Santa Rosalia, or Why Are There So Many Kinds of Animals?" and "The Paradox of the Plankton."[63]

"Hutchinson's 'Concluding Remarks' is . . . certainly one of the least explicit titles in the history of population ecology, masking . . . one of the most highly touted and disputed productions of that discipline," wrote Robert McIntosh.[64] It was Hutchinson's niche theory, formalized in that paper, that "propelled the pioneer existence on the intellectual frontier" of ecology, as Platil and Rosenzweig wrote in the preface to their book.[65] Their image of Hutchinson and his graduate students spearheading the "colonizing" of ecology is a striking one. Hutchinson's paper, given at the conclusion of a Cold Spring Harbor Symposium on demography, spawned a tremendous amount of new research (including this author's) on the ecological niche and competitive exclusion or the lack thereof in both animals and plants and in many ecosystems, both aquatic and terrestrial. It also engendered critiques and modifications. Hutchinson was by that time the leading American theoretical ecologist. Hutchinson the teacher, especially of graduate students, is also evident in this paper. In an almost unprecedented footnote to the printed paper, he thanked ten participants in his seminar in advanced ecology and wrote: "Anything that is new in the present paper emerged from this seminar and is not to be regarded specifically as an original contribution of the writer."[66] One of those thanked, Robert MacArthur, finished his PhD in 1958 and went on to become a leading mathematical ecologist. It was Hutchinson and MacArthur, wrote Sharon Kingsland, "who helped to make ecology intellectually exciting."[67] Before his early death, MacArthur published many influential papers and books not only on population ecology but also on biogeography. I leave to MacArthur the concluding remarks about his mentor, from an unpublished letter he wrote early in his career.

> It is of course hard to characterize in a few words a great man. . . . I think Hutchinson is, above all, an intellectual. Any fact, no matter how small, and any rational idea, no matter how insignificant is treated with respect. And the people who discuss these facts and ideas are also treated with respect. . . . Perhaps his most significant achievements have been by using procedures of other sciences on ecological problems: He has used his tremendous memory and the methods of the historian to study the problems of the guano deposits; he used the methods of electrical engineering to elucidate the effects of time lags on animal populations; he used the method of radioactive tracers (in a novel way) combined with a knowledge of theoretical hydrodynamics to explain motion of phosphorous in a pond.
>
> I believe a few words about him as a teacher are in order. . . . He never made appointments. Students go into his room to discuss, not to get advice, and are treated as equals no matter how far-fetched their ideas. . . . I believe he has influenced me the most by his genuine enthusiasm about the small, novel ideas we may have. This enthusiasm is the more sincere sort of encouragement. I feel rather foolish having said all this "in print" but I hope it may contribute to the understanding of a remarkably fine man.[68]

Hutchinson continued to work in his office at Yale through 1990. He returned to his native England and died there in 1991.

A SWIMMING GRASSHOPPER.—When I was looking for Hemiptera in a pond here, I knocked a grasshopper into the water. It fell about 18 inches from the bank and commenced to struggle, and I then saw it was getting under the water. When it was well under it began to swim, using its hind legs and its front ones, but not its intermediate ones. It swam back to the bank and climbed up a stem out of the water. I put the insect back into the water and let it swim again. The insect was not apparently much exhausted after its swim. I do not know what species of grasshopper it is so I enclose it for examination. Could you return it if possible. I enclose stamps to cover the postage.—G. E. HUTCHINSON, " Woodlands," Holt, Norfolk. *June 20th.* [The species is *Tetrix bipunctatus.* Neither Dr. Chapman nor I were aware of the fact that a grasshopper would apparently be so much at home in the water.—H.J.T.]

in my day in zoology was hardly a success. The M.A. degree followed automatically three years after the B.A. on payment of a fee. Anyone who wanted anything better could publish enough significant work to get an Sc.D. By the time I had done this there was no point in spending fifty pounds for the degree fee, though the scarlet gown would have been fun.

When I came up to Cambridge the biological eminence of the university was very unevenly distributed among the various departments of what are now often called life sciences. This was indeed true of any group of subjects, as is inevitable in a university. To my certainly prejudiced mind it seems that in the thirty years from about 1890 to 1920, Cambridge was the intellectual capital of the world, much as Berkeley was for a few years during the past decades. Within the period of which I am thinking Russell and Whitehead were writing the *Principia Mathematica*, G. E. Moore the *Principia Ethica*, J. G. Frazer *The Golden Bough*, and Jane Harrison *Themis*; Marshall and later Keynes were establishing modern Western economics and W. W. Skeat providing the basis for modern studies of Middle English; J. J. Thomson and Ernest Rutherford were beginning to dissect the atom and F. Gowland Hopkins doing more than his share of founding biochemistry. Vague indications of all this were apparent throughout my boyhood. By the time I was an undergraduate some of the glory had departed, partly because of the First World War and partly because the rise of the other universities was spreading the most exceptional talent, which might be recognized wherever it appeared, more thinly than before. Nevertheless the tradition was still strong. The greatness, however, was not uniform through the university, being concentrated in Henry VIII's huge foundation of Trinity, in King's, and to some extent in Saint John's.

The role played by the college in undergraduate

education was undefined and very varied. In the natural sciences it was possible never to have any college instruction at all, and in fact, I never did. Most science students had a supervisor, who was either a college fellow or had been appointed from outside the college fellowship to do this work. My father once noticed a student leaving his laboratory before he had finished the exercise and asked him why. He replied he had to go to see his supervisor. My father then discovered that the supervisor was a postgraduate student in mineralogy, and he pointed out to the young man that it might be better to stay and learn from the university lecturer rather than from his student. I was, therefore, warned against all supervision. In contrast, in some arts subjects, one could get all the teaching needed in college, particularly in a large college. There was, I think, always an advantage in having really eminent men around, but I had plenty of ways of meeting them, so that their comparative rarity in Emmanuel did not worry me.

In zoology the great days at Cambridge had been the late nineteenth and very early twentieth centuries, with F. M. Balfour and later Adam Sedgwick, both very distinguished embryologists. By the time I came up no one of first-class distinction was apparent to the beginning undergraduate; and the group of students to which I belonged was rather impatient with their teachers. The professor, J. Stanley Gardiner, had been a leader in the exploration of the coral reefs of the Indian Ocean. His expeditions had done a great deal to make known not only the marine biology but also the terrestrial life of the Maldives, Laccadives, Seychelles, and Aldabra. In 1931, he published a very good book on coral reefs, still read with profit by modern workers.[2] He was, however, odd

2. J. Stanley Gardiner, *Coral Reefs and Atolls; Being a Course of Lectures Delivered at the Lowell Institute at Boston, February 1930* (London: Macmillan, 1931).

and uncertain and an adept at putting one at dis-ease. He could not inspire students, at least in my time, and I have heard him called a "very bad man." Yet when he came to recommend me to Yale he wrote a superb and most understanding letter. The three more senior lecturers were L. A. Borradaile, F. A. Potts, and J. T. Saunders. They cooperated in that excellent textbook *The Invertebrata*.[3] Borradaile was a classical morphologist, interested primarily in the body plan of arthropods. He was immensely learned and very generous to anyone who really wanted to learn from him, but by the time that I knew him he seemed somewhat disillusioned intellectually and was said to spend more time on his stamp collection than on zoology. What I learned from him and the files of notes that he lent me was invaluable on the two occasions when I was called in as intellectual midwife at the birth, or at least recognition, of the two major new groups of living Crustacea to become known in this century, the Mystacocarida of Pennak and Zinn and the Cephalocarida of Sanders.

Potts had done very nice work on parasitic Crustacea but again had lost interest to some extent. Saunders taught in the main invertebrate laboratory and gave a hydrobiology course to part II students. He did some of the best, indeed, for a time the only critical work on the ecological role of pH, but he drifted into academic administration and ended as the principal of Ibadan University.

Vertebrate zoology was taught by Hans Gadow, a marvelous relic of the nineteenth century who walked in from Shelford, four or five miles, to give his lectures. He began the course by writing in transliterated Greek the first verse of the first chapter of Saint John's Gospel. He

3. L. A. Borradaile, F. A. Potts, L. E. S. Eastham, and J. T. Saunders, *The Invertebrata* (Cambridge: Cambridge University Press, 1932).

had pronounced orthogenetic views of evolution, in which he clearly felt the Logos operating. As a good Lutheran he once nailed to the door of the zoology laboratory a set of theses on the homologies of ear ossicles, but I never heard that anyone took up the challenge.

F. Balfour Brown taught entomology. A committed Scot with a passion for water beetles, he gave a good straightforward course; not a powerful intellect, he is dear to me for having introduced me to the Hebrides.

Friends, particularly Bill Thorpe, with more inside experience of the department than I could have had, tell me that much of the tension that we felt was a psychological aftermath of the First World War. This fits in with my general impression that the university had lost some of the early-twentieth-century glory before I came up. British intellectual society, however excellent it was in other ways, clearly had no completely effective mechanism for absorbing the shock, as well as the loss of life, resulting from the war.

More exciting people appeared when one got to part II. H. Munro Fox, with a reputation as the best-dressed man in British zoology, had done beautiful work on the lunar periodicity of breeding in sea urchins and later on chlorocruorin and other animal pigments. He was also much interested in the passage of animals through the Suez Canal. He soon left Cambridge for a chair in Birmingham and later in London.

James Gray was certainly the most generally distinguished man in the department at the time. He was tall, long-legged, and was believed, I think falsely, to have served in 1914–18 in a camel corps. He was working on ciliary movement and gave a very good, skeptical course on experimental cytology. He tended to disbelieve in any structure not visible in intact living cells, all the rest of cytology being a study of artifacts. At this time such a

restriction greatly reduced the range of the subject. Fortunately, S. T. P. Strangeways in the pathology department
was demonstrating mitotic figures in living fibroblast cells;
his films, in fact, may still be used in teaching. This demonstration was very helpful if one took Gray a little bit more
seriously than he intended. It is also interesting to recall
his worry that the histones of the sperm head were among
the simplest of proteins, quite unfit to be the stuff Mendelian factors are made of. Gray succeeded Gardiner as
professor in 1937 and at once began turning the department
into one of the outstanding centers of experimental
zoology in the world.

The last teacher in zoology that I would mention was
to me the most important, G. P. Bidder. He was an eccen-
tric of considerable wealth, largely nocturnal, with beard,
moustaches, and cloak in truly operatic fashion. He gave
a course on sponges, or rather on some aspects of the
calcareous sponges and any related ideas that came into
his head. This disorganized piece of teaching was the most
important that I ever had. The ideas that he implanted
kept on working and growing and still do so when trans-
planted. His great theme was the ecological significance
of size as such and the complete difference of living in
aquatic worlds in which Brownian movement, viscosity, or
inertia was the dominant force. This concept is not only
extremely interesting in itself but provides a model of ways
of classifying the spaces in which organisms live, by using
properties that are not intuitively obvious.

As well as the teaching faculty, and Bidder was, I
fancy, only an honorary member, there was an important
zoological museum. The curator, C. Forster Cooper, the
discoverer of *Baluchitherium*, the largest known terrestrial
mammal, fossil or living, was a charming and able man who
had fallen out with Gardiner and was obviously glad to be
called to the directorship of the Natural History Museum

in South Kensington. The other full-time member of the museum staff was Hugh Scott, a meticulous taxonomist working mainly with beetles, who had been involved with the *Fauna Hawaiensis* and with the terrestrial insects of the Seychelles. He was obviously, but very quietly, in love with the diversity of the insect world and was a good, sweet-tempered man. His entomological learning was extraordinary and he was also an amateur archaeologist and collector of old English ironwork. We became great friends though he must have been thirty years older than I was. I spent many happy hours helping arrange Hemiptera in the museum. Quite recently his successor, Dr. John Smart, showed me a couple of death's-head hawkmoths, *Acherontia atropos*, preserved in a cabinet drawer, with a long and characteristic label by Hugh Scott that identified them as having appeared in George III's bedroom late in his career.

The physiology department was housed in a building much newer than the rather ramshackle zoological laboratory. The professor, J. N. Langley, who had done important work on the autonomic nervous system, seemed, like Gardiner, to have mislayed his earlier eminence. I was in fact advised not to take his elementary lectures, so I signed up only for the lab, which was half histology and half nerve and muscle frog physiology. Instead of Langley's course I therefore took Adrian's Nervous System in my first term, followed by Hartridge on special senses, Marshall on the physiology of reproduction, and a few other unimportant offerings of which I have almost no recollection.

Langley had one engaging quality. If there was good ice on a large field, flooded for the purpose, south of Coe Fen, he put up a notice indicating that the laboratory was closed on account of the inclemency of the weather. This was a treat that only happened in some years.

Adrian's course was a tremendously exciting experience. The audience was immense, including every intended medical student in the university, and had to be held in the examination schools after the first meeting, no other available room being big enough. Adrian always strode in rapidly, a few minutes late, removing an immense pair of fur gloves. He never wore an overcoat even at the height of winter. He gave a lively summary of all the contemporary research on neurons and on the mammalian spinal cord and brain, beautifully arranged and clearly delivered. He always used yellow chalk, the dust from which supplied admirably macabre makeup for an impersonation of a Jacksonian epileptic going into a fit. At the beginning of the course he recommended Bertrand Russell's *Analysis of Mind* as the best treatment of the mind-body problem, adding that it had nothing to do with psychoanalysis. His death, while this book was being written, removes the last person whose formal lectures I attended at Cambridge.

The other exciting member of the department at the time was Barcroft, but I do not think he took any part in the part I teaching, and though I had met him on and off during my school days, I did not really get to know him until much later.

Marshall, who taught what I imagine was the first course in the physiology of reproduction given in Britain, was a member of the agriculture faculty and not of the department of physiology. He spoke with a pronounced lisp, "In the pretheding leththure we dithcuthed the phenomenon of ptheudopregnanthy in the bitth." To those of us in zoology the excitement of Marshall's course lay in its comparative aspects. The bitch might become pseudopregnant but some other female mammals could not. Women ovulated spontaneously, but rabbits at least needed to be stroked. In Adrian's treatment of the nervous system the hominid or medical emphasis was not only practical but intellectually reasonable, since we after all did have the best one available. Marshall, coming from a non-medical area, provided something comparative that was specially informative, though of course the subject itself had an immense inherent fascination. Of the other teachers of physiology I will only mention, as an antiquarian footnote, Dr. Shore, who had worked with Maray in Paris, and whom we all believed had kept the horse happy with continual wisps of hay as the gauge was being slid into its heart in the first *in vivo* measurement of intracardial pressure. An Englishman should, of course, have understood horses.

Though when I took part I, biochemistry was still part of physiology, the development of a separate department was fast going ahead. This department *in statu nascendi* was the great intellectual center of Cambridge biology in the 1920s. As was the tradition in most science departments, the lectures of the introductory course were given by the professor, in this case F. Gowland Hopkins. He was able to impart an aura of authority even when, as once happened, he tried twice to put the structural formula of tryptophan on the board and then gave up with the remark, "I knew what this was when I elucidated it in 1911."

Hopkins, in 1906 and 1907, had done experiments that indicated the essential nature of some minor constituents of an unknown kind in the diet of growing rats. The results had been mentioned in lectures at Guy's Hospital, but illness had prevented their publication. Funk's classic paper on the substance, now vitamin B_1 or thiamin, which was derived from the outer coat of rice seeds and prevented the polyneuritis produced in pigeons by a diet of polished rice, appeared in 1911. Immediately after, in 1912, Hopkins published the results of a new set of experiments confirming those done five or six years earlier. In Cambridge we always regarded Hopkins, I think justifiably, as a codiscoverer of vitamins with Funk, as indeed did the Nobel committee.[4] Some time between 1912 and 1914 I had been taken with my cousin Bill Stewart to see Dr. H. H. Dale (later Sir Henry Dale, president of the Royal Society), at the Wellcome Physiological Research Laboratory and was there shown

4. F. Gowland Hopkins, Feeding Experiments Illustrating the Importance of Accessory Factors in Normal Dieteries, *J. Physiol.* 44:425–60, 1912. Hopkins writes (p. 425) that the "particular experiments . . . were undertaken to put upon a more quantitative basis results which I obtained as far back as 1906–1907." He adds a note that publication was delayed on account of illness but that the "results of experiments made at this time were summarized in lectures delivered at Guy's Hospital in June 1909." It would be interesting to know if any notes exist, taken by someone in his audience. Hopkins also writes in the same paper (p. 449): "Convinced of the importance of accurate diet factors by my own earlier observations, I ventured, in an address delivered in November 1906, to make the following remarks. 'But further, no animal can live upon a mixture of pure protein, fat, and carbohydrate and even when the necessary inorganic material is carefully supplied the animal still cannot flourish.'" This remark is quoted from F. G. Hopkins, The Analyst and the Medical Man, *The Analyst* 31:385–97, 1906, which was based on an otherwise dull lecture given to the Society of Public Analysts on 7 November 1906.

Funk's results were published, from the Lister Institute, as: C. Funk, On the Chemical Nature of the Substance which Cures Polyneuritis in Birds Induced by a Diet of Polished Rice, *J. Physiol.* 43:395–400, 1911.

Funk and Hopkins were certainly on friendly terms, as is clear in S. L. Becker, Butter Makes them Grow, *Conn. Agri. Exp. Stat. Bull.* 767, 1977, which gives an excellent account of the early work on nutrition by T. B. Osborn and L. B. Mendel.

It had, of course, been obvious for a couple of centuries that the lack of something in diets could cause scurvy, and comparable knowledge had existed in Japan with regard to beri-beri. Perhaps one should not try to say vitamins were discovered at a specific time and by certain workers, but rather that a knowledge of such substances started growing rather rapidly in the second decade of the twentieth century.

FIGURE 8   Sir Frederick Gowland Hopkins O. M., F.R.S. (Barnet Woolf, originally published in *Brighter Biochemistry* 1925; by permission of the Department of Biochemistry, Cambridge University).

experiments on the cure of polyneuritis in pigeons by extracts of the material removed in polishing rice. These were presumably an early confirmation of Funk's work. When I was a student Hopkins was best known for his recent discovery of glutathione (figure 8).

The elementary laboratory in biochemistry was run by Sydney Cole, who was I think really more interested in golf than in science. He and his wife would in fact have fitted, as minor characters, into a silent film of a Wodehouse novel. During the First World War he was worried by what seemed to him the calamitous scarcity of blood charcoal, then widely used as a sorbent in certain analytical and preparative processes. He wrote a letter to *Nature* about it.[5] This gave rise, I suspect in the fertile mind of M. A. Rushton, to:

> Sydney Cole
> Would sell his soul
> For half a bowl
> Of blood charcoal,

which in the perfection of its sound rhyme, set against varying eye rhymes, has a certain fascination for the student of minor verse. Cole liked to improve on the work of his predecessors, so that most of the standard procedures appeared, in the laboratory manual that he wrote for the course, as Cole's modification of——'s method. All the apparatus was set up ahead of time by the lab boy, as such attendants were usually called, so that even the most thick-headed aspirant to an MB degree could complete his experiment. As a result the course was entirely useless as an exercise in biochemical technique. It had for me, however, an extraordinary fringe benefit, namely that one of

5. S. W. Cole, The Preparation of "Blood Charcoal," *Nature* 99:226, 1917.

the demonstrators was Joseph Needham; so having let the apparatus complete Cole's modification of this, that, or the other, I heard about the work being done in chemical embryology and many other things.

J. B. S. Haldane, then reader in biochemistry, was active in his experiments on salt metabolism, which on occasion involved running round and round Market Hill, a crowded flat area in the middle of the town, with a bottle to be filled for analysis in the subterranean men's room, which on another occasion had been used as Tutankhamun's tomb in a mock excavation. Haldane was also beginning his classical work on the mathematical theory of natural selection, which is a major foundation stone of modern population genetics and evolutionary theory.

Though I do not think he ever worked on enzyme kinetics himself, he gave an excellent course on enzymes which later turned into a book. Once when I was the only student to appear on time for a lecture in this course, he began in a deadpan manner, "When John Hunter found himself having to address an audience of this size, he was accustomed to bring a skeleton into the room so that he could address them as gentlemen."

I soon got to know another biochemist, Robin Hill, as he was an Emmanuel man who sought me out as soon as I had settled in. Although officially working on photosynthesis, which of course ultimately led to the Hill effect,[6] he was devoting a great deal of time to calculations for a fish-eye lens that would photograph the entire sky and so, by the use of two cameras synchronized electrically, permit the triangulation of lightning flashes. Hill was also much

6. R. Hill, Oxygen Evolved by Isolated Chloroplasts, *Nature* 139:881–82, 1937. Hill found that chloroplasts suspended in a medium containing ferric oxalate produced oxygen when illuminated, the ferric iron becoming reduced to ferrous. This was the beginning of modern work on the nature of the reactions occurring in photosynthesis.

interested in the preparation of permanent mineral pigments for artists painting in oils. Mention of Needham, Haldane, and Hill indicates the extraordinary range of intellectual interests that centered around biochemistry, but it does not exhaust the list of significant people. R. A. Peters soon left for Oxford and I did not really meet him until much later in life. Marjorie Stevenson was founding the comparative biochemistry of bacterial metabolism.

In spite of the great strength of the nascent department of biochemistry, it is possible that the most important biochemical work done in Cambridge about the time that I was up was the discovery of cytochrome by David Keilin, an entomologist interested in parasitic fly larvae, working in the Molteno Institute of Parasitology. He was a friend of E. J. Bles, and I came to know him quite well. I think he put me on to Delcourt's work on what we should now call introgression, to which I shall return on a later page. He took such evolutionary studies rather lightly and evidently felt that Delcourt, in attempting to raise *Drosophila* axenically on a medium of known composition, had gone on to do something really serious and worthwhile.

One further name, that of the Honorable Huia Onslow, must be added to the list of Cambridge biochemists who influenced me. He perforce worked as an amateur at home, but though I never met him personally, he has haunted my life in a most curious and beneficial way. The son of Lord Onslow, governor general of New Zealand, and named for an extraordinary extinct bird, the Huia or *Heteralocha acutirostris* (Gould), Onslow had had a serious diving accident as a student and did all his scientific work in a more or less recumbent position. This included a very important analysis of apparent blending inheritance of yellow ground color in the currant moth, *Abraxas grossulariata*. By applying a quantitative colorimetric method, he showed the inheritance to be Mendelian. He did very early biochemical work on the genetics of melanin production in rabbits and also an extraordinary study of irridescent physical colors in insects.[7] When I was admitted as a fellow of the Royal Entomological Society in 1923 Onslow's death had just been announced at the meeting, as his name was coming up for election. Many years later, Sir James Gray (in a letter of 7 February 1966), wrote "D[oncaster] introduced me to Onslow who was the first biochemist I ever met— a terribly nice and modest man with superb courage." Onslow's work on *A. grossulariata* has been most useful to me when in recent years I have tried to give an interpretation of the fantastic range of variation in this moth. Onslow's wife, Muriel Wheldale, was also a distinguished and very early biochemical geneticist, working on anthocyanins. The Cambridge interest in such matters, insofar as it was not an obvious next step, probably derived something from Bateson.

Genetics was taught twice a week, at five o'clock in the Michaelmas term, by R. C. Punnett, who had been Bateson's close collaborator and aid. He was a mild man with an overdominant wife who had been a major tennis player. Her opponents must have been terrified. Punnett had fine collections of Chinese porcelain and Japanese

7. H. Onslow, A Contribution to Our Knowledge of the Chemistry of Coat-Colour in Animals and Dominant and Recessive Whiteness, *Proc. Roy. Soc. Lond.* 89 (B): 36–58, 1915. In this paper Sydney Cole is thanked for "his invaluable suggestions and help." Helping Onslow in this work may have been Cole's greatest contribution to science. H. Onslow, The Inheritance of Wing Colour in Lepidoptera. I. *Abraxas grossulariata* var. *lutea* (Cockerell), *J. Genetics* 8:209–58, 1919; H. Onslow, On a Periodic Structure in Many Insect Scales and the Cause of their Iridescent Colours, *Phil. Trans. Roy. Soc.* 211 (B): 1–74, 1923.

prints in a delightful house backed by an experimental garden, and he devoted himself largely to the genetics of sweet peas. The Punnetts gave Sunday lunches with superb wine to an incongruous set of students, half biological intellectuals, half athletes, all I think men. Only about half a dozen students took Punnett's course. It was given in the large zoology lecture theater. All but one of us sat in the front row. My friend Ivor Montagu, on the very rare occasions that he attended, sat in the back row. Punnett was a strong believer in the presence and absence theory of dominant and recessive factors (genes were not yet respectable in Cambridge); I used to argue, in retrospect interminably, that his defense was circular. He seemed to enjoy these arguments. The chromosome theory was still widely debated. Bateson was usually skeptical, though I know he accepted it for about a fortnight shortly before his death. Punnett tended to be more receptive to the idea. One evening the high point of the course arrived unexpectedly; Punnett came in demurely and then announced that he had just finished all the calculations of linkage of the various characters he had studied in the sweet pea and that indeed there were as many linkage groups as haploid chromosomes. The chromosome theory had worked for a plant as well as an animal and therefore might be reasonably expected to be of general validity.

I have said nothing about chemistry, which I also took for part I. I remember little of the teaching except Fenton writing with both hands at once on the board, and an organic lab run by the dourest Scot I have ever had the misfortune to encounter. His scientific interests were supposed to be the synthesis of more and more poisonous gases for military purposes. He may have been misjudged; as a fellow of King's perhaps he was an interesting and humane man. Anyhow,

> It's certain that some of the ultimate things
> Are hidden from even the fellows of King's.[8]

I have little to say of botany either except to express, as so many others have done, gratitude, admiration, and affection for H. Gilbert Carter. He was a great knower of plants; his labels in the Botanic Gardens told one what the Latin, Greek, Hebrew, and Persian poets had called them. On field trips, he insisted that the leaf of each species studied, unless he knew it was poisonous, had to be tasted to supply a new set of taxonomic characters. He ran a private seminar on Virgil and was the only person whom I knew who really understood my passion for the folklore of *Mandragora*, which he grew.

In general, the relations between botany and zoology were much less satisfactory than between zoology and biochemistry. Tansley was mainly concerned with psychoanalysis when I was taking botany for part I of the tripos and without H. Gilbert Carter I should have learned nothing from the course.

---

8. I suspect this appeared in *Punch* during the First World War when some fellows of King's were much criticized for their pacifism.

CHAPTER FIVE

EMMANUEL : LEARNING

A VERY IMPORTANT PART OF CAMBRIDGE LIFE consisted of
the various college and university societies. Many of these
were of fundamental importance in our education. The
Cambridge University Science Club elected only senior
students of outstanding ability, mostly in their third year.
The meetings began with the junior member being called
on to "eat the whale," which was a sardine on toast. The
club then turned to the more serious business of a paper
and discussion, usually at a very high level. It was commonly believed that one in three of the members of the
club would ultimately become fellows of the Royal Society.
I never addressed the club, being ill the day of my paper
in my last year. Fortunately, it had been written out and
was read for me. It dealt with the problem of local variation
and subspecies.

Every college had at least one science society of some
sort. In Emmanuel, in those days, there were two, the
Science Club, which elected only scholars, and the Science
Society for the rest. Robin Hill called on me to introduce
me to the Science Club. He was incredibly shy and
unaudible, but I did gather we were going to a meeting

at which he was speaking. When the room was hushed he presented a brilliant but unresolved discussion of the problem of why art appears to involve an element of illusion. At a later meeting he gave, under the title "Cumulonimbus," a beautiful account of his work with the fish-eye lens on thunderstorms. Though I must have attended some dozen meetings, these are the only ones that I can remember.

The Cambridge Natural History Society was an old-established body consisting of both professionals and amateurs, some associated with the university and others with the town. It was a very healthy organization in my day. The most enthusiastic and prominent amateur member from the town was William Farren. D. G. Lillie had in fact drawn a caricature (figure 9) of the society showing three members, my uncle Arthur Shipley, as president, C. G. Lamb, an academic engineer who was deeply interested in flies, particularly from the islands of the Southern Hemisphere, and Farren, then secretary, sitting round a table looking at a small stuffed bird, the first smoking a cigar and the second a self-rolled cigarette. Farren had a taxidermist's shop on Regent's Street in Cambridge, later converted into Farren the Furrier, that I had known from childhood, flattening my nose against the window whenever I could persuade nurse or mother to pass that way. Farren had been an industrious collector of moths and later a very successful bird photographer, and during my undergraduate days he became a great friend. I had got interested in the geographical variation of the ermine or stoat, and he pointed out to me that in East Anglia, where these animals rarely became even partially white in the winter, if extensively whitened individuals were found, they were always females. Later Hamilton reported the same phenomenon in the North American *Mustela*

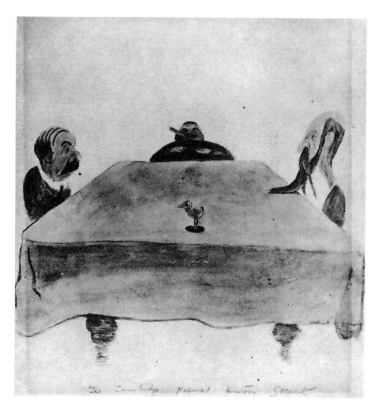

FIGURE 9   D. G. Lillie, *The Cambridge Natural History Society*. The
broad figure presiding is A. E. Shipley; on his right is C. G. Lamb,
and on his left William Farren (by courtesy of the director of the
Zoological Museum of Cambridge University and of Dr. John
Smart).

frenata, in parts of New York State where whitening is inconstant.[1] Quite recently Pamela Parker and I reviewed the matter, concluding that this may be an example of a rather obscure and little-known phenomenon that also occurs in a few insects. In any region of sufficient environmental instability, if there is a gene adapting the species to one set of environmental conditions and its allele adapting the species to the opposite condition, it would be advantageous for the gene to be dominant in one sex and recessive in the other, so that as long as both alleles are present, even if one is quite rare, both will be phenotypically visible in the population. When the environment changes selection can get to work immediately without waiting for rare homozygous recessives to turn up. When Farren showed me the stoat skins in his shop over half a century ago, I had no idea that I was looking at the result of a recondite and minor but quite interesting evolutionary principle.

Farren was president of the society during part of the time I was up and Michael Perkins senior secretary, a post to which I later was elected.

The most celebrated, or perhaps notorious, meeting of the society was one at which Paul Kammerer gave an account of experiments that he believed demonstrated the inheritance of acquired characters. This meeting is a central episode of Arthur Koestler's *The Case of the Midwife Toad*; all my recollections of it were communicated to him.[2]

Kammerer's case was generally supposed to rest most

1. W. J. Hamilton, The Weasels of New York, *Amer. Midl. Natural.* 14:289–344, 1933; G. E. Hutchinson and P. J. Parker, Sexual Dimorphism in the Winter Whitening of the Stoat *Mustela erminea*. To appear in Mammal Notes in *J. Zool.*

2. A. Koestler, *The Case of the Midwife Toad* (New York: Random House, 1971).

critically on his claim that if the midwife toad, *Alytes obstetricans*, is reared under very hot humid conditions, it can be converted into an animal breeding in water like other amphibians rather than one breeding on land. This change involves the development of a ridged horny nuptial pad on the front leg of the male, which though present in most other frogs and toads, is lacking in ordinary fully terrestrial *A. obstetricans*. Kammerer found minute indications of the pad in the second aquatic generation and a clear small pad in the third, which became more pronounced in the fourth and fifth generations reared in water. Much of Kammerer's material disappeared in the First World War, but a fifth generation specimen, showing the pad, survived and was photographed in Vienna for J. H. Quastel in 1922. The pad is clearly ridged. The same specimen was shown in Cambridge at a conversazione of the Natural History Society. I had some organizational duties and did not see it when it was taken out of its jar and placed under a binocular microscope. J. B. S. Haldane did examine it and remarked on the ridges, though just before his death he could not remember the incident. Later what was supposed to be the same specimen was found by Dr. G. K. Noble, a very distinguished student of the Amphibia, to have no true nuptial pad but rather a dark mark apparently produced by the injection of India ink.[3] Six weeks after this discovery Kammerer committed suicide.

I believe with Koestler and so far as I know with everyone who met Kammerer in Cambridge, including his hosts Mr. and Mrs. E. J. Bles, that Kammerer was indeed honest. I do not, however, think that the experiments indicate what he thought they did. Three reviewers

3. G. K. Noble, Kammerer's Alytes, *Nature* (London) 118:209–10, 1926.

of Koestler's book, an anonymous writer in the *Times Literary Supplement*, Stephen Gould in *Science*, and Lorraine Larison in the *American Scientist* all point out, after studying Kammerer's original accounts, that the investigation began with a very large number of specimens, among which there was considerable mortality.[4] There is also evidence of rudimentary pads occurring very rarely in nature. The original stocks may well have been genetically heterogenous, and very intense natural selection could have taken place in Kammerer's vivaria, leading to conventional Darwinian selection for the pad in the course of the four or five generations of the experiments. That this possibility was never mentioned, as far as I can remember, in all the discussion of Kammerer's work that I heard and took part in at the time, is a measure of the extraordinary change in attitude to evolutionary processes that has taken place during the past half century.

The doctoring of a specimen of *Alytes*, which Koestler suggests must have taken place shortly before Noble's visit to Vienna, was believed by Dr. Hans Przibram, director of the Biologische Versuchsanstalt where Kammerer worked, to have been done by someone, possibly with anti-Semitic motives, who wanted to discredit Kammerer. Koestler gives some reason to suspect that the suicide was as much due to the failure of a love affair as to the apparent destruction of a scientific reputation.

4. Individual Paradigms and Population Paradigms (review of Koestler), *Times Literary Supplement* 70:1309–10, 1971. This very interesting review points out that the rise of Mendelian genetics displaced the Galtonian attitude to inheritance in populations treated statistically by an emphasis on individual genomes. The population paradigm did not return strongly till the 1930s.

S. Gould, Zealous Advocates (review of Koestler), *Science* 176:623–25, 1972. This gives in greatest detail the neodarwinian interpretation of Kammerer's work on *Alytes*.

L. L. Larison (review of Koestler), *Amer. Scient.* 60:644, 1972.

shows the wing cases open, disclosing the yellow color of the front part of the body; in the other European species this area is black, so there is no doubt about the identity of the insect. No other really good figure of *N. maculata* or, for that matter, of any other species, appears for about five hundred years.

While we speculated interminably on phylogeny, little theoretical biology of a modern kind existed. Bidder's work came closest to what we now seek in biological theory, but although I treasured it, some years passed before I was able to put it to use. I had been familiar with D'Arcy Thompson's great work "On Growth and Form" since it came out, as the author was a friend of my father's and we had the book at home.[19] Though as a piece of literature I knew it to be important, I did not see how to reconcile what I learned from it with what I knew of genetics and evolutionary biology, till much later in my career.

My own interest in theoretical constructs in biology seems to have arisen in a peculiar way. The issue of vitalism versus mechanism was hotly debated, mainly in terms of the German and American work on experimental embryology. I read avidly and more or less at random in the area without finding anything theoretically satisfying, though Harrison's work on the development of symmetries in amphibian embryos fascinated me, as it still does. The mechanistic argument that ultimate reduction to physics would always be possible seemed to postulate a false simplicity, the vitalist argument that some special property was needed to produce living beings seemed to imply a naive belief that if one could not think of a mechanistic explanation no such explanation exists. Two areas seemed to suggest profitable ways of proceeding.

19. D'A. W. Thompson, *On Growth and Form* (Cambridge: Cambridge University Press, 1917).

One way was what is now called parapsychology, which we then called psychical research. Cambridge had earlier been an important center for this sort of study; the celebrated and partly inexplicable Italian medium Eusapia Palladino had stayed there and was known to have cheated when playing croquet, though perhaps not when in full trance. When I was up, there was little activity of this sort in the university, though the subject was not dead. I had bought William McDougall's *Body and Mind* at the end of 1923 and found it both a systematic and a stimulating account of the psychophysical problem as understood at the time:[20] at a very basic level I suspect we now have mainly gained in humility rather than understanding, in spite of vast progress in our knowledge of the physiology of the brain.

My grandmother died in August 1924 and as she wished to be buried at Culgaith my parents asked me to go north to discuss the matter with the local parson while they were concerned with the necessary arrangements in Cambridge. When I got to Culgaith I learned that the tenants in our house had been sitting in the drawing room on the evening that my grandmother had her fatal stroke, and that two of them believed they had seen an elderly lady in black walk through the room and disappear. As far as I could learn, this had been discussed well before they knew of my grandmother's death. There is, of course, an enormous literature on such phantasms, which are hard to treat statistically, though they are very convincing when one is brought up against a case personally.

McDougall's book had prepared me to take psychical phenomena seriously, and the incident at Culgaith reinforced such an attitude. The few experiments that I have

20. W. McDougall, *Body and Mind, A History and Defense of Animism*, 5th ed. (London: Methuen, 1920).

3 *Limnology*

Astonishing Microcosms

David M. Post and David W. Schindler

The great difficulty of limnology is that many quite disparate things must be done at once.
—G. Evelyn Hutchinson (1963)

Hutchinson discovered limnology relatively late in his development as a biologist; he had already left graduate school and completed a stint as a lecturer when he was offered the opportunity to study the chemical properties of South African vleis—shallow lakes. As he put it, "At last, I had found what I wanted to do."[1] Clearly, the abundant opportunities for comparative natural history combined with a perspective that encouraged integration of physical, chemical, and biological properties of systems was deeply appealing to Hutchinson. With interests ranging from *The Biogeochemistry of Vertebrate Excretion* and phosphorous cycling in small lakes to the ecological interactions and evolutionary consequences of organisms interacting within lakes, Hutchinson was able, like few others of his or later generations, to bridge the great intellectual span of limnology.[2] In his four-volume series *A Treatise on Limnology,* Hutchinson demonstrated his genius for thoroughly absorbing knowledge from all corners of limnology and combining seemingly disparate facts into a more general understanding of the working of lakes. Hutchinson was drawn to limnology, in part, because the lake, as a microcosm, provided a unifying framework for exploring nature from various points of view. In his 1963 essay "The Prospect Before Us" Hutchinson wrote: "We have in lakes units that are satisfactorily unitary from many points of view. The lake as a microcosm was the first important theoretical construct to develop in American limnology. The idea has become a guiding principle of limnologists everywhere and will continue to give intellectual content to our science."[3] Hutchinson excelled in the integration of his and others' empirical observations to produce a new, and often unique, understanding of the working of the natural world. More than anything, his work as a limnologist provided him with the intuition required to synthesize the large quantities of facts he had accumulated. Hutchinson's best-known works are primarily found in this anthology within the chapter covering his contributions to ecological theory, but those papers, with little doubt, would be poorer if not for the empirical perspective Hutchinson derived from decades of research on lakes. For example, Hutchinson's development of the n-dimensional niche was clearly influenced by his studies of aquatic bugs in the family Corixidae (water boatmen).[4] Hutchinson had a broad cross section of experience in lakes from which to draw. He studied mesotrophic and eutrophic lakes in Connecticut, high alpine lakes in Ladakh, North India, and pan lakes in Africa. He also learned techniques from the founding fathers of North American limnology, Edward Birge and Chauncey Juday, while studying lakes in northern Wisconsin. The challenges of working in lakes, where direct observation is difficult but the interacting influences of physical, chemical, and biological processes are obvious, provided an important context in which generalization could emerge.

Here we provide six papers and two excerpts from *A Treatise on Limnology* that represent a small fraction of Hutchinson's research on, and thinking about, lakes. In his paper "Limnological Studies at High Altitudes in Ladak," Hutchinson provided a brief picture of the physical, chemical, and biological limnology of high alpine lakes in India in relation to other high alpine lakes.[5] Although short, this paper highlights Hutchinson's early capacity to approach lakes from multiple perspectives.

In "Chemical Stratification and Lake Morphology," Hutchinson used measurements of alkalinity and lake morphology in Linsley Pond to explore competing hypotheses for the movement of water in the hypolimnion (bottom water) of a stratified lake.[6] His data were consistent with horizontal rather than vertical movement of water within the hypolimnion, suggesting that the hypolimnion was effectively a closed system.

In his 1941 paper "Limnological Studies in Connecticut, IV: The Mechanisms of Intermediary Metabolism in Stratified Lakes," Hutchinson used this observation of horizontal move-

ment and the shape of alkalinity-depth profiles to deduce that the main source of alkalinity to the hypolimnion of a lake was in the sediments.[7] Chemical methods of the day were not sensitive enough to study the mechanisms of alkalinity generation, but they provided pivotal evidence that led to the realization forty years later that lakes acidified by polluted precipitation could at least partially recover as the result of sulfate reduction and denitrification.[8]

By the mid-1940s it was clear that phosphorus played an important role in regulating phytoplankton biomass and community succession. Yet, paradoxically, biologically available phosphorus was often at levels too low to measure during periods of high phytoplankton production late in the summer. By adding radioactive phosphorus to the surface waters of Linsley Pond, Hutchinson and Bowen demonstrated the vertical flux of phosphorus from the epilimnion to the hypolimnion and provided evidence for the rapid turnover of phosphorus in the open water of lakes.[9] Measuring rates of phosphorus cycling in lake waters remains a challenging methodological issue, as turnover rates regulate primary and secondary production.[10] It has also been crucial for our understanding of eutrophication, one of the foremost problems in freshwater ecosystems. The study of eutrophication interested Hutchinson greatly, as evidenced by his historical account of the topic at the start of the 1967 International Symposium on Eutrophication and by his many papers on phosphorus in Linsley Pond.[11]

Two other papers reprinted here are more synthetic and together provide a glimpse into Hutchinson's musings about the inner workings of lakes and lakes as model systems for understanding natural systems. In "The History of a Lake," Hutchinson discussed the substitution of space for time and paleoecological methods for studying the history of lakes.[12] In summarizing several years of research by multiple collaborators at two lakes near Yale, East Twin Lake and Linsley Pond, Hutchinson outlined the value of integrating multiple perspectives to produce fundamentally new understandings of lakes.

"The Lacustrine Microcosm Reconsidered" derives from an address given by Hutchinson at the opening of the Center for Limnology at the University of Wisconsin.[13] In this address, Hutchinson revisited the theme of a lake as a microcosm, first introduced by Forbes, by using observations from lakes to discuss species coexistence, rarity, and niche differentiation among the plankton—themes that permeate many of Hutchinson's most influential papers.[14] At the end of this essay, Hutchinson remarked on the surprising diversity found at all levels of biological organization and its potential in maintaining the "harmony" of natural systems, foreshadowing later research on the role of complexity and biodiversity in maintaining stability.[15]

Finally, excerpts from *A Treatise on Limnology* provide an introduction to Hutchinson's tremendous ability to gather and synthesize facts.[16] The four-volume set of the *Treatise* arguably best represents this ability and remains an important reference for modern limnology. As Gordon Riley noted, Hutchinson believed "that every fact should be useful, but no fact is useful until it is used."[17] This belief was put to test in the *Treatise*. In the first volume, Hutchinson discussed the geological origin and physical and chemical properties of lakes, a truly impressive scope for any one author to cover.[18] In the second volume, Hutchinson began the enormous task of cataloging the diversity of and interactions among the inhabitants of lakes, for in lakes we find representatives of every group of organisms inhabiting the earth.[19] It would take Hutchinson a full three volumes (volumes 2, 3, and 4) and a great portion of his life to cover the biology of lakes.

From volume 1 we provide Hutchinson's opening remarks and his summary of the phosphorus cycle in lakes, which occupied so much of his attention over his career.[20] Volume 2 contains another central area of research for Hutchinson—interactions among the plankton. In volume 2, Hutchinson offered treatments of the hydromechanics of plankton, the nature and distribution of phytoplankton and zooplankton, seasonal (successional) changes in plankton assemblages, horizontal and vertical migration of pelagic organisms, trophic interactions, and cyclic changes in the morphology of zooplankton (cyclomorphosis). While some of the ideas and observations outlined by Hutchinson have not stood the test of time (for example, much of the early work on cyclomorphosis focuses on the effects of temperature, while modern research has clearly identified predation as the central cause of these often striking morphological changes), the enormous catalog of questions spurred by the *Treatise* inspired generations of biologists to explore the wonders of the biology of lakes. Many limnologists still reach for their well-worn copies of the *Treatise* when working on a lecture, digging up background for a grant, or simply searching for inspiration.

136   N A T U R E   JULY 22, 1933

and experiments on this blood are in progress at Cambridge. It is hoped that a full account will be published shortly in "Discovery Reports".

ALEC H. LAURIE.

"Discovery" Investigations,
52 Queen Anne's Chambers,
London, S.W.1.
June 1.

Limnological Studies at High Altitudes in Ladak

IN his studies of the Alpine lakes of Central Europe, Pesta[1] has characterised the smaller lakes at the highest altitudes as pan-oligotrophic. Such lakes, despite their small size, show a very high oxygen content in the hypolimnion and have a poorly developed bottom fauna and plankton. Recently, during the work of the Yale North India Expedition under the leadership of Dr. H. de Terra, it has been possible to make limnological studies on a number of lakes in Ladak lying at altitudes between 4,267 m. and 5,274 m. In waters more than 10 m. deep at

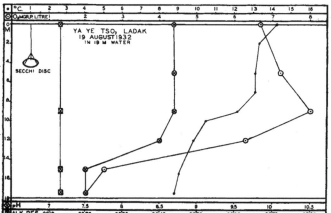

FIG. 1. Thermal and chemical stratification in Yaye Tso, Ladak.

such altitudes the surface temperature in August and the beginning of September lay between 11·73° C. and 14·19° C., that is within the ranges recorded for the pan-oligotrophic type by Pesta.

Two types of basin are to be found in the region studied. The first type comprises the tectonic depressions containing the Panggong Tso, Tso Moriri and Pangur Tso. The first two of these, though lacking outlets and therefore rich in accumulated electrolytes, are comparable to normal oligotrophic lakes. Pangur Tso is very shallow, is not thermally stratified and appears to be entirely carpeted with flowering plants.

The second class of basin is of glacial origin and includes a number of small lakes in kars and kettle holes. Of these, Yaye Tso, a freshwater lake lying at 4,686 m. and draining into the Indus just south of Mahe, was most intensively studied. It is clear from Fig. 1 that this lake has a quite well-developed eutrophic oxygen curve. Calculations based on a bathymetric map made from two lines of soundings indicate that on August 11, 1932, the lake had an actual oxygen deficit of 1·1 mgm. per square cm. of hypolimnion surface. This figure lies within the

limits of those derived for some of the lakes of north-eastern Wisconsin from the data of Juday and Birge.[2]

In two other small lakes, Khsagar Tso and Mitpal Tso, considerable oxygen deficiencies were also observed in the hypolimnion. Since the water is brackish and no bathymetric maps are available for these lakes, it is not possible to use the chemical data obtained so completely as in the case of Yaye Tso.

A fourth glacial lake, Ororotse Tso, lies at 5,274 m., and is apparently the highest lake yet studied limnologically. When visited on July 11, 1932, Ororotse Tso was still covered with ice save at the extreme edge and it seems doubtful if it ever becomes entirely clear. The water was probably being disturbed by convection currents from the margin, but a bottom sample from 14 m. showed a slight oxygen deficit. The lake contains, moreover, considerable amounts of plankton and has a quantitatively rich bottom fauna, approximately 8,900 animal organisms, all chironomid larvæ, occurring per square metre in 14 m. of water. This number is considerably greater than the estimate for the Yaye Tso, in which lake there were approximately 4,800 organisms per square metre, mostly chironomid larvæ and tubificid worms. Such figures may be compared with that for the oligotrophic Tso Moriri, in which but 450 organisms per square metre, all chironomid larvæ, were found at a depth of 50 m.

Since the area studied is largely semi-desert, the Ladak lakes might be expected to resemble those of other semi-arid regions. This expectation is to some extent fulfilled in that lakes in such regions frequently lack outlets and accumulate large quantities of electrolytes. The Ladak lakes, however, differ markedly from those of other semi-arid regions[3] in that they show no trace of a paratrophic condition. It appears, therefore, that the smaller high altitude lakes of Ladak have more in common with the less extreme eutrophic or mesotrophic lakes of the lowlands of Europe and North America than they have with the pan-oligotrophic lakes of the Alps or with the lakes of less mountainous semi-arid regions.

I have great pleasure in expressing my thanks to Dr. de Terra for continual opportunities for prosecuting these studies and for his never-failing help and encouragement in the field. A full account of the limnological work of the expedition will be published as soon as taxonomic studies have been completed by various specialists.

G. EVELYN HUTCHINSON
(Biologist, Yale North Indian
Expedition).

Osborn Zoological Laboratory.
Yale University,
New Haven, Conn.
May 19.

[1] *Die Binnengewässer*, **8**. "Der Hochgebirgsee der Alpen", Stuttgart, 1929.
[2] *Trans. Wisc. Acad. Sci. Art. Lett.*, **27** ; 1932.
[3] Decksbach, *Verh. Int. Ver. theor. angew. Limnol.*, **2** ; 1924. Stanković, *ibid.*, **5** ; 1931. Hutchinson, Pickford and Schuurman, *Arch. Hydrobiol.*, **24** ; 1932.

VOL. 24, 1938   *GEOLOGY: G. E. HUTCHINSON*   63

This study demonstrated that colchicine is effective in the production of cytogenetically changed cells. The process affects the mitotic divisions; hence, it is a study of independent cells as structural and functional units. These units divide mitotically at a given time and do so independently of the activity of the adjacent cells. The writer was not concerned with a study of entire tissues influenced by colchicine treatment. The exact rôle played by colchicine is essentially the inhibition of the mitotic spindle which prevents separation of the daughter nuclei, and cell plate formation, with the subsequent division into two cells. The failure of the reagent to interfere with the process of chromosome formation by longitudinal equational divisions, shows a specificity of a high degree for inhibition of certain phases of cell division and the apparent promotion of other phases of the mitotic process.

[1] Allen, E., *Anat. Rec. Abst.*, **67**, 49 (1936).

[2] Mr. E. L. Lahr gave the writer some helpful suggestions in the initiation of this investigation.

[3] Kostoff, D., *Genetica*, **12**, 33–118 (1930).

[4] Eigsti, O. J., *Manuscript Thesis*, University of Illinois (1935).

[5] Eigsti, O. J., *Bot. Gaz.*, **98**, 363–369 (1936).

[6] Nemec, B., *Jahr. Wis. Bot.*, **39**, 645–730 (1904).

[7] Winkler, H., *Zeits. Bot.*, **8**, 417–531 (1916).

[8] Jorgenson, C. A., *Jour. Genet.*, **19**, 133–211 (1928).

[9] Randolph, L. F., these PROCEEDINGS, **18**, 222–229 (1932).

[10] Bakeslee, A. F., *Compt. Rend.*, **205**, 476 (1937).

[11] Blakeslee, A. F., *Jour. Hered.*, **28**, 392–411 (1937).

[12] Nebel, B. R., *Biol. Bull.*, **73**, 351–352 (1937).

CHEMICAL STRATIFICATION AND LAKE MORPHOLOGY

By G. Evelyn Hutchinson

OSBORN ZOÖLOGICAL LABORATORY, YALE UNIVERSITY

Communicated January 11, 1938

Data collected during an extended investigation of Linsley Pond, a small eutropic lake (area 94,400 m.[2], max. depth 14.8 m., mean depth 6.7 m.) near North Branford, Connecticut, throw considerable light on the nature of the water-movements in the hypolimnia of thermally stratified lakes. That such movements occur is clear from the rise in concentration of substances, that can only have been derived from the bottom mud, at distances from the latter far exceeding those over which molecular diffusion could be effective. The nature of such water movements has been a matter of discussion, three main hypotheses having been advanced in recent years. Birge,[1] Thienemann[2] and more explicitly Grote,[3] have regarded

88   *Limnology*

64   *GEOLOGY: G. E. HUTCHINSON*   Proc. N. A. S.

wind-generated turbulence as the most important agency in the transport of heat and dissolved material throughout the lake. This hypothesis implies that the hypolimnion is not really a closed system and that considerable vertical tubulent diffusion occurs. McEwen,[4] while admitting the fundamental importance of turbulence, believes that at least in the upper hypolimnion and in the epilimnion, a system of vertical convection currents generated by surface cooling, is of great importance. Alsterberg,[5] on the other hand, regards the hypolimnion as a closed system, in which water movements are confined to thin horizontal laminae, such movements having a negligible vertical component. This hypothesis appears to be the only one consistent with the remarkable optical microstratification discovered by Whitney.[6] It is perhaps unfortunate that in previous investigations most attention has been focused on temperature, in part controlled by direct radiation, and on oxygen, the most reactive of all dissolved substances in a lake, and in the present case known to be primarily controlled (Riley *unpublished*) by consumption and production in the free water. Evidence based on the distribution of alkalinity (bicarbonate ion, determined by titration with methyl orange as indicator) is given below in support of Alsterberg's hypothesis, though the results obtained on the oxygen deficit in Linsley Pond do not permit us to follow this author in all his conclusions.

Exclusion of Vertical Turbulent Transport.—In figure 1, the solid lines represent the variation, with depth, of alkalinity, the curves being based on the means of determinations made weekly, at one meter intervals, during three successive five-week periods, throughout the summer of 1937. It will be observed that during the period of investigation alkalinity tends to rise, but that the rise is much greater in the lower hypolimnion than in the epilimnion. Mineral analyses made at the end of September indicate that in the epilimnion sufficient calcium and magnesium exists to balance the bicarbonate, but that the concentration of these substances hardly varies with depth. In the hypolimnion, the excess bicarbonate is apparently balanced by considerable quantities of ammonia and iron, the latter presumably in the ferrous condition. This iron can only have come from the bottom and the large amounts of ammonia appear to indicate a like origin for the volatile alkali.

At any depth (y), the rate of change $\dfrac{\partial \theta}{\partial t}$ in any property (θ) with time due to turbulent diffusion is given by

$$\frac{\partial \theta}{\partial t} = \mu^2 \frac{\partial^2 \theta}{\partial y^2},$$

where (μ^2) is a virtual diffusion coefficient, the coefficient of turbulence, which is always positive. If the second derivative with respect to depth be negative, the first derivative with respect to time must be negative also.

If the three curves be compared, it appears that they show a marked similarity in form, showing points of inflection at the same depths. (Cf. also figure 2.) These points of inflection are such that in the lower water,

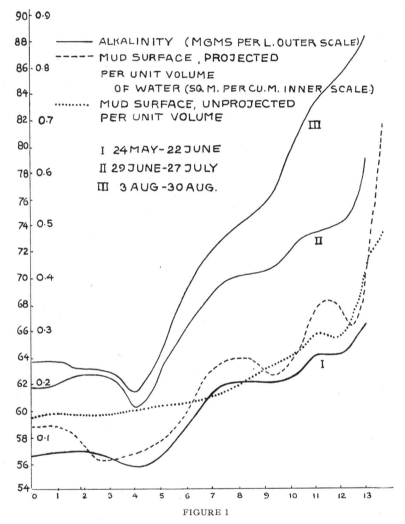

FIGURE 1

Solid lines, variation of alkalinity (HCO₃) with depth (mean values over successive five week periods). Broken line, ratio of projection of mud surface to volume. Dotted line, ratio of mud surface to volume.

the second derivative with respect to depth is persistently negative at 8 m. and 11 m. If the rise in alkalinity were due to turbulent movement bringing ferrous and ammonium bicarbonate up from a constantly re-

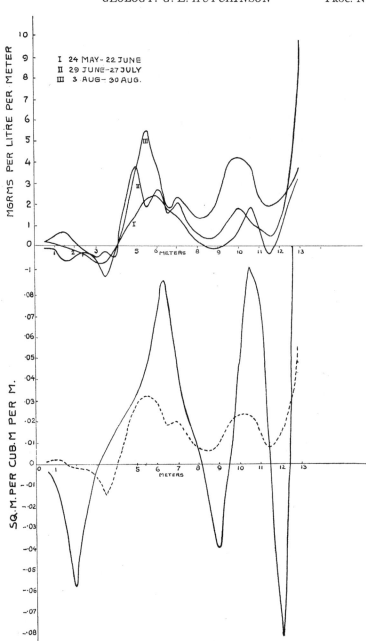

FIGURE 2

Upper curves, first derivative of alkalinity curves in figure 1. Lower curve, solid, first derivative of projected mud surface curve; broken, mean of the three upper curves.

Vol. 24, 1938   *GEOLOGY: G. E. HUTCHINSON*   67

generated and highly concentrated microzone on the mud, it is clear that the persistence of sections of the curve with a negative second derivative would be quite inconsistent with the continued accumulation of bicarbonate at all levels of the hypolimnion as a whole. A similar situation is apparent in the individual curves relating alkalinity to depth during periods of rapid rise in alkalinity. Such curves are less regular than the mean curves of figure 1, doubtless owing to uneven horizontal distribution at any given time, but regions with a negative second derivative can always be found, though not invariably at the same depths as in the smoothed curves. The unaveraged data, therefore, likewise give ample evidence of the occurence of considerable rises in alkalinity in layers of water where the turbulence hypothesis would, as in the case of the average curves, indicate that the alkalinity should be decreasing.

Particular significance is to be attached to the occurrence of a zone with a negative second derivative at 11 m., because, if McEwen's theory be accepted, this region should show the effects of almost pure turbulence uncomplicated by the type of convection current that this author has postulated. While it is hoped in the future to examine the implications of this very complex theory in greater detail, the correlation indicated below strongly suggests that it is inapplicable in the present case.

Morphological Concomitants of Horizontal Streaming.—As Alsterberg himself pointed out, if the hypolimnion be regarded as closed, and if its water movements are practically exclusively horizontal, a relationship should exist between the chemical characters of any layer of free water, and the amount of mud surface to which the edges of the layer are exposed. In attempting such a correlation it must be remembered that a steeply sloping lake bottom will probably neither receive nor retain as much mud as one that is more horizontal. A practical investigation of this problem would be extremely difficult, but as a first approximation it has been assumed that the chemically active mud is proportional not to the actual area of the lake bottom, but to this area projected on to a horizontal plane.

The broken line in figure 1 represents a curve drawn by plotting at 0.5 m., 1.5 m. and at succeeding meter intervals, the ratio of the area between successive one meter contours on a bathymetric map of the lake, to the volume enclosed between the planes represented by these contours. A comparison of this curve with the alkalinity curves of the same figure indicates that a relationship of the type demanded by Alsterberg's horizontal current hypotheses does exist. This is further brought out when the derived curves in figure 2 are examined. Owing to the large amount of steep lake bottom, particularly on the western side of the basin, the relationship is obscured when the ratio of contour length to area is taken as the measure of mud to water at any depth (dotted line in figure 1). The existence of any such relationship, even though it involve an empirical but quite objective correction, speaks very strongly in favor of the hypothesis of horizontal movement, and this, taken in conjunction with the

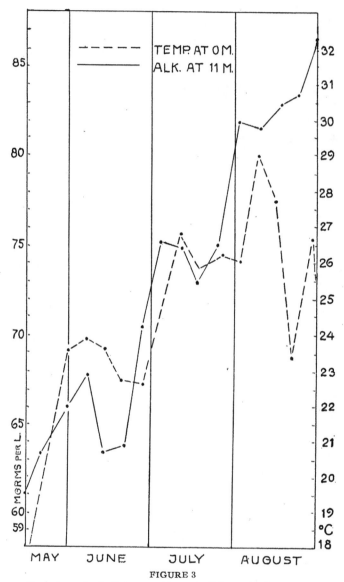

FIGURE 3

Variation of alkalinity at 11 m. (solid), and of temperature at 0 m. (broken), with time.

evidence given above, makes any other hypothesis of hypolimnetic water movements most improbable.

Temporal Variation in Alkalinity.—Conclusions based on comparison of the details of curves showing the variation of alkalinity with time at any given depth, are, for various technical and other reasons, less satisfactory than those based on the depth distribution curves for any given day. Although the significance of certain features in such curves is doubtful, the solid line in figure 3, representing the alkalinity at 11 m. from the end of May to the end of August, and typical of the events of all depths, indicates marked discontinuities in the rise of alkalinity during this period. The cause of the minima before the sudden rises is at present obscure, the two large rises (22 June–6 July and 27 July–3 August) certainly lie outside experimental error, and since the high values reached are in each case maintained almost unchanged for a week, they must be regarded as real and general throughout the water layer. These two large rises, and a third discontinuous rise beginning on the 24 August, appear to be initiated by temperature minima *at the surface* (broken line). Examination of the density curves at the time of such minima indicated that the stability of the top two meters of the lake is reduced at such times and rises more suddenly between two and three meters than at times of temperature maxima. This suggests that though the rise in alkalinity cannot be due to vertical turbulent diffusion, the best conditions for such a rise are provided by the existence of a well defined and freely circulating epilimnion of measurable thickness below which the stability rises sharply. Preliminary experiments with small scale models indicate that horizontal streaming of the type postulated can be produced in a lower, more dense layer when a less dense layer of water lies upon it, and when the boundary between the two is made to oscillate by an air current which causes turbulent motion in the upper layer, but no mixing of the upper, less dense water with the denser water at the bottom of the tank. It is hoped to be able to investigate the nature of such currents in models under more favorable conditions in the near future.

Intensive work in 1937 was made possible by a grant from the BACHE FUND, of the NATIONAL ACADEMY OF SCIENCES, here gratefully acknowledged. My thanks are also due to Professor A. E. Parr, for the loan of apparatus, to Mr. H. J. Turner and Miss Anne Wollack for the care they have taken over analyses entrusted to them, to Dr. G. A. Riley for unpublished morphometric data and to Dr. Riley and Mr. E. S. Deevey for help in the field.

[1] Birge, E. A., *Trans. Wisc. Acad. Sci. Arts. Lett.*, **18**, 341 (1916).

[2] Thienemann, A., Die Binnengewässer, **4**, (1928).

[3] Grote, A., Die Binnengewässer, **14**, *Arch. Hydrobiol.*, **29** (3), 410 (1935).

[4] McEwen, G. F., *Bull. Scripps. Instn. Oceanogr. Tech.*, **2**, 197 (1929).

[5] Alsterberg, G., *Bot. Notiser*, **255** (1927).

[6] Whitney, L. V., *Science*, **85**, 224 (1937).

LIMNOLOGICAL STUDIES IN CONNECTICUT

IV. THE MECHANISMS OF INTERMEDIARY METABOLISM IN STRATIFIED LAKES

G. EVELYN HUTCHINSON

Osborn Zoological Laboratory
Yale University

TABLE OF CONTENTS

THE MECHANISMS OF INTERMEDIARY METABOLISM
IN STRATIFIED LAKES

INTRODUCTION

THE FUNDAMENTAL problem to be considered in the present paper is raised by the work of Juday and Birge (1931), who, unable to find clear evidence in a number of lakes of the depletion of phosphate in the epilimnion during summer stagnation, write "in some of the . . . lakes the quantity of soluble phosphate was maintained or was even increased somewhat during the growing season in spite of the fact that these bodies of water sustain a relatively large growth of phytoplankton." This observation suggests that our knowledge of the detailed movements of nutrients and other substances dissolved in the water of lakes is inadequate. A considerable body of information is available as to the total quantity of various important substances present in lakes. Observations on the oxygen deficit and various studies of the photosynthetic and katabolic activity of the plankton have given some information, often, however, of a very relative nature, as to the total metabolism of lakes. The intermediary aspect of metabolism, to continue the analogy with the individual organism, is extremely little known.

The method to be adopted in the present attempt to elucidate this problem may be briefly outlined in the following general terms. Before any biological activities can be considered, it is necessary to have an understanding of the purely physical movements of substances in a thermally stratified lake. The most easily treated type of disturbance causing transport of material is vertical turbulence, whereby conservative properties (insofar as they do not affect density) are transmitted through the water-mass according to laws of the same forms as those expressing the conduction of heat through a solid, but in which the coefficient of conduction is replaced by a virtual diffusion coefficient, the coefficient of turbulence or eddy-conductivity coefficient. Such vertical turbulence is the result of horizontal currents, generated by the wind. It is first necessary to examine the competence of turbulence to act as the agent producing observed effects, and insofar as this competence can be established, to estimate the value of the virtual diffusion coefficient involved. In certain cases it becomes apparent that vertical turbulent transport is inadequate to explain observed changes in properties. Knowing the distribution of the property in question, its rate of change, and the magnitude of the coefficient of turbulence, it is possible to estimate what part of the observed change is due to vertical turbulent mixing and what part to other causes. Such other causes are primarily the horizontal movements of water from the mud-water interface towards the middle of the lake, movements which, by generating vertical turbulence, are also indirectly responsible for the vertical exchanges. This procedure is the reverse of that usually adopted in oceanography but is probably a far more convenient one for the study of small lakes than one founded primarily on direct observation of horizontal current systems. The first part of the paper, in which turbulence coefficients are evaluated, is based on a study of vertical series of temperatures. The second part in which residuals due to non-turbulent transport are obtained, is based primarily on a study of the bicarbonate content of a single lake, the concentration of this ion being regarded as conservative in the free water in the lake in question. In the third part the theory of water-movements evolved in the first two parts is applied qualitatively to elucidate the behavior of the phosphorus concentration, a highly non-conservative property of great biological interest. A number of related topics, of minor importance to the development of the general theory, but of considerable interest in themselves, are discussed in their appropriate places throughout the paper.

My best thanks are due to the Bache Fund of the National Academy of Sciences for generous financial support of my investigation, and to the Sheffield Fund of Yale University for a grant for special apparatus. I am also grateful to Dr. E. A. Birge and Dr. C. Juday for a table of weekly mean temperatures of Lake Mendota, which has been of the greatest value in the study of turbulence, to my former students and collaborators, Dr. G. A. Riley and Dr. E. S. Deevey, Jr., for invaluable help throughout the investigation, to Miss Anne Wollack and Mr. H. J. Turner who have expended much care in the analyses entrusted to them, and to the property holders on the shores of Linsley Pond, particularly Mr. C. S. Sargent and Mr. Alexander Day, who have taken an encouraging interest in the investigation. I should also like to express my debt to Dr. G. F. McEwen's paper "A Mathematical Theory of the Vertical Distribution of Temperature and Salinity in Water" (1929). The present contribution could never have been made without the stimulus supplied by McEwen's ingenious treatment, even though the conclusions that I have reached differ greatly from his. I am well aware that since my own studies have led me to theorize somewhat extensively, time may show that my speculations are unfounded. I can only hope that they will prove as great an impulse to some future investigator as McEwen's theory has been to me.

TEMPERATURE AS AN INDICATOR OF TURBULENCE

THE DATA

1.1.1. A series of fourteen sets of temperatures taken in the morning with a reversing thermometer in the center of Linsley Pond (Fig. 1), at one meter intervals, from the surface to a depth of 14 m., every week from the beginning of June to the end of August, 1937, and a similar series of eight

Ecological Monographs
Vol. 11, No. 1

Fig. 1. Linsley Pond, from the northern shore.

sets taken in the center of Lake Quassapaug during July and August, 1938, constitute the new data to be analyzed. Of these, the Linsley series, owing to its greater completeness, is by far the more important. A description of both localities has been given by Riley (1939) in an earlier paper. Equally valuable in the initial stages of the investigation is the table of mean weekly temperatures in Lake Mendota prepared by Dr. E. A. Birge and used by McEwen (1929) in his analysis of the heating mechanism of lakes.

1.1.2. In all series of single temperature readings, irregular variations are observed, which are undoubtedly primarily due to the piling up of the epilimnion down wind and the consequent distortion of the isotherms. Such irregularities, while probably minimal in series from properly chosen stations, cannot be entirely avoided, and vitiate any attempt to consider the events relating to a single date. Accordingly, in the analysis of the Connecticut data, it has been found necessary to use only mean values for each depth, the mean value of the temperature, $\bar{\theta}$, at any depth z, being taken as the independent variable to which is related the mean rate of change of temperature with respect to time at that depth, $\left(\dfrac{\partial \theta}{\partial t}\right)$.

To obtain the latter quantity, the data relating temperature to time have been fitted to a straight line, the slope of which is taken to be the rate required.

This does not imply that the rate of heating is constant at any depth, but merely that such a procedure is the simplest and probably only practicable method of obtaining a mean rate of change. In treating the data from the Connecticut lakes, the correlation coefficient of temperature and time was determined as a measure of the significance of the relationship observed. In one or two cases, the correlation is low enough to indicate that accidental disturbances are playing so great a part in determining the observed temperatures that at such depths the rate of change calculated from the regression equation is of little significance. In the case of the Mendota data, the rise in temperature throughout the heating period is, owing to the very large number of readings averaged in a single entry, very regular, and treatment relating to single weekly sets is possible in some cases, but there are irregularities in others that make valid conclusions difficult to draw, and for this reason and in order to compare the results with those from the Connecticut lakes, it has seemed advisable to treat the Mendota data from the middle of June to the middle of August in the same way as the Linsley and Quassapaug data were treated, except that calculations of the correlation coefficient were deemed unnecessary.

To avoid unnecessary zeros and to permit easy comparison with McEwen's results, the unit of depth is taken as one meter, of time as one month (considered as consisting of thirty days).

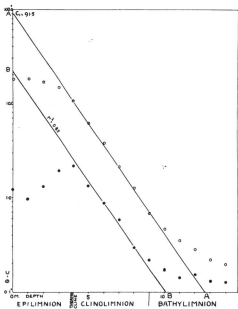

FIG. 2. Linsley Pond. Analysis of thermal conditions by procedure 1; mean data June-August, 1937. A-A', line of $(\theta - C) = C_1 e^{-az}$, $C = 6.82$, $C_1 = 91.5$, $a = 0.533$; observed values of temperature less 6.822 enclosed by open circles. B-B', lines of $\dfrac{\partial \theta}{\partial t} = \mu^2 C_1 a^2 e^{-az}$, where $\mu^2 = 0.85$; observed values of the rate of change of temperature with time entered as solid circles.

OBSERVED VARIATION OF $\dfrac{\partial \theta}{\partial t}$ WITH Z

1.2. The solid circles in Figs. 2-7 indicate the observed values of the rate of change of temperature with respect to time, plotted, following McEwen, on the logarithmic axis of semilogarithmic paper, against depth. It will be immediately apparent that three regions can be distinguished in the lake differing in the nature of the variation of $\dfrac{\partial \theta}{\partial t}$ with depth.

(a) In the upper part of the lake, 0–4 m., in Linsley Pond, 0–6 m., in Lake Quassapaug, 0–9 m., in Lake Mendota, $\dfrac{\partial \theta}{\partial t}$ is relatively high and either may fall slightly with increasing depth, or, if late summer values are included in the calculation, may rise somewhat with depth. The region corresponds to the *epilimnion*.

(b) In the middle part of the lake, 4-9 m., in Linsley, 8–14 m., in Quassapaug, and 9–16 m., in Mendota, the value of $\dfrac{\partial \theta}{\partial t}$ falls exponentially with increasing depth, giving a distribution of points tending to lie on a straight line, at least in Mendota and Linsley, though more irregular in the case of

Quassapaug (vide 1.12). This region is here referred to as the *clinolimnion*, and is limited on its upper surface by the thermocline, treated as a plane; much of the thermocline region, as usually conceived, may be included in the clinolimnion.

(c) In the lower part of the lake, below 9 m., in Linsley, below 14 m. in Quassapaug, and below 16 m. in Mendota, the values of $\dfrac{\partial \theta}{\partial t}$ are considerably higher than would be expected from extrapolation of the exponential fall in the clinolimnion and tend to become approximately constant at a value little if any lower than the lowest value at the bottom of the clinolimnion. This region is here termed the *bathylimnion*.[1] The hypolimnion is regarded in the present paper as comprising the clinolimnion and bathylimnion; the new terms are intended to be used only when problems of heating, turbulence and stability are under discussion.

DETERMINATION OF THE CLINOLIMNETIC COEFFICIENT
OF TURBULENCE

1.3.1. Considering first the clinolimnion and assuming that the transmission of heat through this region is entirely by vertical turbulent mixing, and that density may be taken as constant and unity (Schmidt, 1925),

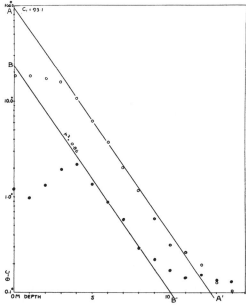

FIG. 3. Linsley Pond. Analysis of thermal conditions by procedure 2; mean data June-August, 1937. $C = 6.91$, $C_1 = 93.1$, $a = 0.540$, $\mu^2 = 0.86$. Otherwise as in Figure 2.

[1] It is not a permanent layer and must not be confused with the monimolimnion of Findenegg, which, however, may well be persistent development of the bathylimnion.

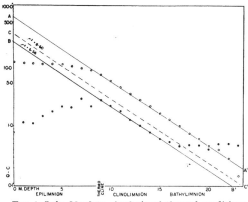

FIG. 4. Lake Mendota. Analysis of thermal conditions by procedure 2; mean data for the period from the third week in June to the second week in August. $C = 11.91$, $C_1 = 63.95$, $a = 0.246$, $\mu^2 = 6.55$. The broken line C-C' represents the calculated value for the rate of change in the hypolimnion if McEwen's procedure, using all points below 12 m., in the calculation of the coefficient of turbulence ($\mu^2 = 8.60$), is employed. Note that the observed values of $\dfrac{\partial\theta}{\partial t}$ are practically constant in the bathylimnion and show no tendency to fall on a line parallel with A-A' below 16 m.

$$\frac{\partial\theta}{\partial t} = \mu^2 \frac{\partial^2\theta}{\partial z^2} \quad \ldots\ldots\ldots\ldots (1)$$

where μ^2 is the coefficient of turbulence, here considered as constant, having the dimensions of $\dfrac{L^2}{T}$, i.e. (following McEwen), square meters per month. In oceanography the symbol A, in C.G.S. units, that is, cm^2 per second, is used; to permit easy comparison with McEwen's results and to avoid unnecessary zeros the former notation is more convenient throughout the greater part of the present section, but in paragraph 1.8.2. it is necessary to use C.G.S. units; both notations therefore have been used in tabulating results ($A = \mu^2 \times 3.858 \times 10^{-3}$). If the rate of change of temperature with time at a series of depths and the variation of temperature with depth in the same region be known, it is theoretically possible to obtain the coefficient by obtaining the second derivative directly. This, however, involves two mechanical differentiations of the temperature curve, which are undesirable, not merely because of the errors introduced in the operations, but because any errors in the original data are apt to produce inflection points in the temperature curve. Such inflection points cause artificial maxima and minima in the curve of the first derivative and changes of sign in the second. It is therefore preferable to use a more indirect treatment, in which the points of the temperature curve are fitted by a function, the second derivative of which is easily obtained. The fit of the observed points, however good, does not necessarily give any

information as to the mechanism of heating in the lake. It merely provides the best available method of obtaining a second derivative and is to be regarded as a quasi-fit (Riley 1939a).

1.3.2. McEwen has pointed out that it is possible to fit the hypolimnetic temperatures of lakes to an expression of the form

$$(\theta - C) = C_1 e^{-az} \quad \ldots\ldots\ldots\ldots (2)$$

where C, C_1 and a are constants to be determined empirically.

From (2)
$$\frac{\partial^2\theta}{\partial z^2} = C_1 a^2 e^{-az} \quad \ldots\ldots\ldots (3)$$

hence
$$\mu^2 = \frac{\partial\theta}{\partial t} \cdot \frac{1}{C_1 a^2 e^{-az}} \quad \ldots\ldots\ldots (4)$$

The problem, therefore, is the determination of the constants in equation (2). According to the findings of the present investigation, this should only be permissible in the clinolimnion (procedure 1), but since the divergence from the fit in the bathylimnion is small and since the addition of more values to the rather meagre number available in the clinolimnion of the lakes under discussion appears to add greatly to the accuracy with which the fitting can be performed, all hypolimnetic values of θ, except the lowest, may be used in the process (procedure 2). The two procedures give essentially the same mean results in the clinolimnion; in working out an example in 1.3.4. procedure 1 has been used for the sake of brevity, but unless otherwise stated all values of the coefficient used in the discussion are those obtained by procedure 2.

1.3.3. Subtract the temperature at each depth, z_{n+1}, from that at unit depth above, z_n. Call the difference, considered as positive, $\Delta\theta$. For the permissible depths (procedure 1) or throughout the whole hypolimnion (procedure 2), construct a table:

z θ $\Delta\theta$ $\theta.\Delta\theta$ $(\Delta\theta)^2$

and add each column. It is now possible, by the least square method, to fit the observed values of θ and $\Delta\theta$ to a straight line of the form

$$\theta = C + b\,\Delta\theta \quad \ldots\ldots\ldots\ldots (5)$$

From equation (2), since for any value
$$\Delta\theta_n = C_1\,[e^{-na} - e^{-(n+1)a}]$$
$$= C_1\,e^{-na}\,(1 - e^{-a})$$
$$= (\theta_n - C)\,(1 - e^{-a})$$
$$\theta = C + \frac{\Delta\theta}{1 - e^{-a}} \quad \ldots\ldots\ldots (6)$$

This is of the same form as (5); C is the same in both equations and
$$e^{-a} = 1 - \frac{1}{b} \quad \ldots\ldots\ldots (7)$$

Having obtained C and a in this manner, the best value of C_1 may be obtained by summing the values of $(\theta - C)$ and dividing by the sum of the values of e^{-az}.

$$C_1 = \frac{\Sigma(\theta - C)}{\Sigma e^{-az}} \quad \ldots\ldots\ldots (8)$$

1.3.4.   EXAMPLE (PROCEDURE 1)

Linsley Pond. June 1-August 31, 1937

z	$\bar{\theta}$	$\Delta\bar{\theta}$	$\bar{\theta}\Delta\bar{\theta}$	$(\Delta\bar{\theta})^2$
4	17.683	4.512	79.78	20.358
5	13.171	2.496	32.87	6.230
6	10.675	1.699	18.14	2.887
7	8.984	0.884	7.94	0.781
8	8.100	0.591	4.79	0.349
9	7.509	0.223	1.67	0.050
	$=: 66.122$	$= 10.405$	$= 145.20$	$= 30.655$

Using the usual formulae for fitting by least squares:

$$p.C + b. \Sigma\Delta\bar{\theta} = \Sigma\bar{\theta}$$
$$\Sigma\Delta\bar{\theta}.C + b. \Sigma(\Delta\bar{\theta})^2 = \Sigma\bar{\theta}.\Delta\bar{\theta}$$

where p is the number of observations.

$$6\,C + 10.405\,b = 66.122$$
$$10.405\,C + 30.655\,b = 145.20$$

whence C = 6.82 and b = 2.42.

From (7) $e^{-a} = 1 - \dfrac{1}{2.42} = 0.587$

From Hayashi (1921)

$$a = 0.533$$

From (8) with observations at six levels

$$C_1 = \cfrac{\overset{9}{\underset{4}{\Sigma}}\ \bar{\theta}\ -6 \times 6.82}{\overset{9}{\underset{4}{\Sigma}}e^{-az}}$$

From Hayashi, where a 0.533.

	az	e^{-az}
4	2.12	0.119
5	2.67	0.070
6	3.20	0.041
7	3.73	0.024
8	4.26	0.014
9	4.79	0.008
		0.276

$$C_1 = \frac{66.12 - 6 \times 6.82}{0.276}$$
$$= 91.5$$

Knowing the constants C_1 and a, the coefficient of turbulence can now be calculated from equation (4)

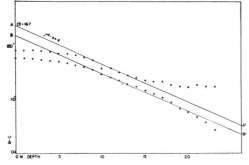

FIG. 5. Lake Mendota. Analysis of thermal conditions for the mean first week in June by procedure 2. C = 10.83, C_1 = 15.85, a = 0.163, μ^2 = 54.8.

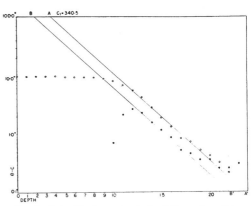

FIG. 6. Lake Mendota. Analysis of thermal conditions for the mean fourth week in August by procedure 2. C = 12.49, C_1 = 340.5, a = 0.345, μ^2 = 4.86.

for all clinolimnetic levels, the values obtained being set out in Table 1. In the same table values for the same region are set out using procedure 2. The results of the calculations by procedure 2 for Lake Mendota (mean) are set out in Table 2. Mean clinolimnetic values for the different weeks of the Mendota chart[2] calculated by procedure 2, are given in Table 3. The results obtained by treatment of the inadequate data from Lake Quassapaug are discussed in 1.12; they raise certain special problems of interest but are too irregular to be of much significance in establishing general principles.

[2] Values of $\dfrac{\partial\theta}{\partial t}$ are obtained by the method of moments used by McEwen.

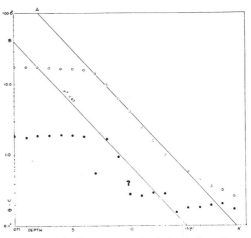

FIG. 7. Lake Quassapaug. Analysis of thermal conditions by procedure 2; mean data July-August, 1938. C = 6.75, C_1 = 232, a = 0.404, μ^2 = 1.07. The ten meter value for $\dfrac{\partial\theta}{\partial t}$ has been rejected in calculating the coefficient of turbulence as the correlation of temperature with time at that level is insignificant (r = 0.418); this point is marked with a query (?).

In both Linsley and Quassapaug the values of A indicate that the observed coefficient is only about three times the coefficient of molecular heat transport (0.12×10^{-2} in C.G.S. units, Grote 1934); the turbulence coefficients in these lakes therefore are approximations, including effects due to molecular conduction.

TABLE 1. Analysis of thermal condition of the clinolimnion of Linsley Pond. (Mean values June-August, 1937.)

Depth m.	$\bar{\theta}$ °C	$\dfrac{\partial \theta}{\partial t}$ °C per month	μ^2, m^2 per month Proc. 1 C = 6.82 C$_1$ =91.5 a = 0.533	μ^2, m^2 per month Proc. 2 C = 6.91 C$_1$ =93.1 a = 0.540	Stability $\dfrac{d\sigma_{\bar{\theta}}}{dz}$ grms per cm^3 per cm.	K = A.$\dfrac{d\sigma_{\bar{\theta}}}{dz}$
4	17.683	2.205	0.72	0.71	7.7×10^{-6}	2.55×10^{-8}
5	13.171	1.367	0.76	0.75	3.4×10^{-6}	1.13×10^{-8}
6	10.675	0.884	0.83	0.83	2.2×10^{-6}	0.73×10^{-8}
7	8.984	0.595	0.95	0.96	0.84×10^{-6}	0.28×10^{-8}
8	8.100	0.297	0.81	0.82	0.32×10^{-6}	0.11×10^{-8}
9	7.509	0.222	1.02	1.09	0.24×10^{-6}	0.08×10^{-8}

Mean μ^2 = 0.85, 0.86
Mean A = 0.86 x 0.3852 x 10^{-2} = 0.331 x 10^{-2}

TABLE 2. Analysis of thermal conditions in the clinolimnion of Lake Mendota (mean data for mid-June to mid-August).

Depth z. m.	$\bar{\theta}$ °C	$\dfrac{\partial \theta}{\partial t}$ °C per month	μ^2, m^2 per month Proc. 2 C =11.91 C$_1$ = 63.95 a = 0.246	Stability $\dfrac{d\sigma_{\bar{\theta}}}{dz}$ grms. per cm^3 per cm.	K = A.$\dfrac{d\sigma_{\bar{\theta}}}{dz}$
9	18.83	2.65	6.27	2.6×10^{-6}	6.56×10^{-8}
10	17.43	2.17	6.54	2.2×10^{-6}	5.55×10^{-8}
11	16.21	1.80	6.96	1.8×10^{-6}	4.54×10^{-8}
12	15.24	1.27	6.31	1.22×10^{-6}	3.08×10^{-8}
13	14.58	1.01	6.41	0.85×10^{-6}	2.14×10^{-8}
14	13.99	0.79	6.42	0.73×10^{-6}	1.84×10^{-8}
15	13.49	0.62	6.36	0.52×10^{-6}	1.31×10^{-8}
16	13.14	0.54	7.15	0.37×10^{-6}	0.94×10^{-8}

Mean μ^2 = 6.55 m^2 per month.
Mean A = 2.52×10^{-2} cm^2 per sec.

CRITERION OF VALIDITY

1.4. In the above treatment it has been assumed that equation (2) gives a sufficiently accurate picture of the variation of temperature with depth in the hypolimnion to permit the use of equation (4), a convenient expression for the coefficient of turbulence. This seems to be justified by the accuracy with which the points representing ($\bar{\theta} - C$) or in the case of figures 5-7, ($\theta - C$) are fitted by the graphs of equation (2), plotted on semi-logarithmic paper. In applying equation (1) it is moreover necessary to assume that over a considerable range of depths the coefficient of turbulence is essentially constant. If this assumption is correct, the values of $\dfrac{\partial \theta}{\partial t}$, i.e. $\mu^2 C_1 a^2 e^{-az}$, should, in any region where turbulence is the sole mechanism of heat transport, not merely fall on a straight line, but on a straight line parallel to that representing equation (2), when plotted against depth on the same semi-logarithmic paper. Inspection of the graphs indicates that, when the best data are used, this is in fact the case in the

TABLE 4. Analysis of thermal conditions in the clinolimnion of Lake Quassapaug. Mean values July-August, 1938. (10 m. level rejected in computing μ^2 on account of the low value of $r_{\theta t}$.)

z m.	$\bar{\theta}$ °C	$\dfrac{\partial \theta}{\partial t}$ °C per month	$r_{\theta t}$	μ^2
8	16.21	1.68	0.912	1.12
9	12.95	0.92	0.856	0.99
10	10.80	0.28	0.418	(0.42)
11	9.20	0.31	0.831	0.71
12	8.28	0.29	0.890	1.00
13	7.74	0.28	0.907	1.44
14	7.48	0.15	0.816	1.13
			mean	1.07

TABLE 3. Lake Mendota (data from Dr. E. A. Birge). Variation of coefficient of turbulence and stability in clinolimnion with time. McEwen's values of μ^2 are given for comparison.

Week	Range of clinolimnion	μ^2, m^2 per month	μ^2, m^2 per month McEwen	A, cm^2 per second	Stability 12 m. grms per cm^3 per cm.	Stability 15 m. grms per cm^3 per cm.	K, 12 m.	K, 15 m.
1st June	9-15	54.8	84	21.1×10^{-2}	0.38×10^{-6}	0.26×10^{-6}	8.02×10^{-8}	5.49×10^{-8}
2nd June	9-15	22.4	27.6	8.64×10^{-2}	0.59×10^{-6}	0.33×10^{-6}	5.08×10^{-8}	2.84×10^{-8}
3rd June	9-16	8.02	15.2	3.09×10^{-2}	0.85×10^{-6}	0.43×10^{-6}	2.63×10^{-8}	1.33×10^{-8}
4th June	9-15	9.26	11.1	3.57×10^{-2}	0.78×10^{-6}	0.46×10^{-6}	2.79×10^{-8}	1.64×10^{-8}
1st July	9-15	13.80*	9.2	5.32×10^{-2}	1.03×10^{-6}	0.44×10^{-6}	5.48×10^{-8}	2.34×10^{-8}
2nd July	9-16	8.27	11.4	3.19×10^{-2}	1.09×10^{-6}	0.64×10^{-6}	3.48×10^{-8}	2.04×10^{-8}
3rd July	9-16	8.01	9.5	3.09×10^{-2}	1.16×10^{-6}	0.52×10^{-6}	3.59×10^{-8}	1.61×10^{-8}
4th July	9-16	4.92	7.9	1.90×10^{-2}	1.54×10^{-6}	0.70×10^{-6}	2.93×10^{-8}	1.32×10^{-8}
1st August	10-16?	3.60	4.5	1.39×10^{-2}	1.68×10^{-6}	0.71×10^{-6}	2.04×10^{-8}	0.99×10^{-8}
2nd August	11-17	4.07	5.6	1.57×10^{-2}	1.79×10^{-6}	0.81×10^{-6}	2.81×10^{-8}	1.27×10^{-8}
3rd August	12-18	3.67	6.4	1.41×10^{-2}	2.02×10^{-6}	0.91×10^{-6}	2.84×10^{-8}	1.28×10^{-8}
4th August	12-18	4.86	6.2	1.87×10^{-2}	2.40×10^{-6}	0.83×10^{-6}	4.99×10^{-8}	1.55×10^{-8}

* Poor fit on semilogarithmic paper.

clinolimnion. Such a criterion was pointed out by McEwen. When it is remembered that the two series of points are obtained in any given case by quite independent treatments, the parallellism shown, for instance in the clinolimnion in Figure 4, is very striking.

CRITIQUE OF MCEWEN'S THEORY

1.5.1. McEwen obtained his values of the coefficient of turbulence in Lake Mendota by summing the separate values of $\dfrac{\partial \theta}{\partial t}$ and dividing by the sum of the separate values of $C_1 a^2 e^{-az}$, *over the entire range of depths below 12 m.*, thus including those depths, between 16 m. and 23 m., here regarded as bathylimnetic, where there is no trace of an exponential fall in $\dfrac{\partial \theta}{\partial t}$ with increasing depth. This procedure clearly violates the criterion of validity given in the preceding section, and leads to somewhat higher values of the coefficient of turbulence than those that have been obtained in the present work. In Figure 4 the dotted line represents the calculated rate of change of temperature with respect to time, due to turbulence, when McEwen's procedure is adopted. At all depths above 17 m. the water heats more slowly than would be expected, and it is necessary to explain the negative residual differences between the observed and calculated values of $\dfrac{\partial \theta}{\partial t}$. These residuals form the basis of the major part of McEwen's theory. It is supposed that cooling, largely by evaporation at the surface, reduces the temperature of discrete masses of water below the mean surface temperature, thus setting up differences of density. After a variable amount of cooling, the cooled elements are assumed to descend to levels appropriate to their new densities. This movement generates upward countercurrents of colder water and so causes a reduction in temperature at all those levels at which residuals are observed. The frequency of different classes of density difference is assumed to follow the normal probability distribution, and an elaborate expression is given, based on that assumption, to which the vertical series of residuals can be fitted. The theory is apparently accepted without comment by Grote (1936), but has been criticized by Behrens (1937) on general grounds. McEwen's choice of 12 m. as the limit beyond which the convection currents are undetectible, and in consequence as the limit below which no significant residuals due to the cold upward stream occur, appears to be quite arbitrary. Examination of the mean Mendota graph indicates that the events at 12 m. are essentially of the same kind as the events at 9 m.; inclusion of all depths up to nine meters, in estimating the coefficient of turbulence, would, however, lower the residuals considerably. Conversely, if the 9 m. level be excluded, so should all depths down to 15 m. be excluded. Moreover, below 16 m. the criterion for the validity of the

assumption of a constant value of the coefficient of turbulence, necessary for the application of McEwen's theory, does not hold; if these depths are rejected, in line with the point of view of the present paper, there are no residuals to form the basis of McEwen's theory.

1.5.2. As the basis of a further examination of the theory in so far as it is founded on the existence of residuals of the kind discussed in the preceding paragraph, three hypotheses as to the values of $\dfrac{\partial \theta}{\partial t}$ in the bathylimnion may be offered.

(a) The apparent constancy of $\dfrac{\partial \theta}{\partial t}$ in Lake Mendota below 16 m. is due to observational error. This appears to be McEwen's position, being implied in his use of a mean value of the coefficient of turbulence. But since the apparent constancy is more clearly marked in the mean curve than in individual curves for separate weeks, such an assumption may be rejected, particularly when it is remembered that several hundred observations are represented by each of the points representing $\dfrac{\partial \theta}{\partial t}$ in Figure 4. The fact that similar phenomena occur in both the Connecticut lakes studied would also appear to exclude the hypothesis.

(b) A cooling mechanism exists through the whole vertical extent of the lake, becoming less and less effective with increasing depth. Such a cooling mechanism could only be the one suggested by McEwen. On the basis of this hypothesis McEwen's theory would be valid, his application of the theory to Lake Mendota invalid. The hypothesis indeed introduces an element of uncertainty into any application. An approximation to the coefficient of turbulence could be obtained from the values of $\dfrac{\partial \theta}{\partial t}$ and $C_1 a^2 e^{-az}$ for the greatest depth available, but there would be no way of telling how much the very large value (in the case of the mean Mendota data of Figure 4, at 23 m. the value would be $\mu^2 = 35.4$) and with it the estimates of the residuals, fell short of the true values. On the basis of this hypothesis the marked exponential form of the fall of $\dfrac{\partial \theta}{\partial t}$ in the clinolimnion must be regarded as an accidental departure from a more complex type of distribution, and the criterion of validity of 1.4. is meaningless.

(c) Turbulence changes in the bathylimnion (1.10.1) or some other heating mechanism (1.10-1.11.) exists in this region. All bathylimnetic estimates of the coefficient of turbulence are therefore irrelevant in calculating the theoretical estimate of heating in the clinolimnion. The exponential fall in the latter region may therefore be taken as satisfying the criterion of validity, indicating that in the region in question only heat transport by vertical turbulence with an essentially constant value for the coefficient is taking place. This is the position of the present paper. The first hypothesis (a) clearly cannot be reconciled with the facts. The second (b) though logically consistent is very complex; and though not in conflict with the observational data, receives no support from them. In the face of the third (c), much simpler hypothesis, it must be regarded as untenable, falling to the razor of William of Occam, with the whole theory of convectional cooling of the hypolimnion.

CONDITIONS IN THE EPILIMNION

1.6.1. The method given in 1.3.2. for the determination of the coefficient of turbulence is only applicable to the clinolimnion. McEwen attempted to apply a modification of the method to the epilimnion, but his procedure does not appear to be valid. Since in the temperature curves of practically all stratified lakes there is a well-marked inflection point defining the thermocline, there is clearly a region in the epilimnion where $\dfrac{\partial^2 \theta}{\partial z^2}$ is negative, and where uniform turbulence produces a fall in temperature. McEwen's treatment on the other hand assumes a second exponential temperature curve set above that in the hypolimnion and entirely unlike the conditions actually found in any lake. All that can be said of the epilimnion is that the conditions are complex and that turbulence is undoubtedly high. A method for relative estimation of its magnitude in a series of lakes is given by Grote (1934) and is discussed in 1.12; no absolute estimate seems at present possible.

1.6.2. Although McEwen's theory is here regarded as definitely inapplicable to the hypolimnion, the possibility of its application to the extreme superficial layers of a lake, layers in which a very small thermal gradient is developed, must not be forgotten, in case some method be devised that would permit a detailed treatment of the thermal phenomena of the epilimnion. This possibility is suggested by an experiment performed by McEwen, in a concrete tank five feet deep, with a fall in temperature at midday from 26.74° C. at the surface to 25.10° C. at the bottom. Analysis of the rate of change of temperature with time during the day in this tank suggested that the convectional system postulated played a part in the heat exchange of the different water-layers. It must, however, be pointed out that in McEwen's treatment of events in the tank, it has been assumed, presumably correctly, that the water was free from turbulence, whereas it is clear that the epilimnion of a lake would be highly turbulent. The chief difficulty in applying the results of the experiment to the upper layers of a lake with a thermal gradient comparable to that in the tank lies in conceiving of the uninterrupted descent of discrete masses of water, through a layer as highly disturbed as the upper few meters of a lake, let alone of the open ocean, for the treatment of which the theory was originally elaborated. No information as to the size of the supposed discrete masses is given, save that they are of more than molecular dimensions.

EFFECT OF SOLAR RADIATION ON THE CLINOLIMNION

1.7. It has seemed desirable to estimate, from the data of Birge and Juday (1929), the rate of heating due to solar radiation at and below 9 m. in Lake Mendota, though it is well known from the classical work of these authors, that such heating is very small. Using a mean value for radiation penetrating the lake surface, of 15,825 cals. per cm.² per month for the period mid-June to mid-August, and a percentile transmission of 45 (mean sun), 4.55 cal. per cm.² reach the 9 m. level per month, and being absorbed at this level at the rate of 2.5 cals. per cm.² per meter, cause a rise in temperature of 0.025° C. per month. Since the observed value of $\dfrac{\partial \theta}{\partial t}$ at this level is 2.65° C. per month, it is clear that an error of about one per cent is introduced by neglecting direct heating by radiation; at all succeeding levels the error will be less. The mean value of the coefficent of turbulence computed by procedure (2) for the whole clinolimnion is 6.54 if solar radiation be taken into account, and 6.55 if it be not; such a difference is insignificant.

VERTICAL CONSTANCY OF TURBULENCE IN THE CLINOLIMNION

1.8. Given constant mixing forces, the value of the coefficient of vertical turbulence is inversely proportional to the vertical stability. Yoshimura (1936, 1938) who has studied the coefficient of turbulence in Japanese lakes, but has so far not published his detailed conclusions, believes that this coefficient has a minimum value in the thermocline and increases in the deeper water of the lake. Such a belief is natural in view of the relationship just stated. A similar position is adopted by Grote (1934) on purely theoretical grounds. McEwen on the other hand makes no comment on his assumption of a uniform coefficient throughout the hypolimnion in spite of manifest changes in stability. Consideration of either the values derived from the mean Mendota, or from the Linsley data, or from any of the individual weekly series from the former lake, shows very definitely that the stability changes greatly with depth throughout the vertical extent of the clinolimnion, in which region the coefficient of turbulence is essentially constant. Putting the coefficient into C.G.S. units we can write, since A is proportional to the reciprocal of the stability,

$$A = K \cdot \frac{dz}{d\sigma_0}$$

or

$$K = \frac{d\sigma_0}{dz} \cdot A \quad \ldots\ldots\ldots\ldots (9)$$

K may be taken as a measure of the mixing forces producing turbulent exchange, and is a function of the horizontal current velocity and the vertical gradient of such velocities. Assuming that the criterion of validity is satisfied, K is clearly proportional to the stability throughout the clinolimnion. On this assumption values have been computed by converting μ^2 to C.G.S. units (A) and multiplying by the stabilities throughout the clinolimnion. In the case of Mendota the values of stability were determined by mechanical differentiation of a density curve based on temperature alone. In the case of Linsley Pond, the chemical component of stability has been allowed for approximately, on the basis of an increase in density of 0.0000018 per milligram per litre HCO_3.[3] The generalization that the coefficient of turbulence is minimal in the more stable layers is clearly erroneous insofar as the clinolimnion, defined by the condition of validity, is concerned. Moreover, if, as seems probable, the coefficient is essentially constant (1.10.3.) throughout the whole hypolimnion, the generalization is essentially false for all depths below the thermocline. The problem raised in this paragraph can only be discussed in terms of the horizontal currents which produce turbulence. The existence of such currents is implicit in all discussions of the turbulence problem, and in the second part of the present paper their implied existence is used to explain the chemical changes in the free water of the hypolimnion. Such currents certainly have velocities varying with depth. It is probable that important information as to their variation in velocity can be obtained from the constancy of the coefficient of turbulence with decreasing stability and decreasing mixing forces.

TEMPORAL VARIATION OF THE COEFFICIENT OF TURBULENCE

1.9. In marked contrast to the constancy of the mean values of the coefficient of turbulence in the clinolimnion with decreasing stability and increasing depth, there is a very marked decline in the value of the coefficient at all levels with increasing stability as the season advances (Table 3). In Figure 8 the individual values of $\mu^2 = \dfrac{\partial \theta}{\partial t} \cdot \dfrac{1}{C_1 a^2 e^{-az}}$, at 12 and 15 m., are plotted against the reciprocal of the stability for all the weeks from the first in June to the last in August. The theoretically preferable mean values of the coefficient have not been used, to avoid overemphasis of the accuracy of the mean. The two depths were chosen as the highest and lowest that can be regarded as clinolimnetic throughout the entire period. In the same diagram the individual mean (mid-June to mid-August) values for the clinolimnion (Table 2) are indicated by the solid circles, which form an essentially horizontal line across the graph.

[3] This figure is based on determinations with a 50 cc. picnometer on 13 m. filtered water taken on June 13, 1939, and having an alkalinity of 63.1 mgrms. per litre. Surface water of June 23, having an alkalinity of 58.6 mgrms. per litre gave a value of 0.0000014 per mgrm. HCO_3 in excess of distilled water, at 25.0°. The higher value of the 13 m. water is probably due to differential accumulation of other substances (organic matter, silicate, etc.); since the correction is applied to hypolimnetic water, this higher value has been used.

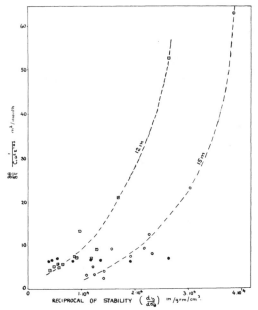

Fig. 8. Lake Mendota. The relationship between the value of $\dfrac{\partial \theta}{\partial t} \cdot \dfrac{1}{C_1 a^2 e^{-az}}$ and the reciprocal of stability at 12 m. (open squares) and 15 m. (open circles) for the various weekly mean temperatures for the period June-August, and between the same quantities at each depth between 9 and 16 m. (solid circles) for the mid-June to mid-August mean data.

The broken curves, which are fitted by eye and intended primarily to aid in the study of the diagram, would be straight lines if the mixing forces were constant throughout the period considered. Their form indicates that actually the mixing forces decline considerably during the progress of stratification. This is also apparent from Table 3, but after the middle of June the change is irregular, small, and perhaps insignificant.

CONDITIONS IN THE BATHYLIMNION

1.10.1. In the bathylimnion, the criterion of validity of 1.4. is no longer satisfied. In the mean Mendota graph (Figure 4) it is apparent that the rate of change of temperature with respect to time is essentially constant throughout the region. Two hypotheses have to be considered in the treatment of the region; the first is that turbulence is variable in the bathylimnion, the second that turbulence remains constant throughout practically the whole of the hypolimnion, but that other heating mechanisms play a part in the deeper water. A third hypothesis, that turbulence varies and that other mechanisms exist, is too complex to be considered unless the first two fail. Two equations, other than (1), give expressions for the coefficient of turbulence at any point. When turbulence is variable, the more general form of (1) is (Schmidt 1925)

$$\frac{\partial \theta}{\partial t} = \frac{d(\mu^2)}{dz} \cdot \frac{\partial \theta}{\partial z} + \mu^2 \frac{\partial^2 \theta}{\partial z^2} \quad \dots (10)$$

while in all cases the transport equation (Schmidt 1925)

$$\frac{d\Theta}{dt} = -\mu^2 \frac{\partial \theta}{\partial t} \quad \dots\dots\dots\dots (11)$$

is valid, where $\dfrac{d\Theta}{dt}$ is the rate of total heat transport across unit area. Equation (11), on account of its simplicity, has been used by many workers, including some interested in the transport of biologically active materials (e.g. Redfield, Smith, and Ketchum, 1937). It suffers, however, from the very grave defect that the value of μ^2 at any level is dependent on all exchanges, due to turbulence or to any other mechanism, beyond that level. Although, as will be shown in paragraph 1.10.3., this equation is of great utility in solving the present problem, its use naturally brings up the question of heating mechanisms other than vertical turbulent exchange. It is therefore appropriate to investigate briefly the value of the much more intractible second order differential equation (10) on the basis of the mean Mendota figures, which, here as elsewhere, may be used as a norm to which conditions in other lakes may be referred.

1.10.2. Assuming that the temperature curve may be expressed by equation (2), at least almost to the bottom of the lake, and that the rate of change of temperature with respect to time is constant throughout the bathylimnion, i.e.,

$$\frac{\partial \theta}{\partial t} = b = \text{constant}$$

equation (10) takes the form

$$b = \mu^2 C_1 a^2 e^{-az} - \frac{d\mu^2}{dz} C_1 a e^{-az} \quad \dots (12)$$

Now putting

$$\frac{b}{C_1 a} = k$$

and writing in operational form, this becomes
$$(D-a)\,\mu^2 = -k e^{az}$$
The solution of this is
$$\mu^2 = (Q - kz)\,e^{az} \quad \dots\dots\dots\dots (13)$$
This belongs to a class of solutions[4] of (10) which has no physcial meaning below a certain depth, the turbulence coefficient decreasing more and more rapidly, and becoming zero at $z = \dfrac{Q}{k}$. Now if it be assumed that the variation of the coefficient is continuous across the boundary of the clinolimnion and

---

[4] Putting $\dfrac{\partial \theta}{\partial t}$ = constant, it is easy to see, by differentiating (10) with respect to depth, that, provided the temperature curve has negative first and third derivatives with respect to depth, which is presumably the normal condition in holomictic lakes, the coefficient of turbulence can vary, given suitable forms of $\theta = f(z)$, so that its first and second derivatives with respect to depth are both negative, or both positive, or the first is positive, the second negative; the case in which the first is negative, the second positive is excluded. Equation (12) belongs to the first of these classes.

bathylimnion, the constant of integration Q can be determined by putting μ^2 equal to the clinolimnetic value when the depth z is taken as the depth of the boundary. In this way it can be shown, using the mean Mendota data, that the coefficient of turbulence would assume meaningless negative values below 20.2 m. The hypothesis of purely turbulent heating in the bathylimnion therefore does not seem a fruitful one, though it may be urged that the impasse just reached is implicit in the assumptions rather than in the observational data.

1.10.3. Turning now to equation (11), the values of the coefficient may first be computed by summing the heat income per unit of time and area, of each unit layer of a column in the deepest part of the lake. This quantity, when divided by the rate of change of temperature with depth, the minus sign being omitted, gives a set of apparent values for the coefficient of turbulence. This has been done for the mean Mendota data in Table 5, $\dfrac{\partial \theta}{\partial z}$ being taken as $-C_1 a e^{-az}$. In Table 6 the same procedure has been applied to the Linsley data, but in this case, as the fit of the temperature curve to equation (2) is far from perfect, values have been calculated both from the expression $-C_1 a e^{-az}$ and from values of $\dfrac{\partial \theta}{\partial z}$ obtained directly from the temperature curve by means of the Richards-Roope tangentmeter (Richards and Roope, 1930). In the case of the Mendota data the depth of the lake was taken as 24 m., the figure used by Birge (1916a), and the rate of change of temperature at 24 m. was taken to be the bathylimnetic mean of 0.476° C. per month. In the case of Linsley Pond the whole 13-14.8 m. stratum was taken to heat at the rate observed at 14 m. Uncertainty as to what maximum depth should be used in the case of Mendota does not invalidate the final treatment of the data. It will be observed that at the top of the clinolimnion, the values of the coefficient of turbulence approach those obtained from the simple second order equation (1), but that with increasing depth they diverge from the clinolimnetic values accepted in Tables 2 and 3. Since, at least in Figure 4, the criterion of validity is so clearly satisfied, it is certain that these increasing apparent values are erroneous. If now, at each level, the rate $\dfrac{d\Theta'}{dt}$ at which heat is received through unit area in unit time be calculated from the clinolimnetic coefficient previously obtained, i.e. 6.55 m²/month in Mendota, 0.86 m²/month in Linsley, it will be observed that there is a well defined tendency for the difference between these calculated rates of heat income and the observed rates to be constant, as is indeed implicit on the exponential arrangement. In Mendota this is very marked in the whole clinolimnion, the mean difference being 184.1 cals. per cm² per month. In Linsley Pond there is great divergence between the two methods of calculation at 4 and 5 m., but from

6 to 9 m. fair constancy obtains, with mean values of 62.4 cals. per cm² per month if the derivative be calculated and 58.9 cals. per cm² per month if it be obtained by mechanical differentiation. The most reasonable conclusion to be drawn is that heat is entering the bathylimnion by non-turbulent mechanisms at a rate of 184 cals. per cm² per month in the case of Lake Mendota and of about 60 cals. per cm² per month in the case of Linsley. Now let it be assumed that the coefficient of turbulence is constant throughout the entire hypolimnion. Using the values of $\mu^2 C_1 a^2 e^{-az}$ for each bathylimnetic level, it is possible

TABLE 5. Computation of the mean rate of heating not due to vertical turbulent mixing, below one square centimeter over the deepest water (taken as 24 m.) of Lake Mendota (mid-June to mid-August), as the mean of the clinolimnetic residuals (mean of column 6) or as the sum of bathylimnetic residuals on the assumption of constant hypolimnetic turbulence (sum of column 9).

(1) Depth z m	(2) $\dfrac{\partial \bar{\theta}}{\partial z}$ °C per m. or cals per m³ per m x 10^{-6}	(3) $\dfrac{d\Theta}{dt}$ cals. per cm² per month or month x 10^{-4}	(4) Apparent coefficient of turbulence $\dfrac{d\Theta}{dt} \cdot \dfrac{\partial z}{\partial \bar{\theta}}$ m² per month	(5) $\dfrac{d\Theta'}{dt}$ $=\mu^2 C_1 ae^{-az}$ where $\mu^2 = 6.55$ cals per cm² per month	(6) Residual heating $\dfrac{d\Theta}{dt} - \dfrac{d\Theta'}{dt}$ cals per cm² per month	(7) $\mu^2 C_1 a^2 e^{-az}$ assuming constant turbulence $\mu^2 = 6.55$ °C per month	(8) $\dfrac{\partial\theta_{nt}}{\partial t} = \dfrac{\partial\theta}{\partial t} - (7)$ °C per month	(9) Rate of non-turbulent heating for unit layer below depth z cals. per cm² per m. per month
9	−1.72	1309.9	7.61	1126.0	183.9			
10	−1.34	1069.0	7.96	880.0	189.0			
11	−1.05	870.8	8.29	688.0	182.8			
12	−0.822	717.2	8.73	538.0	179.2			
13	−0.642	602.9	9.39	420.3	182.6			
14	−0.500	512.7	10.25	327.5	185.2			
15	−0.393	442.5	11.26	256.2	186.3			
16	−0.307	384.7	12.52	201.1	183.6	0.495	0.045	6.1
					mean 184.1			
17	−0.241	334.5	13.88			0.388	0.077	13.2
18	−0.187	286.8	15.33			0.302	0.187	21.5
19	−0.146	238.4	16.33			0.236	0.243	23.4
20	−0.115	193.9	16.86			0.185	0.225	29.7
21	−0.0897	147.3	16.43			0.145	0.368	38.0
22	−0.0708	96.3	13.60			0.114	0.392	38.7
23	—	—	—			0.089	0.383	39.6
(24)	—	—	—			0.068	0.408	Sum 210.2

TABLE 6. Computation of mean non-turbulent heating in the bathylimnion of Linsley Pond.

(1) Depth m.	(2) $-\dfrac{\partial\bar\theta}{\partial z}$ $=C_a e^{-az}$ From graph °C per m	(3) From $\textcircled{2}$	(4) $\dfrac{d\Theta}{dt}$ From $\textcircled{3}$ cals per cm² per month	(5) $\dfrac{d\Theta}{dt}$ From (2) cals per cm² per month	(6) $\dfrac{d\Theta'}{dt}$ From (3) cals per cm² per month	(7) $\mu^2 C_1 a^2 e^{-az}$ assuming constant turbulence °C per mo.	(8) $\dfrac{\partial\theta_{nt}}{\partial t}=\dfrac{\partial\theta}{\partial t}-(7)$ °C per month	(9) non-turbulent rate of heating of unit layer below depth, cals per cm² per m. per month
4	5.79	4.9	447.2	−29.8	50.4			
5	3.47	4.2	334.8	52.8	− 5.2			
6	1.966	1.93	232.2	72.9	75.8			
7	1.145	1.30	158.3	59.9	53.1			
8	0.609	0.88	113.7	59.5	42.4			
9	0.377	0.28	87.6	57.2	64.4	0.175	−0.047	6.1
	mean (6–9m)			62.4	58.9			
10						0.100	0.075	7.9
11						0.058	0.083	10.1
12						0.034	0.120	11.7
13						0.018	0.114	11.6
14						0.012	0.117	21.1
								(to 14.8 m)
								Sum 68.5

to calculate the rate of turbulent and non-turbulent[5]

$\left(\dfrac{\partial\theta_{nt}}{\partial t}\right)$ change of temperature of each level. The total rate of non-turbulent income of heat into the bathylimnion can be calculated by summing the mean non-turbulent rate of change for each layer, and is estimated as 210.1 cal. per cm² per month in the case of Mendota, 68.5 cals. per cm² per month in the case of Linsley Pond. These figures indicate that while the estimates of the turbulent heating based on assuming the clinolimnetic coefficient to hold are perhaps a little too high, the coefficient of turbulence cannot be very different in the bathylimnion from what it is in the clinolimnion. Similar phenomena also occur in Lake Quassapaug. The values of the residuals obtained when the heating below any clinolimnetic level as calculated is subtracted from the value observed are fairly constant between 11 and 14 m., the mean value being 72.0 cals. per cm² per month: the values derived from the 8-10 m. levels are low, probably due to the unreliability of the data. The value calculated on the assumption of constant turbulence

[5] Turbulent and non-turbulent here refer only to *vertical* turbulent exchanges.

in the whole hypolimnion is 76.8 cals. per cm^3 per month; considering the nature of the data the agreement is remarkable. It is important to note that the quantity of heat brought into the bathylimnion of this lake by mechanisms other than vertical turbulent mixing is of the same order of magnitude as in Linsley Pond, in spite of the lesser depth of the latter. Per unit depth or volume, therefore, the non-turbulent mechanism is best developed in Mendota, least in Quassapaug.

1.10.4. In the use of equation (11) in the preceding paragraph, it has been assumed that the whole of the heat taken up by a column of water, below any given horizontal surface, has entered through the area of that surface defining the top of the column. It has been shown that, considering a column in the deepest part of the lake, heat enters the base of the column, in the bathylimnion, by some agency other than vertical turbulent transport. In the clinolimnion, since the criterion of validity of 1.4 is satisfied by the distribution of rates of change, the effect of such an agency is negligible. It remains to be considered whether this heat has been transmitted by vertical turbulent mixing, through some area of a horizontal plane at depth z_1, lying above the top of the bathylimnion, lateral to the deepest part of the lake, in relatively shallow water. Such a transmission of heat would imply that some of the heat passing through the area a_1 arrives, after lateral displacement, in more centrally situated deep water. This type of transmission is assumed in the calculation of heat budgets by the Wojeikoff-Birge method, while the budget (Θ) used in the preceding paragraph is essentially a Forelian budget. Birge (1916) has shown conclusively that the Forelian mode of estimation is erroneous, in that it leads to impossibly high figures in certain cases. The rate of change of the heat budget below a given level can be calculated not merely on a Forelian basis, as in the last paragraph, but on a Birgean basis ($\dfrac{d\Theta_b}{dt}$) also. This is done by multiplying the mean rate of change for any unit layer by the volume of the layer, summing upwards to any desired depth z_1, and dividing by a_1, the area at that depth. This gives a rate of change based on the assumption that all the heat present in the lake below any level z_1 has been evenly distributed over the area a_1 in its passage through that level; some form of lateral heat transmission is necessarily assumed. It is now possible, by the use of equation (11), to determine anew the coefficient of turbulence in the clinolimnion. The results of such a procedure for Linsley Pond are set out in Table 7, together with the apparent values of the coefficient obtained in the preceding paragraph on a Forelian basis, and the values of $\dfrac{\partial\theta}{\partial t} \cdot \dfrac{1}{C_1 a^2 e^{-az}}$ obtained by procedure (2) in Table 1. It will be observed, that though the estimates in the second column show an increasing trend, this is much less marked than is the trend

TABLE 7. Estimates of Coefficient of Turbulence. Figures in parentheses obtained by mechanical differentiation.

	$\dfrac{d\Theta}{dt} \cdot \dfrac{\partial z}{\partial \bar{\theta}}$	$\dfrac{d\Theta_b}{dt} \cdot \dfrac{\partial z}{\partial \bar{\theta}}$	$\dfrac{\partial\theta}{\partial t} \cdot \dfrac{1}{C_1 a^2 e^{-az}}$
4	0.77 (0.91)	0.66 (0.78)	0.71
5	0.96 (0.79)	0.66 (0.54)	0.75
6	1.18 (1.20)	0.68 (0.69)	0.83
7	1.38 (1.22)	0.72 (0.63)	0.96
8	1.70 (1.29)	0.88 (0.70)	0.82
9	2.32 (3.13)	1.21 (1.63)	1.09
mean	1.66 (1.70)	0.80 (0.83)	0.86

in the first column. As an estimate of the coefficient of turbulence, the mean of the second column is less good than is the mean of the third column where the trend is even less well marked, though these two means are almost identical. Comparison of the second and third columns indicates that in the upper part of the clinolimnion, the value of $-\dfrac{d\Theta_b}{dt} \cdot \dfrac{\partial z}{\partial \bar{\theta}}$ is too low to give a reliable coefficient of turbulence, thus implying that part of the heat brought through the horizontal planes of this region is lost by absorption in the mud; this is also implicit in the exponential arrangement of $\dfrac{\partial\theta}{\partial t}$ in the clinolimnion. The nine-meter level on the other hand may not be truly clinolimnetic, but the difference between 0.96 at 7 m. and 1.09 at 9 m. in the third column is not great enough to warrant excluding the latter value in the clinolimnetic estimate of the coefficient of turbulence. If the estimates of the second column are continued to greater depths they increase, but, as has been shown in the previous section, it is very doubtful if pure turbulence can be evoked to explain heating at depths below the clinolimnion. In the next paragraphs the possibility of another mechanism of heating is suggested which is particularly significant as it explains the fact that the residuals apparently due to non-turbulent heating are greatest in the deepest water, and are best demonstrated in the most eutrophic lake of the three under consideration.

ROLE OF DENSITY CURRENTS

1.11.1. It has long been known, from the observations of Buchanan (1879) and very many subsequent observers, that heating of lakes can occur under an ice cover, and that the bottom temperatures resulting from such heating may rise above 4.0° C.[6] This is known to occur in Lake Mendota (Birge and Juday, 1911, data for March 6, 1909) as well as in a number of other lakes, not involved in the present discussion; a concise summary of the relevant literature has been presented by Yoshimura (1936a). Apart from crenogenic meromixis, two principal factors play a part in the phenomenon. Firstly, heating of water by

[6] Buchanan's example may be a meromictic lake.

solar radiation at the surface, particularly in shallow water, sets up a thermal density current as the temperature of the cold water approaches 4.0° C. Secondly, the passage of heat stored in the mud in the summer into water cooled to below 4° C. similarly will cause a density current carrying heat towards the deeper part of the lake. These two factors were first considered quantitatively by Birge, Juday and March (1928). In order to explain the rise in temperature above 4.0° C. it is necessary to suppose that a chemical as well as a thermal increase in density in the water immediately over the mud is involved. This is clearly shown in Table 8, giving the temperatures, alkalinities and oxygen contents of Linsley Pond, before and after the breaking of the ice in the spring of 1936. It is extremely doubtful if there was a full circulation in the spring of 1936. The very high bottom temperature and alkalinity observed at the end of winter stagnation in this year were undoubtedly due to the exceptionally hard winter and the resultant persistent ice sheet. Subsequent years, in which the ice has never lasted as long as in 1935-36, have shown similar phenomena but not on such a marked scale. It is highly probable that the initiation of biogenic meromixis takes place after exceptionally severe winters when the spring circulation is incomplete and when almost all the accumulation of dissolved substances of a long winter stagnation can be added to the accumulation developed in the succeeding summer.

TABLE 8. Thermal and chemical stratification of Linsley Pond after about 3 months under ice (freezing occurred between December 14 and 27).

		0 m.	5 m.	8 m.	12 m	14 m.	14.5 m.
Dec. 14, 1935	Temp.	3.50	3.50	3.50	3.51		3.51
No ice.	HCO₃	68.6	69.1	68.3	70.0		69.5
	O₂	10.37	9.89	10.05	9.89		9.76
Mar. 7, 1936	Temp.	1.68	4.12	4.21	4.21	4.70	
50 cms. ice	HCO₃	8.9	72.5	76.9	76.8	84.9	
	O₂	12.13	5.08	3.22	3.09	—	
Mar. 21, 1936	Temp.	4.83	4.58	4.43	4.76		4.73
Almost no ice	HCO₃	56.0	58.0	72.7	75.4		85.4
	O₂	8.50	8.19	4.61	3.15		0.10
Mar. 28, 1936	Temp.	7.68	6.48	5.42	5.21		5.15
Ice free	HCO₃	60.6	62.7	62.8	64.3		73.0
	O₂	9.55	7.56	8.10	7.86		4.17

1.11.2. Alsterberg (1930, 1931) has pointed out that the concept of profile-bound density currents need not be limited to winter stagnation. Since the mud is colder than the water in summer, the thermal density current, as Alsterberg points out, is a cold descending current. If, however, sufficient dissolved material passes into the water immediately over the mud of the slope of the basin, this dissolved material may cause the water to flow down the slope to a level at which the free water has a lower temperature, but not a greater density, than the water of the current; in this way heat can be brought down to the level in question. In view of the fact that chemical density currents must clearly be involved in the winter heat-

ing of all holomictic lakes in which the bottom water rises above 4.0° C., it seems probable that part of the heat delivered in summer to the lowest part of the bathylimnion of Linsley Pond and Lake Mendota is also carried by such currents. In the case of Lake Quassapaug, where, in spite of the striking development of the bathylimnion, an increase in alkalinity and possibly in phosphate is not clearly indicated below 15 m. until the end of August, conductivity studies might well demonstrate an earlier bathylimnetic increase in total concentration. The greatest objection to accepting the hypothesis that density currents are entirely responsible for the formation of the bathylimnion is derived from an examination of the early weeks of heating in Lake Mendota (e.g., Fig. 5), where the region appears to be developing very rapidly. For the earliest date (first week in June) for which a reliable value of the clinolimnetic coefficient can be obtained, $\dfrac{\partial \theta_{nt}}{\partial t}$ has the following values:

18 m.	0.55°C. per month
20 m.	0.83
22 m.	1.08
23 m.	1.18.

These seem rather high for a period so early in chemical stratification. No other non-turbulent mechanism[7] can be suggested and it is possible that early in the summer in Lake Mendota the vertical variation of the coefficient is more marked than at later dates or in the other two lakes. Figure 9 indicates diagrammatically the mechanisms of heating believed to occur in a stratified lake.

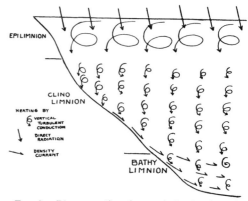

FIG. 9. Diagrammatic scheme of the heating of a lake by radiation, turbulent conduction, and density currents according to the theory developed in the present paper.

[7] Biological oxidation must produce some heating. A rise of 0.1° per month, however, corresponds to 100 cals. per litre per month, i.e., to an oxygen uptake of about 20 cc. per litre per month or 28.6 mgrms. per litre per month. Reference to the value of the non-turbulent rate of change shows that 0.1°C. is a reasonable figure, but as is indicated in 2.5.2, a metabolic rate of the order implied is about five times too great in the case of Linsley Pond. Cooling by the upward current displaced by density currents will be negligible in the clinolimnion on account of its great volume compared with the bathylimnion.

GROTE'S REGIONAL MODULUS OF TURBULENCE

1.12. The Quassapaug data have been little used in the preceding discussion as they are clearly inadequate, owing to the limited number of dates on which readings were taken, and the considerable accidental oscillations of temperature observed at some levels. At 10 m. in particular, the correlation of temperature with time is only 0.418, which, with eight sets of observations, is entirely insignificant statistically. Such fluctuations in the thermocline are to be expected in a fairly large sharply stratified lake. Omitting the ten-meter level, a fair exponential fall is obtained between eight and fourteen meters, below which depth the bathylimnion is well defined. The seven meter level shows a remarkably low value of $\dfrac{\partial \theta}{\partial t}$ which, however, seems to be reliable. Taking the clinolimnion as lying between eight and fourteen meters the coefficient of turbulence may be estimated as 1.07, which is not significantly higher than that in Linsley Pond. Both lakes have essentially the same range of hypolimnetic temperatures, and differ in this respect, as in their clinolimnetic coefficients of turbulence, most strikingly from Lake Mendota. Now Grote (1934) has pointed out, that making certain reasonable assumptions, the square of the depth of the thermocline, i.e., of the thickness of the epilimnion, can be used as a measure of the turbulence of the latter so that in a geographically and climatically uniform lake-district, a modulus of turbulence can be established, based on some lake taken as unity. Taking the thermocline as at four meters in Linsley Pond and eight meters in Quassapaug, we find that though the clinolimnetic coefficients of turbulence are practically identical, the epilimnetic modulus in Quassapaug is four times that in Linsley Pond, taken as unity. This epilimnetic difference is to be expected in view of the very much greater area of the former lake (1.72 km^2 as opposed to 0.094 km^2); that the clinolimnetic coefficients should bear no apparent relation to the area is more remarkable.

COMPARISON WITH THE RESULTS OF PREVIOUS INVESTIGATORS

1.13. In his classical work on turbulent transport, Schmidt (1925) gives values for A in the Lunzer Untersee, calculated from the total heat income below a given isobath by means of equation (11). The computations were made for 5 m., 10 m., 15 m., and 20 m. In April there is a more or less constant value of A = 2 at least down to 15 m. In July the value at 5 m. had fallen to 0.18; at 20 m. to 0.01. The deep water summer values are clearly of the same order of magnitude as in the clinolimnion of Lake Mendota (A = 0.0252). The present investigation was well advanced before Schmidt's work could be consulted, and for this reason McEwen's more elaborate treatment was taken as a convenient starting point. The more detailed analysis made possible by McEwen's treatment of equation (1) has, however, amply justified the present approach. The only other estimate of the turbulence coefficient in any lake is that made for Lake Balaton by Defant (1932), who used a most ingenious method based on the discrepancy of the observed period of the longitudinal seiche and the period calculated by classical methods. The resulting value of the coefficient of turbulence (i.e., the coefficient of eddy viscosity), A = 45, seems extraordinarily high, though within the range of values possible in the ocean.

ALKALINITY AS AN INDICATOR OF WATER-MOVEMENTS AND OF THE METABOLIC CONDITION OF THE HYPOLIMNION

THE DATA

2.1. Along with the temperatures taken in Linsley Pond and Lake Quassapaug, a series of determinations of the alkalinity by methyl orange titration was obtained. Lake Quassapaug contained too little bicarbonate to be of any great interest in the present study, and the analysis of the data is restricted to that from Linsley Pond. Saunders' method (1927) of using a standard prepared by saturating distilled water with carbon dioxide was employed in all titrations. The results are expressed as milligrams of HCO_1 per litre.

THE FORM OF THE ALKALINITY CURVE

2.2.1. In a preliminary communication (Hutchinson, 1938) it was pointed out that when the mean alkalinity, at any depth in Linsley Pond over a period long enough to avoid accidental disturbances due to distortion of the stratification, is plotted against depth, the resulting curve bears an evident relationship to the morphometry of the lake basin. This is most clearly seen when the morphometry is expressed as the variation of mud surface per unit volume of water at any level; the curve showing the mean radius of the lake is less instructive (Fig. 10), but indicates essentially the same relationship. Owing to an error in the method of calculating the relation of the area

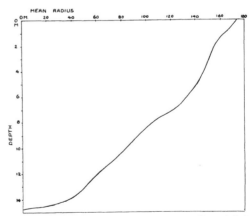

Fig. 10. Linsley Pond. The mean radius plotted against depth, or generalized profile. Note the well-developed shelf at 7 m. and the suggestion of a second shelf at about 11 m.

at any depth of the mud surface to the volume at that depth, it was supposed that this only held when the area was projected onto a plane surface. Actually the ratio of the projected element of area to the corresponding element of volume is practically the same as the ratio of the actual element of area to the element of volume. In Figure 11 the mean alkalinity for the whole period from the beginning of June to the end of August 1937 is plotted against depth, as is the ratio of the element of area of mud surface to element of volume at that depth. The latter curve was obtained by drawing a curve relating area to depth, differentiating by means of the Richards-Roope (1930) tangentmeter and dividing the value of the derivative by the area at each depth

$$\frac{dA}{dz} \cdot \frac{1}{A} = \frac{dA}{dz} \cdot \frac{dz}{dV} = \frac{dA}{dV}.$$

Interpolation of half-meter points has been used and is probably justified, though the finer form exhibited as the result of this process (e.g., the minimum at 8.5 m.) cannot be shown by the alkalinity curve based

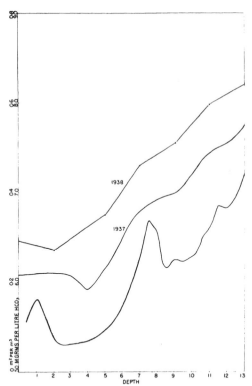

FIG. 11. Linsley Pond. Lowest curve, the *morphometric curve*, showing the variation of $\frac{dA}{dV}$ with depth. Middle curve variation of mean bicarbonate content in milligrams per litre with depth, for June-August, 1937. Topmost curve, the same for 1938. Note that in spite of the large intervals in the 1938 curve, the variation in slope in the hypolimnion is similar in both years.

on points one meter apart. The correspondence of the form of the curves is immediately apparent. In 1938, the work on other lakes, particularly Lake Quassapaug, made such an extensive series of observations in Linsley Pond impossible. A series of determinations, for the most part at two-meter intervals, was made. The plot of the mean of this series shows that the same phenomenon was occurring in both years, though its exact nature is less evident in the less extensive 1938 series than in the 1937 series. The significance of the epilimnetic differences in the two years is uncertain but not relevant to the present discussion. The relationship observed in the first year is therefore a general one, characteristic of the lake. As was pointed out in the preliminary communication, the only reasonable explanation of the relationship is that the bicarbonate originates in the bottom mud, and is carried into the free water by horizontal currents, probably of variable direction, which are sufficiently rapid to permit the development of a pattern in the alkalinity distribution in spite of the slight turbulent mixing in a vertical direction necessarily postulated for the distribution of heat. Further analysis indicates, however, that the problem of the alkalinity curve is more complex than at first appeared, though nothing has yet been discovered that casts doubt on this primary hypothesis.

2.2.2. When the morphometric relationship of the alkalinity curve had been established it was natural to inquire whether any similar distributions had been observed by previous investigators, and to attempt to obtain information from other Connecticut lakes that could be used in confirming the observations made in Linsley Pond. The results of this inquiry have so far been disappointing, as it would seem that certain special conditions must be satisfied before the relationship can become apparent. Soft water lakes do not contain enough bicarbonate to give any reliable variation of alkalinity with depth; conductivity determinations might provide interesting data in such cases. In the majority of hard-water lakes, containing far more bicarbonate than does Linsley Pond, there is a tendency for the alkalinity to fall, in the upper hypolimnion at least, during the course of stagnation. This is apparent in some of the data tabulated by Birge and Juday (1911) for Lake Mendota, and was observed by Dr. Deevey in the hard water lakes examined in Connecticut and New York State in the hope of throwing further light on the present problem. This fall is possibly due largely to bacterial precipitation of calcium carbonate, though Williams and McCoy (1934) who have investigated the matter, were unable to obtain definite proof of this. Any such disturbance naturally renders the morphometric effect unobservable. It is, moreover, obviously necessary for the development of the effect that the lake should be productive enough to produce adequate carbon dioxide in the mud, deep enough to be well stratified, small enough for the mud to have an appreciable influence on the free water and of such a shape that some conspicuous morphological feature can be easily recognized in the alkalinity curve if the relationship

G. EVELYN HUTCHINSON

Ecological Monographs
Vol. 11, No. 1

holds. It is evident that Linsley Pond is exceptional in the coincidence of all these requirements. The Schleinsee described by Einsele and Vetter (1938) also fulfills these requirements and probably exhibits comparable phenomena, but the number of points on the alkalinity curve and the published information on the morphometry of this lake, though suggestive, are not sufficient to permit a final conclusion.

THE RATE OF CHANGE OF ALKALINITY WITH TIME

2.3.1. Since the alkalinity is uniform throughout the entire range of depths at the time of the spring turn-over, the determinations for any one date give a rough idea of the speed at which bicarbonate has accumulated at any depth since the over-turn, and it is important to remember that the characteristic pattern, strikingly shown in the mean curves, is really an expression of differences in the rate of accumulation at different levels. Though the curves give unequivocal evidence of the general distribution of rates of increase, they cannot be used in further analysis. Even in Linsley Pond there appear to be occasions on which a part of the bicarbonate can be lost at any level, though the amount so lost is small and technical error cannot be altogether excluded. A few irregularities of this sort near the beginning or end of the series may produce a great difference in the slope of the best fitting line but little difference in the mean for that level. In the preliminary communication it was suggested that rapid rises in bicarbonate content tend to occur at a time when the stability of the epilimnion is low; the observations made in 1938 cast some doubt on this supposition. It is, however, certain that the rise in bicarbonate content does not occur at a constant rate; apart from the embroidery previously attributed to stability changes, there is a clear general tendency for the curve relating alkalinity and time at any depth to have a sigmoid form. Since the gradient is steepest at the bottom, sampling errors will be greater at the bottom and not in the thermocline and upper clinolimnion, as is the case with temperature readings. Computation of mean rates from the available data, by the method employed in 1.1.2. is therefore unfortunately somewhat more questionable than is the case with temperature; the resulting rates are likely to be the least reliable near the bottom, rather than in the metalimnion. Such mean rates, must, however, be used and in general show the same trends as do the means, except near the bottom, where wide departures occur, as at 11 m. and 13 m.; here the mean rates, based on a more complex procedure than the simple means, must be regarded as suspect.

2.3.2. Considering only the clinolimnion between 4 m. and 12 m. over which the clinolimnetic turbulence coefficient may be applied ($\mu^2 = 0.86$) it is possible to calculate what part of the observed change in alkalinity is due to turbulence (Fig. 13). To do this it is necessary to obtain the second derivative of the curve relating mean alkalinity to depth. This was done mechanically, by means of the Richards-Roope tangentmeter, the first derivative curve being obtained and a second differentiation carried out on it. The

values, for the different depths, of $\dfrac{\partial^2[HCO_3]}{\partial z^2}$, in milligrams per litre per meter per meter are then multiplied by 0.86 to give the calculated rate of change due to turbulence; subtraction from the observed values gives the non-turbulent rate of change $\dfrac{\partial[HCO_3]_{nt}}{\partial t}$.

Since the observed turbulence coefficient is not very large in comparison with the coefficient of molecular heat conduction, some error is certainly introduced by neglect of molecular diffusion. The procedure also introduces other opportunities for small errors. The fall in the value of the non-turbulent rate at seven meters would not appear if a very slight alteration in the drawing of the curve relating mean alkalinity to depth were made prior to the first differentiation; such an alteration might be justified, as the position of the inflection point of the curve is to a large extent arbitrary. In spite of this irregularity there is a very evident step-like region in the curve between 6 and 8 m. corresponding to the maximum in the morphometric curve between 7 and 8 m. Both curves undoubtedly rise steeply beyond nine meters. The general superposition of the curve for the non-turbulent rate of change on that for the observed total clearly indicates the greater importance of horizontal and possibly profile-bound currents as opposed to vertical mixing in determining the chemical structure of the lake.

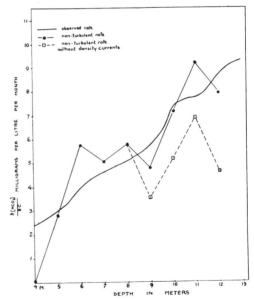

FIG. 12. Linsley Pond. Smooth unmarked curve, the mean rate of increase in bicarbonate concentration with time plotted against depth for 1937. Circles joined by unbroken lines, the non-turbulent rate of increase plotted against time. Squares joined by broken lines, the non-turbulent rate less the supposed contribution of density currents.

2.3.3. Assuming now the validity of the argument presented in 1.10.3., we may extend the process into the bathylimnion, calculating first the rate of change due to turbulent mixing, and from the observed value obtain the residual $\dfrac{\partial[HCO_3]_{nt}}{\partial t}$. In view of the uncertainties inherent in the fit of the temperature curve, this procedure can hardly be valid below 12 m. According to the argument presented in 1.11.2. part of the rise in alkalinity at all bathylimnetic levels is due to density currents running down the slope of the lake basin to levels appropriate to their densities. Assuming that the non-turbulent rise in temperature is due to such currents, a rate of change of 0.1° C. in unit time may be taken as equivalent to a rate of decrease in density of 0.000005 in unit time. Using the figure given in 1.8. for the difference in density between the water of Linsley Pond and distilled water, this rate of decrease is balanced by an increase of 2.8 mgrms. per litre of HCO_3 in unit time, so that it is possible to compute the amount of bicarbonate brought in by the density currents. Subtracting this from the non-turbulent rate gives the rate at which the concentration of bicarbonate, derived from the level in question, rises. In view of the fact that the density probably depends on substances other than and varying independently of bicarbonate, no great accuracy can be attached to the final estimate of the output from the mud at each level of the bathylimnion, but the results presented in Table 9 and shown graphically in Figure 12 are probably of the correct order of magnitude. The complete analysis described above is given in Table 9.

ESTIMATE OF BICARBONATE OUTPUT FROM UNIT
MUD SURFACE

2.4. A further treatment of the alkalinity curve is now possible by estimating the rate at which bicarbonate must leave the mud to cause the observed non-turbulent rate of change at each level. Below 9 m. the residuals, after subtracting the contributions due to density currents, should be used if the latter

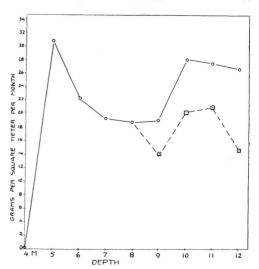

FIG. 13. Linsley Pond. Circles joined by unbroken lines, the rate of output of bicarbonate from unit area of mud surface in 1937, uncorrected for the effect of density currents. Squares joined by broken lines, the same less the increase in alkalinity due to density currents.

exist; as some doubt may be felt as to their reality, both the total non-turbulent rates and the rates from which the contribution of density currents have been subtracted are used. These, when multiplied by the reciprocal of the morphometric factor, $\dfrac{dA}{dV}$, used in preparing the lower curve of Figure 11, give the required values in grams per square meter per month. There is clearly a tendency to a bimodal distribution with respect to depth (Fig. 13). If the density current hypothesis be accepted the upper mode is clearly the best developed; the fall between 7 and 9 m. may in part represent bicarbonate lost from these layers and delivered to the lower layers by the

TABLE 9. Analysis of Rate of Change of Bicarbonate Concentration (all figures in mgrms. per litre per month unless otherwise stated).

	(1) $\dfrac{\partial^2[HCO_3]}{\partial z^2}$ mgrms. per litre per m. per m.	(2) $\mu^2 \dfrac{\partial^2[HCO_3]}{\partial z^2}$ =turbulent rate of change	(3) $\dfrac{\partial[HCO_3]}{\partial t}$ =observed rate of change	(4) (3) − (2) = $\dfrac{\partial[HCO_3]nt}{\partial t}$	(5) $\dfrac{\partial\theta_{nt}}{\partial t}$ °C per month	(6) (5) x 27.5 rate of change of HCO_3 due to density currents	(7) (5) − (6) rate of change of HCO_3 due to horizontal movements	(8) $\dfrac{dA}{dV}$ m² per m³	(9) $(4)\cdot\dfrac{dV}{dA}$ =(4) ÷ (8) estimates of rate of output into horizontal current grms. per m² per month	(10) $(7)\cdot\dfrac{dV}{dA}$ =(7) ÷ (8)
4....	2.80	2.41	2.42	0.01				0.0685	0.15	
5....	0.29	0.25	3.03	2.78				0.0904	30.8	
6....	−2.05	−1.76	4.00	5.76				0.1461	22.4	
7....	−0.49	−0.42	4.60	5.02				0.2571	19.5	
8....	−0.80	−0.69	5.09	5.78				0.3069	18.8	
9....	+1.05	+0.90	5.70	4.80	0.047	1.3	3.5	0.2507	19.1	14.2
10....	+0.09	+0.08	7.28	7.20	0.075	2.0	5.2	0.2558	28.1	20.3
11....	−1.78	−1.53	7.70	9.23	0.083	2.3	6.9	0.3230	28.5	21.4
12....	+0.96	+0.83	8.77	7.94	0.120	3.3	4.6	0.3651	24.6	12.6

density currents. If this is accepted there is probably a progressive decrease from 5 to 12 m. in the rate of removal of bicarbonate from the mud surface into the lake water. This progressive decrease may well be an expression of the vertical variation of the velocity of horizontal currents transporting the substance in question.

CARBON DIOXIDE ANOMALY

2.5.1. A very important corollary to the fact that the greater part of the variation in the bicarbonate content of the hypolimnion is to be accounted for by horizontal movements of water from the mud-water interface towards the center of the lake, may now be tentatively discussed. Ohle (1934) has considered the quantity by which the observed free CO_2 is in excess or in defect of the amount expected from the actual oxygen deficit (δO_2), a respiratory quotient of unity being assumed. This anomaly may be called the free carbon dioxide anomaly (ΔCO_2). It is clear, however, that the free carbon dioxide represents only a part of the metabolic carbon dioxide produced in the hypolimnion, since half of the carbon dioxide present in the bicarbonates of non-volatile bases and the whole of the bicarbonates present in equilibrium with the ammonium ion is to be regarded as metabolic. The metabolic carbon dioxide anomaly ($\Delta'CO_2$) is therefore probably of greater interest than the anomaly considered by Ohle. One series of carbon dioxide determinations only is available for Linsley Pond, made on September 20, 1937. In order to determine the metabolic CO_2, the increase in alkalinity since the lake was of uniform alkalinity, 143 days previously, on April 30, 1937, has been computed. Data for volatile alkali are not available for September 20, but figures for October 2, 155 days after the initial date, are available and have been multiplied by 0.923 to correct approximately for the later date. The data derived in this way are set out in Table 10 and Figure 14. It will be observed that while the original data for carbon dioxide show a vertical distribu-

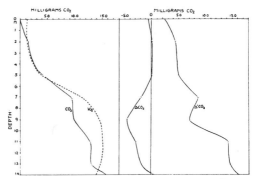

Fig. 14. Linsley Pond. Left, free CO_2 (solid) and CO_2 equivalent of oxygen deficit; right, free carbon dioxide anomaly (ΔCO_2) and metabolic CO_2 anomaly ($\Delta' CO_2$), all for September 20, 1937.

2.5.2. Since the only probable source of the excess carbon dioxide represented by $\Delta'CO_2$ is the anaerobic metabolism of the mud, and since the observations made by Riley and confirmed less extensively in the present investigation show that the oxygen consumption of the water of the hypolimnion in a dark bottle *in situ* is sufficient to account for the oxygen deficit, we can clearly divide the metabolic changes that affect the free water of the hypolimnion into two categories. One category comprises changes taking place in the water *in situ* and may be called *hydrometabolism;* the other category, comprising changes in the most superficial layer of the mud or at the mud-water interface, insofar as such changes affect the free water, may be termed *pelometabolism.* Substances, the concentration of which is primarily due to hydrometabolism, will show no morphometrically determined inflections on the curve of vertical distribution; this is most conspicuously true of the oxygen curve. Substances whose concentration is primarily determined by the pelometabolism, will show such inflections, as in the case of the curve of vertical distribution of alkalinity. Since in succeeding sections of the present work the great importance of pelometabolic changes is emphasized, it is of interest to attempt to obtain some measure of the relative magnitudes of the two metabolic categories. The mean value of the metabolic carbon dioxide anomaly on September 20 for the whole hypolimnion, below 4 m., duly weighted for the volume of each layer, is 7.98 mgrms. per litre. This corresponds to an anaerobic production of pelometabolic carbon dioxide of 1.68 mgrms. per litre per month. Though this estimate is liable to error in that neither is information available as to the departures from a linear relation between the anomaly and time or from a value of unity for the respiratory quotient, nor has loss of CO_2 by turbulent transport been considered, it may be taken as an approximate and probably minimal estimate of the pelometabolism of the hypolimnion. Estimation of the hydrometabolism is somewhat more difficult. Since it appears (Riley, 1940) that the low and relatively constant oxygen concentrations observed in the hypolimnion during

TABLE 10

(1)	(2)	(3)	(4)	(5)	(6)	(7)	(8)	(9)	(10)
Depth	HCO_3 incr April 30–Sept. 20	Volatile alkali (0.923 x 2 Oct. data)	½CO_2 of non-vol. alkali= ½(2)– (3)] x 44 — 61	CO_2 of vol. alkali (3) x 44 — 61	Free CO_2	Metabolic CO_2 (4)+ (5)+ (6)	δO_2 x1.375	ΔCO_2 (6)– (8)	$\Delta' CO_2$ (7)– (8)
0	9.4	0	3.4	0	0.07	3.5	1.3	—1.2	2.2
2	11.9	0	4.3	0	1.5	5.8	1.4	0.1	4.4
5	12.6	0.8	4.3	0.6	3.9	8.8	3.8	0.1	5.0
7	25.1	3.6	7.8	2.6	9.5	19.9	11.3	—1.8	8.6
9	27.0	5.4	7.8	3.9	9.7	21.4	14.2	—4.5	7.2
11	36.6	9.3	9.9	6.7	12.7	29.3	15.1	—2.4	14.2
13	36.2	10.2	9.4	7.4	12.9	29.7	14.8	—1.9	14.9
14	35.1	(10.2)	9.0	7.4	13.6	30.0	13.8	0.2	16.2

tion comparable to that of the bicarbonate ion, so reflecting the morphometry of the basin, this step-like distribution is much accentuated in the curve for the metabolic carbon dioxide anomaly.

the latter half of the summer represent a steady state between photosynthesis and oxidative consumption, the oxygen deficit cannot be used in a way comparable to the metabolic anomaly. Riley's experiments indicate that on an average the water of the lake when placed *in situ* in a dark bottle, consumes oxygen at an initial rate of about 4 mgrms. per litre per month.[8] This figure was obtained during the period between October and June and *a priori* might be expected to be too high when applied to water containing very little oxygen. A few data obtained in September, 1937, suggest that this is indeed the case; on the other hand a series obtained at the beginning of August shows higher values. In default of better information, Riley's figure may be provisionally accepted; it is certainly of the correct order of magnitude. 4 mgrms. O_2 correspond to 5.5 mgrms. of carbon dioxide produced per month, assuming again a respiratory quotient of unity. The pelometabolism, as measured by the metabolic CO_2 anomaly for the whole hypolimnion, would therefore appear to constitute about one fourth of the total metabolism as measured by the total rate of CO_2 production of the hypolimnion. This estimate is admittedly very rough, but is of value in showing that, owing primarily to the horizontal current system, the biochemical conditions of the hypolimnion as a whole may be considerably influenced by events taking place in the insignificantly small volume of mud in immediate contact with the water. In any consideration of substances, which unlike carbon dioxide cannot be produced in appreciable quantities in the free water, the importance of the pelometabolic contribution is vastly more important. In large lakes the ratio of the two metabolic categories will be much more in favor of hydrometabolism than in Linsley Pond, and all substances of exclusively pelometabolic origin will be present in lower concentrations. The principle developed in the present section is in fact an attempt to give quantitative expression to one aspect of Strom's (1932) generalization that productivity is a function of the ratio of volume to bottom surface. The mechanism by which the mud affects the free water in a stratified lake is of course entirely incomprehensible without a knowledge of the movement of dissolved substances by horizontal currents.

BASES IN EQUILIBRIUM WITH BICARBONATE

2.6.1. A few observations of the nature and distribution of the bases present in the water of Linsley Pond may not be out of place. Calcium has been determined essentially as recommended by Meloche and Setterquist (1933). Magnesium has been determined by Barnes' (1928) modification of Kolthoff's colorimetric method, using titan yellow after determination of the calcium, the concentration of calcium in the sample being adjusted to correspond to a known concentration in the standards. Total iron has been determined following Standard Methods of Water Analysis (8th ed. American Public Health

[8] This rate is not maintained, so that the oxygen concentration of water isolated at the spring overturn in a dark bottle hung in the hypolimnion never falls as low as that of the free water surrounding the bottle some months later.

Association, 1936), and total manganese by Winkler's method, following Maucha (1932). The titan yellow method for magnesium is very sensitive, not perhaps as accurate as might be desired; on one occasion (0 m. August 22, 1937) it gave results that are obviously too low and have been rejected.

2.6.2. Owing to the defective magnesium determination at the surface for August 22, and the presence of a large amount of decaying Anabaena liberating ammonia, after copper sulphate treatment, on August 5, the series for the surface water is less good than at 5 m. The mean value of the sum of the calcium and magnesium at the latter depth between July 12 and September 16, 1938, was 1.181 milli-equivalents per litre, the bicarbonate content at this depth and time 1.144 milli-equivalents per litre. The corresponding surface values with the defective dates omitted are 1.075 milli-equivalents for calcium and magnesium, 1.049 for bicarbonate. At these depths, therefore,

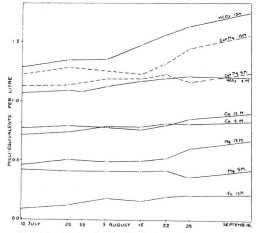

FIG. 15. Linsley Pond. Variation of bicarbonate and calcium and magnesium at 5 and 13 m., and of iron at 13 m. throughout the latter half of stagnation in 1938. (Milli-equivalents per litre.)

there is sufficient of the alkali earth kations to balance the whole of the bicarbonate and with small amounts (3-4%) in equilibrium with other anions (probably sulphate and silicate) as well. At nine meters the combined calcium and magnesium, give a mean of 1.289, the bicarbonate 1.343 milli-eq. per litre. The excess of bicarbonate is even more strongly marked at 13 m. where the mean for the two kations is 1.332, for the bicarbonate 1.490 milli-eq. per litre. The graphical presentation (Fig. 15) of the condition at 5 m. and 13 m. indicates how the excess of bicarbonate over calcium and magnesium develops as stagnation progresses. A few ammonia determinations and numerous data on the iron content show that there are present more than enough ammonium and ferrous ions to balance the bicarbonate under such conditions. Thus on August 5 at 13 m. there was present

Ecological Monographs
Vol. 11, No. 1

Mg	0.484 milli-eq.	HCO₃	1.357 milli-eq.
	per litre	Sum of	per litre
Ca	0.770 ,,	bases	1.499
NH₄	0.074 ,,	Difference	0.142
Fe (as Fe ··)	0.171 ,,		

At this time some of the iron is probably in a form other than ferrous ions, but a small quantity of manganous ion may have been present. It is clear that anions other than the bicarbonate also increase during the stagnation; presumably silicate, which has not been studied in the deep water of the lake, is the most important. In 1937 it was believed (Hutchinson, 1938) that the whole of the rise in the bicarbonate concentration in the deep water was due to ferrous and ammonium bicarbonate, since the 1937 analyses show a deficiency of bases in the deep water; it is clear that the calcium and magnesium determinations of these preliminary studies were too low.

2.6.3. With the exception of sodium, studied only on September 20, 1937, when about 1 mgrm. per litre was present at all levels, which, as is to be expected (Yoshimura, 1933) shows no clear vertical stratification, there is a marked tendency for all the bases present (Ca, Mg, NH₄, Fe, Mn)[9] to increase in concentration in the hypolimnion during stagnation. In the cases of magnesium, ammonia and iron, this increase progresses most rapidly at the bottom, less so at lesser depths. A few representative determinations for the first two substances are presented in Figure 18, the question of iron is discussed in 3.7. Calcium shows a tendency to higher values in the middle water than at the bottom during August and September, becoming regularly stratified only at the end of stag-

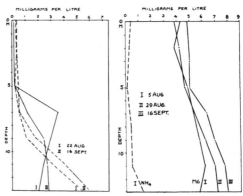

Fig. 17. Linsley Pond. Vertical distributions of manganese (solid) to show the difference in stratification from that exhibited by total iron (broken).

Fig. 18. Linsley Pond. Vertical distributions of magnesium (solid lines) and of ammonia (broken line) in the latter half of stagnation in Linsley Pond, 1938.

nation (Fig. 16). The concentrations at the different levels vary rather irregularly, and though high maximum at 7 m. on August 22 was checked and is certainly not due to technical error, its absence on August 29 may be due to sampling error as it reappears on September 13. No great significance should be attached to the rather meagre data available, but they suggest that calcium passes into solution more easily in the layers poor in iron than the layers where, in what will be defined (3.4) as phase III of stagnation, much ferrous iron is found. The data given in Figure 17 for manganese show very considerable and well-established variations, with an intermediate maximum on August 27 very much like that for calcium. Small maxima of calcium in the upper hypolimnion are recorded in some of Ohle's (1934) analyses of Baltic lakes, while apart from the extraordinary metalimnetic maximum of manganese in Ranu Klundungan in Java, Ruttner (1931) has recorded smaller maxima in intermediate water not unlike those reported above. At present it is merely desirable to call attention to the chemical activity exemplified by the changes in manganese, and to a less extent calcium, concentration, and to point out that if the density current hypothesis of 1.11.2 be accepted, the water at the bottom has gained some of its alkalinity at higher levels, but has subsequently passed over a considerable mud surface that may provide great opportunities for base exchange. It is hoped in the future to obtain more data on the variations of these substances.

COMPARISON WITH THE RESULTS OF OTHER INVESTIGATORS

2.7. The horizontal water-movements of the hypolimnia of stratified lakes have been studied by Wedderburn and Watson (1909), Wedderburn (1910), Möller (1928, 1933), Whipple (1927) and others, but too little information exists to permit any empirical generalizations of value in the present

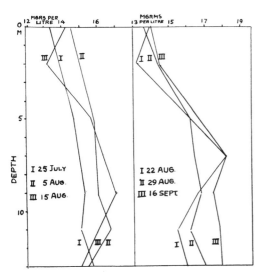

Fig. 16. Linsley Pond. Vertical distributions of calcium.

[9] A slight stratification of sulphate was observed on September 20, 1937 (0 m. 0.75, 2 m. 0.68, 5 m. 1.47, 9 m. 1.64, 13 m. 1.70). On October 13, 1938, 0.05 mgrms. per litre of titanium was detected in 13 m. water; this may have been in suspension.

study, other than to state that the existence of such currents is well established. Defant's (1932) interesting theoretical treatment assumes a steady state resulting from the action of the wind on the lake surface, and deduces two complete circulation systems, one in the epilimnion, one in the hypolimnion. This implies zero horizontal motion (dead water) at a certain depth in both regions. The type of circulation deduced by Defant does not seem to fit in very well with the meagre observations available, which suggest that a steady state is very rarely achieved. Recently Elster (1939) has applied the methods of dynamic analysis used in Oceanography, to a study of the deep-water currents of Lake Constance. In general he finds a decrease in the motion between any two stations in a transverse profile with increasing depth, except near the shores of the lake. A current system of this sort may well be developed only in large lakes, but would fit the observed changes in alkalinity discussed in the present paper. The direct measurements of both Wedderburn and Möller, however, suggest a more irregular type of distribution of velocities. Hypolimnetic water-movements have been postulated by Alsterberg (1927) to explain the distribution of oxygen concentrations, and in particular the very low values sometimes found in the thermocline region. Alsterberg's concepts were in fact the starting point of the treatment given in the present paper. This explanation of the metalimnetic minimum is considered probable by Yoshimura (1939), but Alsterberg's theory cannot be applied to the oxygen of Linsley Pond (cf. Riley, 1940). Alsterberg's whole theoretical position has recently been attacked with considerable vigor by Grote (1936). There is probably much to be said for the position adopted by both investigators; Alsterberg's hydrodynamic scheme (1927, Fig. 6) is certainly an unlikely one, but is not really necessary to a theory of the effect of morphometry on the vertical distribution of dissolved substances. Any series of predominantly horizontal currents would be equally effective. Of all previous investigators, Einsele and Vetter (1938), who have observed very large and rapid changes which they attribute to hypolimnetic water movements in the concentration of iron, phosphorus, silica and other substances in the deeper water of the Schleinsee, perhaps come closer to the position adopted in the present paper. A very brief mention of comparable changes, attributed to water movements, is made by Juday, Birge and Meloche (1938).

THE PHOSPHORUS CYCLE IN LINSLEY POND

THE DATA

3.1.1. The methods used in the determination of phosphorus were those employed by Juday and Birge (1931), oxidation of organic phosphorus being carried out as recommended by Robinson and Kemmerer (1930). Occasionally in the oxidation of total phosphorus the final solution obtained was not as colorless as would be desired, but it is improbable that any appreciable error has been introduced by this. In determining soluble phosphorus, an effort has been made to read below 0.001 mgrms. per litre. Watercolor was at first compensated by Pearsall's (1930) method, using Bismarck Brown; later it was found better to employ the glass screens used in determining the color in a Hellige comparator for this purpose. Owing to the disturbing effect of water color and the difficulty of obtaining perfectly matched and colorless Nessler tubes, it is realized that the determinations of soluble phosphorus at great dilution are not very accurate. When the great variations of total phosphorus summed over the entire lake were first noticed, it seemed possible that some technical source of error was vitiating the experiments, and the need for careful control of the technique was realized. The details of a set of controlled determinations are of sufficient interest to be mentioned briefly. On July 5, 1938, the phosphorus content of the lake was very low, the hypolimnetic values being the lowest ever recorded. A sample of 9 m. water was set aside to be oxidized along with a blank and control of known phosphorus content; all these were oxidized with the series taken a week later. Fortunately one of the greatest rises recorded took place in the interval, the lake gaining over 9 kilos of phosphorus. The 9 m. sample of the previous week gave the correct value (0.004), as did the control (0.010); the blank was completely negative. Since the enormous rise recorded between July 5 and July 12, 1938, was therefore technically above reproach, there is no reason to reject the smaller rises occurring at other times when the control was less rigorous.

3.1.2. Since the main object of the investigation was the construction of a phosphorus balance sheet during stagnation, more attention was given to total phosphorus than to the fractions into which it may be divided. Determinations of *soluble phosphate phosphorus* have, however, been made on occasions at all levels, and in the surface water at very frequent intervals. In the case of the surface water, moreover, determinations of total phosphorus on both unfiltered water and on water after passage through a 35 second membrane filter have been made. The difference between the two determinations is referred to throughout the paper as *sestonic phosphorus*. Subtracting the soluble phosphate phosphorus from the total phosphorus of the membrane filtrate gives the *soluble organic phosphorus*. From the autumn of 1938 to mid-summer of 1939 a series of determinations of soluble phosphate phosphorus in the unfiltered surface water and in a membrane filtrate of the same water has been obtained. The difference between these two determinations, if detectible, gives the amount of *acid-soluble sestonic phosphate phosphorus*. The above scheme of fractionation is identical with that proposed by Ohle (1939); unfortunately his paper and that of Cooper's (1938) were published too late for the technical refinements introduced to be used in the present study.

Ecological Monographs
Vol. 11, No. 1

NATURE AND PROPORTIONS OF THE FORMS OF
PHOSPHORUS PRESENT IN THE
SURFACE WATER

3.2.1. Twenty-three analyses are available of the phosphorus content of the surface waters of Linsley Pond, separated according to the above scheme.

The mean values and standard deviations are set out below:

Seston P ...... 0.0133 mgrms. per litre, = 0.0082
Organic
   soluble P ... 0.0060 mgrms. per litre, σ = 0.0036
Phosphate P .. 0.0017 mgrms. per litre, σ = 0.0019[10]

It will be observed that the mean phosphate phosphorus in the surface waters constitutes but 8.1% of the mean total, the organic soluble 28.6%, the mean seston phosphorus 63.3%. The mean total phosphorus, 0.021 mgrms. per litre, is comparable to the figure of 0.023 mgrms. per litre given by Juday and Birge (1931) for the surface waters of 479 lakes in northeastern Wisconsin. The organic phosphorus of the latter authors corresponds to the sum of the sestonic and organic soluble phosphorus in the present study. The Wisconsin workers found about 88% of their phosphorus in combined form; in the present investigation about 92% is combined in either solid or dissolved form. The general distribution of the phosphorus in the surface waters of Linsley Pond and of the Wisconsin lakes therefore superficially appears to be similar. It is probable that in spite of this superficial resemblance an important difference exists between the distributions of the two mean series. Juday and Birge give five analyses of the phosphorus content of centrifuge plankton; such phosphorus corresponds to the sestonic phosphorus of the present investigation. These five analyses indicate a phosphorus content of from 0.185-0.394% of the dry centrifuge plankton. In the present investigation, with a larger number of observations the comparable determinations range from 0.063-0.775%. Half the Linsley determinations, however, fall into the Wisconsin range; the mean of the five Wisconsin determinations is 0.290%; of the twenty-three Linsley determinations, 0.353%. The two means are clearly closely comparable. If the Linsley sestonic phosphorus be considered as entirely in organic combination and be expressed in terms of the organic seston the latter is found to have a mean phosphorus content of 0.464%. A similar or slightly lower figure may clearly be expected in the case of the Wisconsin lakes. Now according to Birge and Juday (1934) the mean organic sestonic content of these lakes is 1.36 mgrms. per litre. This would therefore correspond to a sestonic phosphorus content of 0.0063 mgrms. per litre, or 27.5% of the mean total phosphorus of the Wisconsin series. The soluble organic would likewise be 0.014 mgrms. per litre or 60.9% of the mean total. It would appear therefore that while in Linsley Pond on an average over 60% of the phosphorus in the epilimnion is in sestonic form, in

the Wisconsin lakes a like proportion is probably in soluble organic form. Juday, Birge, Kemmerer and Robinson (1928) comment on the fact that much of the organic phosphorus present in the Wisconsin lakes appears to be in solution. Both series of analyses contrast markedly with those given for the sea by Redfield, Smith and Ketchum (1937) who find from 70-90% of the phosphorus to be present as phosphate and but 3-6% to be in the seston.

3.2.2. The considerable variation in the phosphorus content of the total seston indicates that factors other than the mere quantity of suspended matter are involved in determining the quantity of sestonic phosphorus. Certain of these factors are amenable to statistical analysis. From twenty-four determinations of sestonic phosphorus (Ps), total seston (Ts), organic seston (Os), seston ash (As) and chlorophyll (C) the following simple correlation coefficients can be computed:

$$r_{Ps.Ts} = 0.560$$
$$r_{Ps.Os} = 0.560$$
$$r_{Ps.C} = 0.527$$
$$r_{Ps.As} = 0.340$$

The first three, being significant to above the 99% level (Fisher, 1934), are regarded as statistically significant, the last is statistically insignificant. The correlation with total seston is not amenable to further analysis. Putting Ps in terms of milligrams per cubic meter to avoid unnecessary zeros, the regression equations corresponding to the second and third correlation coefficients are

$$Ps = 2.81 \ Os + 5.7 \ \dots \dots (14)$$
$$Ps = 0.478 \ C + 6.9 \ \dots \dots (15)$$

In both equations the independent constant corresponds to almost half the mean sestonic phosphorus; neither can therefore be regarded as very satisfactory expressions of the variation of the phosphorus content with the two variables considered independently. The variation of chlorophyll is, in the present series of observations, largely independent of the variation of organic matter

$$r_{Os.C} = 0.335$$

which is statistically insignificant.[11] The partial correlation coefficients

$$r_{PsOs.C} = 0.456$$
$$r_{PsC.Os} = 0.436$$

significant to the 95% level, indicate that a considerable part of the variation of the seston phosphorus with one variable is independent of the variation with the other. The multiple correlation coefficient

$$R_{Ps(C.Os)} = 0.654$$

is a considerable improvement on either simple coefficient. The associated multiple regression equation

$$Ps = 2.07 \ Os + 0.353 \ C + 2.7 \ \dots \dots (16)$$

is likewise an improvement on either of the equations (14) and (15) in that the independent constant is considerably reduced. Figure 19 shows the way in which this equation roughly expresses the major variations in seston phosphorus in 1938-39. Neglect-

---

[10] Phosphate phosphorus data obviously are not normally distributed, owing to the high autumnal values, which include acid-soluble sestonic phosphate phosphorus.

[11] Riley (1939a), using different dates, none of which fell at the height of summer, obtained a significant correlation.

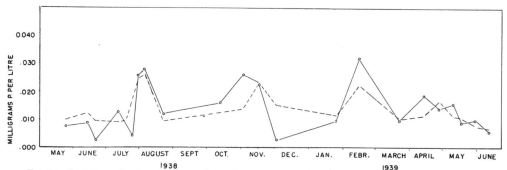

FIG. 19. Variation of seston phosphorus in surface water 1938-1939 observed (solid) and computed (broken) from the multiple regression equation (16).

ing variation due to causes extraneous to either the mass of organic matter or the chlorophyll, and substituting mean values, (16) becomes

$$13.9 = 6.0 + 5.2 + 2.7$$

Very approximately therefore the variations in the phosphorus associated with chlorophyll are as important as variations associated with the whole suspended organic mass. Now since only about 20% of the total organic seston of Linsley Pond is phytoplankton (Riley, 1940), it is clear, for this relation to hold, that the phosphorus content of unit mass of phytoplankton must be greater than that of unit mass of the dead seston or detritus. This conclusion is comparable to that reached by Birge (Birge and Juday, 1934), by more direct methods, relating to the distribution of nitrogen in the planktonic organisms and detritus of the seston.

TOTAL PHOSPHORUS ESTIMATED FOR THE ENTIRE LAKE

3.3.1. From the total phosphorus determinations at individual levels at a station in the center of the lake, the total phosphorus content of the entire lake has been calculated. The mean concentrations of the layers 0-2 m., 2-5 m., 5-7 m., 7-9 m., and 9-11 m. have been taken as the means of the total phosphorus concentrations at the top and bottom of each layer. In the case of the lowest layer, 11-14.8 m., only the 13 m. concentration was used in 1938, the mean of the 12, 13, and 14 m. determinations in 1937. Since there is some reason to believe that the very high values observed in the epilimnion in the latter part of 1938 are due to liberation of abnormal amounts of phosphorus from decomposing algae in the littoral, killed by copper sulphate treatment of the water, the 1937 data will be primarily considered in the analysis of the natural variations in the phosphorus content of the lake. The values for 1938, however, give critical information as to the progress of changes in the hypolimnion. Horizontal variations in all chemical parameters doubtless occur; the experience of most investigators seems to indicate that they are small in lakes of the size of Linsley Pond, and no attempt was made to consider them.

3.3.2. In Figure 20 and Table 11 the estimates of

the total phosphorus content of the lake throughout the thermally stratified period (April to August) in 1937 are presented. The mean of all observations gives a value of 8.98 kilograms of phosphorus in the lake. There is, however, no evidence of constancy of content, nor in this year, of a steady trend either of increase or decrease; the total content rather varies in an irregularly rhythmical manner, from low values of about 6 kilos to high values of about twice that amount. In the earlier part of 1938 similar, and even one more spectacular (3.1.1.), changes occurred; later, after two treatments with copper sulphate, a steady rise took place. Gross technical error having been excluded as an explanation of the observed changes as indicated in 3.1.1., two possibilities present themselves. The increases and decreases might be due to gain of phosphorus through the inlets and loss through the outlet, or it might be due to purely internal changes leading to gain of phosphorus from the mud across the mud water interface, and loss by sedimentation of particles, that had incorporated this phosphorus gained by the free water from the mud. Though the first hypothesis is inherently improbable, as there are cases (Fig. 22) of gains at all levels occur-

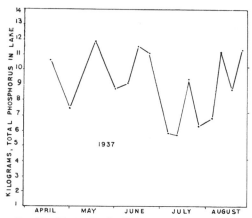

FIG. 20. Linsley Pond. Variation of total phosphorus in the entire lake, 1937.

ring simultaneously under conditions of extreme thermal stratification, when the inlets could hardly affect the hypolimnion, the role of the paralimnion, or environment of the lake, must be considered before an attempt is made to evaluate quantitatively the limnetic or internal changes.

3.3.3. In order to estimate the paralimnetic contribution to the observed changes, estimates of the rate of flow of the two inlets and of the outlet of the lake were made throughout the period of observation, by timing the descent of floating objects in the streams and measuring the mean depth and the width of the streams in the middle of the courses selected. This method is very rough, but the mean flow of the inlets and the outlet, when compared; show that the results must be of the correct order of magnitude. Total phosphorus determinations were made on the same water.

	Water cubic m. per wk.	Total Phosphorus kilos per wk.
Mean income, Inlet 1	22,012	0.354
Mean income, Inlet 2	5,508	0.118
Total mean income	27,520	0.472
Mean loss, outlet	24,271	0.365
Net gain	4,289	0.107

The excess of water entering the lake over that leaving would correspond to an excess of 3.4 cms. per week evaporation over precipitation and surface run-off into the lake; this is perhaps excessive and would indicate that the gain in phosphorus, which is very small, is also probably slightly overestimated. If now the individual weekly gains or losses in phosphorus due to the paralimnion are computed from the individual determinations in the waters of the streams, it will be found that they are never commensurate with the observed changes, and that very large residuals due to purely limnetic events are obtained when the gains and losses due to the streams are subtracted from the gains and losses observed in the lake (Table 11).

3.3.4. Differences between successive weeks show that in 1937 net increases of as much as 4.93 kilos per week and net decreases of as much as 3.18 kilos per week, due entirely to limnetic events, are possible; while in 1938 gross increases, in which the inlets can have played but a small part, of 9 kilos per week occurred. In order to obtain at least a minimal estimate of the mean natural rate of replacement, it may, however, be assumed, as a limiting case, that phosphorus only enters the free water from the mud when the total content is rising, only is sedimented when the total content is falling. The net weekly increments when rises are occurring may then be added and divided by the whole time, giving the lowest estimate for the mean rate of regeneration; the net weekly losses when falls are occurring may similarly be added and divided by the whole time giving the minimum estimate for the mean rate of sedimentation. The values so obtained for 1937 are 1.09 kilos per week for the mean gain by regeneration, 0.98 kilos per week for the mean loss by sedimentation. In spite of the fact that these estimates are minimal

TABLE 11   Estimates of gain and loss of total phosphorus from the entire lake. Linsley Pond, 1937.

Date	Kilograms P in Lake	Rate of gain or loss through inlets and outlets	Mean rate of gain or loss between observations	Gross change per week	Net change per week
April 17	10.60	+0.33			
April 30	7.54	—0.49	—0.08	—1.65	—1.57
May 17	11.79	+0.11	—0.19	+2.04	+2.23
June 1	7.55	—0.11	0.00	—1.98	—1.98
June 8	8.21	(—0.01)	—0.06	+0.66	+0.72
June 15	10.54	(+0.09)	+0.04	+2.33	+2.29
June 22	10.13	+0.19	+0.14	—0.41	—0.55
(June 29) July 6	5.86	+0.64	—0.06	—2.15	—2.09
July 13	5.79	+0.22	+0.43	—0.07	—0.50
July 20	9.28	+0.21	+0.22	+3.49	+3.27
July 27	6.30	+0.18	+0.20	—2.98	—3.18
Aug. 3	6.90	+0.02	+0.10	+0.60	+0.50
Aug. 10	12.05	+0.42	+0.22	+5.15	+4.93
Aug. 17	9.77	+0.04	+0.23	—2.28	—2.51
Aug. 24	12.42	—0.21	—0.09	+2.05	+2.74

they are surprisingly high, implying that the average sojourn of a phosphorus atom in the lake, acted on solely by internal forces, is two months, or that during the entire period of the investigation in 1937 phosphorus was entering and leaving the water to and from the mud fast enough to give two entire replacements of the mean amount of the element present in the water. Actually, owing to the nature of the assumptions involved, it is certain that the time of sojourn must be shorter and the rate of replacement more rapid; events in those individual weeks, when 4 to 9 kilos are lost or gained, suggest that phosphorus metabolism of the lake may actually be four to nine times as rapid as the minimum estimate. The mean minimum estimate corresponds to a mean increase of 0.0011 mgrms. per litre, or approximately 0.001 mgrms. per cm^2 of lake surface, which may be regarded as a sufficiently accurate estimate for the mud surface. The maximum increase observed in 1937 (Aug. 3-10) of 5.15 kilos similarly corresponds to an output of about 0.0035 mgrms. per cm^2 of mud; the maximum in 1938 (July 5-12) of 9.78 (gross) to about 0.007 mgrms. per cm^2 mud surface.

PHASES OF STAGNATION

3.4. Except in so far as the epilimnetic values at the end of the summer may be unduly high, there is nothing to indicate that the vertical distribution of phosphorus in the summer of 1938 is abnormal; the data of both years are therefore used in the present discussion. In order to clarify the relations of phosphorus compounds to other substances, and particularly to emphasize the very important relationship, discovered by Einsele (1936, 1938) between the distribution of phosphorus and iron, it has been found convenient to divide the processes during stagnation at any depth into three phases as follows:

Phase I. The oxygen content falls from the high values of the spring turn-over to about 1 or 2 milligrams per litre; alkalinity starts rising as the oxygen falls; total phosphorus remains low, under 0.03 mgrms. per litre; iron has not been determined but is undoubtedly low.

Phase II. The oxygen content remains low, usually between 1 and 2 mgrms. per litre; alkalinity continues to rise; phosphorus (under 0.03 mgrms. per litre) and iron (under 0.50 mgrms. per litre) low, as in phase I. Manganese, however, may be high.

Phase III. Oxygen content as in phase II or slightly lower; alkalinity rising, but more slowly than in the previous phase; phosphorus and iron rising, ultimately reaching values in excess of 0.10 mgrms. per litre and 5.0 mgrms. per litre respectively.

The essential difference between the first two phases lies in the establishment of a relatively constant low oxygen concentration in phase II; the essential difference between phase II and phase III lies in the rapid rise in phosphorus and iron in phase III. In the first two phases it may be assumed that the redox

potential (irreversible oxygen electrode) is of the order of 0.45-0.50 volt; in phase III it may fall to 0.15 volt (Hutchinson, Deevey and Wollack, 1939). The high iron content of water of any level in phase III is primarily, sometimes exclusively, due to ferrous iron, while the phosphorus is largely present as soluble phosphate (e.g., at the end of August 1938 from 60 to 70% of the total phosphorus is present as soluble phosphate in the bottom layers) in marked contrast to the condition in the well-aerated surface water where the soluble phosphate represents but a small fraction of the total. As Einsele has pointed out, most of the soluble phosphate accumulating in the final phase of stagnation will be precipitated out as ferric phosphate when the ferrous iron is oxidized at the turn-over. The events as they affect one particular level (9 m.) are presented graphically in Figure 21.

VERTICAL DISTRIBUTION OF PHOSPHORUS

3.5.1. Typical distributions for four dates in 1937 are given in Figure 22; more detailed distributions for 1938 in Figures 24-30. In May all depths may be regarded as in phase I. During June and July

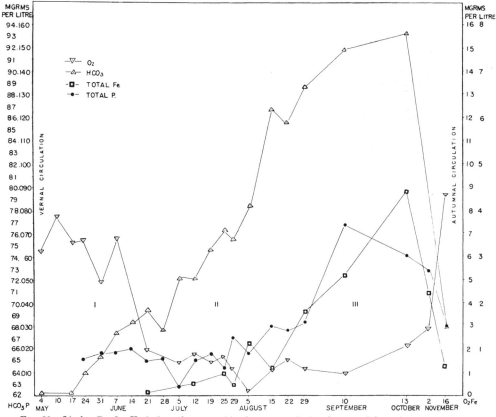

Fig. 21. Linsley Pond. Variation of oxygen, bicarbonate, total phosphorus and iron throughout stagnation in 1938 at 9 m.

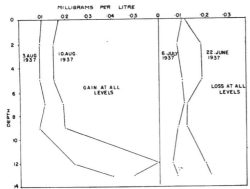

FIG. 22. Linsley Pond. Gain or loss of total phosphorus at all levels between successive determinations in 1937.

the 13 m. water may be regarded as entering phase III, while the rest of the hypolimnion is in phase II. Late in August the 11 and 13 m. levels show phase III fully developed, the 7 and 9 m. levels are still in phase II, the 5 m. level is entering the latter phase. In September phase III is well developed at 9 m., and poorly developed at 7 m. The epilimnion of course is permanently in the early stages of phase I. It is important to note that although great and irregular variations in total phosphorus occur at all depths throughout all phases of stagnation, giving rise to the variations in the total content of the lake discussed in 3.2., there is no tendency to develop any characteristic pattern in the vertical distribution of total phosphorus so long as levels in phases I and II are considered. This is strikingly shown in Figure 23, in

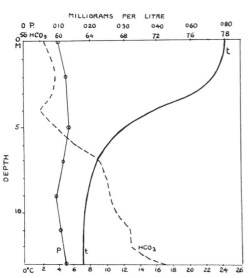

FIG. 23. Linsley Pond. Mean total phosphorus and alkalinity, May 24-July 27, 1937.

which the mean values for the period May 24-July 27, 1937, are presented graphically. During the period in question very considerable changes in the total phosphorus had occurred at all levels, yet the mean curve shows a practically uniform vertical distribution, in marked contrast to the mean alkalinity curve for the same period in which the step-like pattern is already well developed.

3.5.2. Owing to the fact that access of air during filtration precipitates soluble phosphorus in the presence of ferrous salts as ferric phosphate, reliable data for the deeper water could not be obtained (filtration at a low pH might be possible but has not yet been tried). Moreover, it is probable that in the presence of abundant soluble phosphate, even if appreciable amounts of ferrous iron are not present, the filtration

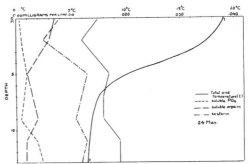

FIG. 24. Linsley Pond. Temperature curve (solid, marked t), total phosphorus (solid), sestonic phosphorus (chain), soluble organic phosphorus (long broken line), and soluble phosphate phosphorus (short broken line) for May 24, 1938.

technique is unsatisfactory owing to adsorption. Nevertheless, a sufficient number of vertical series have been obtained indicating the general nature of the vertical distribution. These series can best be presented in graphical form (Figs. 24-30). Early in the season soluble phosphorus is detectible throughout the lake, and may increase slightly at the bottom. In the latter part of the summer, when all the bottom water is in phase III, practically no soluble phosphorus can be detected in the surface waters but

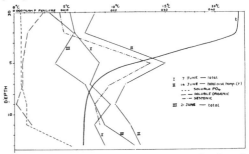

FIG. 25. Linsley Pond. Total phosphorus for June 7 (I), June 14 (II), June 21 (III), and other data for June 14, indicated as in Figure 24. Note the great loss of phosphorus, at 5 m. between June 14 and 21.

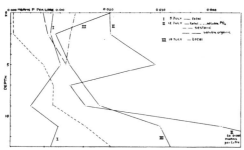

FIG. 26. Linsley Pond. Total phosphorus for July 5 (I), July 12 (II), and July 19 (III),.·other data for July 12, indicated as in Figure 24.

very large amounts are present in the deeper layers of the lake (Fig. 28). The appearance of soluble phosphorus in the hypolimnion is, however, irregular and temporary establishment of phase III at 13 m. is clearly possible, with a subsequent reversal to phase II as between July 19 and 25 when a loss of soluble phosphate, an increase in total phosphate, and a slight rise in iron occurred. In so far as it has been determined, organic soluble phosphorus is more regularly distributed than the other fractions, though on July 12 it is largely responsible for the peak at 5 m. In general the great irregularities that the total curve may exhibit are largely due to the sestonic fraction.

3.5.3. Owing to the rapid variations exhibited by the phosphorus content, no quantitative treatment comparable to that used in analysis of the variations of bicarbonate content is possible. A rough approach to the problem may, however, be made. In Figure 23 the difference between the mean of the three analyses made in the period May 31-June 15 and the last three analyses made in each season (August 10, 17, 24 in 1937; August 29, September 16 and October 13 in 1938) is given. In spite of the great variation in phosphorus content in the upper seven meters throughout the summer, the only significant effect of prolonged stagnation is to increase the phosphorus concentration of the extreme lower layers of the lake. It is therefore reasonable to suppose that while the

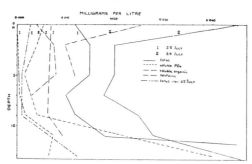

FIG. 27. Linsley Pond. Total, sestonic, soluble organic and soluble phosphate phosphorus for July 25 (I) and July 29 (II), indicated as in Figure 24. Note the enormous increase in sestonic phosphorus at the surface.

regenerative process is dependent primarily on the horizontal current system deduced to explain the observed rates of increase of alkalinity, the cycle proceeds in the epilimnion and the upper part of the hypolimnion in a different way to that in the deepest water. The chemical aspects of this difference have been elucidated by Einsele and are discussed in 3.6.1., but a possible mechanical aspect of the process must be mentioned. Reference to Figures 29 and 30 will show that phase III, though feebly indicated at 7 m., is never properly developed above 9 m. The great increases in phosphorus are confined to the region demarcated as the bathylimnion in the treatment of temperature in 1.2.1. If the speculative interpretation of this region given in 1.10.3. and 2.3.4. be accepted, it seems reasonable to interpret the rise in total phosphorus and total iron in general, and of phosphate and ferrous iron in particular, as due to the effect of density currents carrying a considerable amount of the water present in this region over the highly reductive mud in the depths of the lake. The chemical significance of this process will be clear from the discussion given in 3.6.1.

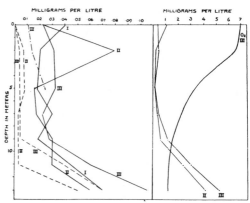

FIG. 28. Linsley Pond. On left, total phosphorus for August 5 (I), August 15 (II), and August 22 (III) and soluble phosphate phosphorus for the second and third of these dates, but with the horizontal scale reduced to one fifth. On right, total iron for the second and third date and oxygen for August 22. Note.the persistence of the stratification of July 29 through August 5, and the subsequent appearance of immense amounts of non-phosphate phosphorus at 2 m. on August 15.

3.5.4. In the epilimnion and in the first two phases of stagnation in the hypolimnion, rapid removal of phosphorus is needed to complete the cycle. The only mechanisms by which phosphorus can be removed from the water of the lake as a whole are the gravitational sedimentation of seston, and the accumulation of seston on aquatic plants in the littoral. The work of Chandler (1937), on the removal of plankton by the vegetation of rivers, suggests that the second of these mechanisms may be effective in the freely circulating epiliminion, but the gravitational sedimentation of seston is doubtless far more important. On account of the high proportion of

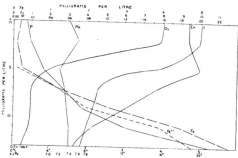

FIG. 29. Temperature (solid, t), total phosphorus (solid, P), oxidation-reduction potential (solid, Eh), oxygen (solid, O₂), total iron (long broken lines), and ferrous iron (short broken lines) for September 16, 1938.

sestonic phosphorus in Linsley, as much as half the mean number of phosphorus atoms may be undergoing sedimentation at any time, but with varying speeds depending on the density and size of the sestonic particles involved. Direct evidence of such sedimentation is not indicated in many of the series. The best instance is provided by the fate of the immense amount of phosphorus that appeared in the *Anabaena* bloom between July 25 and 29. This persisted in the surface water at least until August 5, in spite of the treatment of the lake with copper sulphate on July 30. By August 15, the phosphorus content of the surface layer was more normal but a huge accumulation was found at two meters, presumably in detritus formed as the result of the death of the bloom. A week later, on August 22, a somewhat abnormal amount appeared at five meters, though the two-meter level was still high. By September 17 the five- and seven-meter layers were both high, but the significance of the rise at that latter level is uncertain as the phosphorus concentration increases regularly with depth. It is clear that the descent of the phosphorus observed at this time was much slower than is normal; presumably the density of the detritus involved was abnormally low. In the case of the most remarkable reduction in phosphorus content observed, the loss of 0.022 mgrms. per litre at five meters between June 14 and 21, 90% of the

phosphorus was present in sestonic form at the level in question on the first date. There is no evidence of the reappearance of any of this phosphorus at a lower level on the succeeding visit; losses in fact occurred at all levels. A very large amount of the seston at five meters had therefore apparently fallen to the bottom of the lake or at least below 13 m. in a week. Such a descent implies a rate of fall of 1.14 m. per day, i.e., 4.75 cms. per hour or 0.8 mm. per minute. This seems rather rapid but may well be possible. A somewhat similar but less extensive diminution was observed between July 12 and 19, also at five meters. Here 0.017 mgrms. per litre was lost and in this case only 0.011 mgrms. of sestonic phosphorus was initially present, so part of the rather large amount of soluble (phosphate 0.007, organic

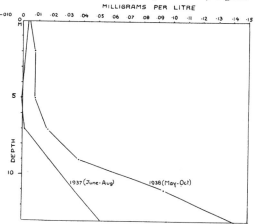

FIG. 31. Linsley Pond. Increment in total phosphorus, at the levels analyzed, occurring between the first three and last three analyses of each season.

soluble 0.011) phosphorus must have been absorbed before sedimentation could occur. There is, however, in this case some indication of a part of the phosphorus reappearing at 7 and 9 m. where rises of 0.003 mgrms. per litre occurred. It is obvious that without a knowledge of the thickness of the band of high concentration initially indicated at five meters, no further analysis is possible. In general, however, it would appear that the only possible mechanism that could produce sedimentation of the magnitude implied by the observed variations is the concentration of detritus and nannoplankton into faecal pellets as the result of the feeding activities of the zooplankton.[12] Such pellets might well have the density and magnitude required to cause the observed changes. Dr. Riley kindly informs me that the phytoplankton (mostly minute green algae) volume at five meters, on June 14, was 0.418 cm³ per m³, on June 21, 0.438 cm³ per m³, the zooplankton on the first date 0.569

[12] Since the above was written, J. Grim (Int. Rev. Hydrobiol. **39**: 193-315, 1939) published data on the rate of loss of diatoms through death and sedimentation in Lake Constance which indicate that the figure given above of 1.14 m. per day is by no means impossibly high even if the material is not concentrated in faecal pellets.

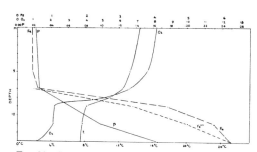

FIG. 30. Linsley Pond. Temperature (solid, t), total phosphorus (solid), oxygen (solid, O₂), total iron (long broken lines), and ferrous iron (short broken lines) for October 13, 1938.

cm³ per m³ and on the second, 0.448 cm³ per m³. An excess in favor of zooplankton in the P:Z ratio is unusual in the lake and may well explain the very great drop at 5 m. on the period involved. By very indirect means Dr. Riley computes that at least 23.5% of the phytoplankton produced at the 5 m. level was eaten between the dates in question. As replacement is fairly rapid this may represent a considerable faeces production. The percentage of phosphorus in the seston would inevitably fall in such a process, but data do not exist to prosecute this phase of the analysis further.

COMPARISON WITH OTHER LAKES

3.6.1. Six sets of determinations of total phosphorus in Lake Quassapaug, made during the summer of 1938, show similar but less rapid changes. The greatest rise took place between July 1 and 15 (Fig. 32) when the total phosphorus of the entire lake increased from 48 kilos to 77 kilos. A slight increase to 92 kilos occurred during the next two weeks, followed by a slow fall throughout August. The rate of output during the maximum rise was about 0.0013 mgrms. per cm² per week, contrasting with the high values of 0.0035-0.007 observed on occasions in Linsley Pond, and very little above the minimum mean estimate for the latter locality. The low rate of output of the mud of Lake Quassapaug, even

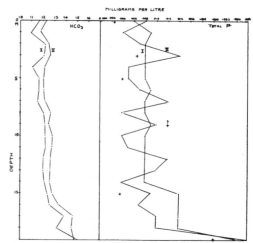

FIG. 33. Lake Quassapaug. Total phosphorus (right) and alkalinity (left) for August 18 and August 25, 1938. Crosses indicate soluble phosphorus determinations for the second date; that at 9 m. is, however, clearly erroneous. Note the extraordinary stratification that developed in the lake.

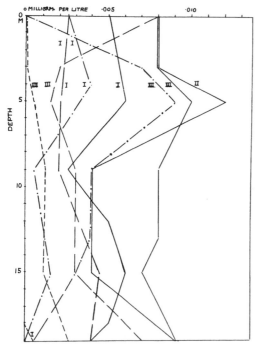

FIG. 32. Lake Quassapaug. Total phosphorus for July 1 (I), July 15 (II), August 11 (III), 1938; sestonic, organic soluble and soluble phosphate phosphorus for the first and third dates. The fractions are indicated as in Figure 24.

at times of relatively rapid rise, is in keeping with the oligotrophic character of the lake. The last two series of determinations, on August 18 and 25 respectively, present a curious problem (Fig. 32). An effort was made to determine total phosphorus at one meter intervals throughout the whole depth of the lake. The first set, obtained on August 18, show a normal distribution, with a slight increase near the bottom of the lake. On August 25, however, although the total content of the lake had changed but little, the total phosphorus had become distributed in a series of bands, reminiscent of the narrow strata often encountered in plankton studies, or of the optical stratification described by Whitney (1937, 1938). A few soluble phosphate determinations show that a considerable amount of the phosphorus is in ionic form; unfortunately at 9 m. there is an excess of soluble over total phosphorus in the recorded figures. This error tends to cast doubt on the whole series, but it is difficult to see how an error presumably increasing the soluble phosphorus, i.e., some contamination, would decrease the total phosphorus at certain levels, as is undoubtedly indicated by the curve for August 25 when compared with conditions a week earlier. The fact that total phosphorus in the lake has changed but little would also favor acceptance of the second series; it is not improbable that the error at 9 m. concerns only this single soluble sample. If the analyses be reliable, they indicate a remarkable transport of phosphorus from one layer to another, and its rapid regeneration *in situ* in the layer to which it is brought. Such an occurrence is presumably possible in a lake in which there is a large amount of zooplankton performing vertical movements and but little phytoplankton, rapid regeneration being known in sea water when zooplankton is decaying (Cooper, 1935). The ratio of zooplankton to phytoplankton is certainly higher in Lake Quassapaug than in the majority of the lakes of Connecticut.

3.6.2. Juday, Birge, Kemmerer and Robinson (1928) give three series of analyses for Trout Lake which exhibit fluctuations very similar to those occurring in the Connecticut lakes. The table showing the organic phosphorus given by Einsele and Vetter for the Schleinsee (1938) also gives clear indications of a rise of about 25% in the total phosphorus content above 9 m. between August 6 and August 19, 1935. Too little data have been published by other workers to permit further comparisons.

RELATIONSHIP OF IRON AND PHOSPHORUS

3.7.1. Einsele (1936, 1938) has made very important studies of the relationship of iron and phosphorus. His main results can be summarized as follows: When a solution of ferrous bicarbonate containing a soluble phosphate is oxidized by aeration, ferric phosphate and ferric hydroxide are precipitated; the phosphate separates before the hydroxide, so that if phosphate is present in excess no hydroxide is formed. The hypolimnetic waters of most eutrophic lakes are rich in both ferrous iron and in phosphate, but the former is usually in excess so that when such waters are aerated ferric hydroxide containing a greater or less amount of phosphate is precipitated. This happens at times of overturn, when a precipitate of ferric hydroxide and ferric phosphate is distributed throughout the lake. In a few weeks this precipitate settles, so that the greater part of the hypolimnetic phosphorus cannot enter into biological circulation. Since iron is precipitated as phosphate before hydroxide is formed, partial oxidation will remove phosphorus and iron differentially, the water being depleted of the former before the latter is wholly removed. This may happen if turbulent mixing occurs in the upper part of the hypolimnion, giving a ratio of Fe:P that decreases with depth. Einsele's further experiments indicate that CO_2 and reducing substances, particularly H_2S, must be considered. If H_2S is added in increasing amounts to a suspension of ferric hydroxide and phosphate in the presence of CO_2 and the absence of oxygen, at first both ferrous bicarbonate and phosphate go into solution, but, with the addition of increasing amounts of the reducing agent, ferrous sulphide is precipitated, leaving the phosphate in solution. It is probable that the conditions in the mud resemble the final state of such an experiment; if the water in contact with the mud is rich in oxygen and a brown oxidative microzone is formed, the conditions in the microzone are to be compared with the initial conditions of the experiment and no phosphate can leave the mud. If, however, the oxygen is reduced in the water, as in phase II of stagnation according to the terminology of the present paper, a time will come when the oxidative microzone will disappear and conditions at the mud-water interface will resemble those of the middle phase of the experiment when ferrous iron goes into solution as bicarbonate and phosphate. In

Linsley Pond, in the mud of which there is an excess of iron over sulphur (Hutchinson and Wollack, 1940) the final condition of Einsele's experiment cannot be realized; reducing substances other than hydrogen sulphide are presumably involved and there is little if any possibility of loss of iron as sulphide in the free water as a final phase in stagnation. It is clear that the ferrous-ferric system is acting in Einsele's experiments as a poising system. Interesting studies could be made by measurement of redox potential in the hypolimnia of lakes rich in decomposing organic matter but relatively poor in iron. Finally Einsele has made a study of the effect of colloidal ferric oxide in adsorbing phosphate from solution. Under the usual conditions found in nature it would appear that, although this process takes place, it is quantitatively unimportant.

3.7.2. The distribution of iron and phosphorus at the end of the summer in Linsley Pond is essentially the same as that found by Einsele, though the absolute amounts of phosphorus are smaller than in the Schleinsee and Stadtsee, and the increase in the phosphorus relative to the iron with increasing depth is less regular than in the two German localities. The data, set out graphically in Figures 29-31, indicate the similarity of the stratification observed in Linsley Pond and in Einsele's localities; while the concomitant increases in iron and phosphorus at 9 m. (Fig. 21) demonstrate the working of Einsele's principle at a single depth most clearly. Some sedimentation of phosphorus doubtless occurs in phase III, but the essential difference between this phase and the previous one must be in the enormous increase in the rate of diffusion of phosphate from the mud once the limitation imposed by the presence of oxidizing ferrous iron in the microzone is removed. Such a condition is likely first to develop in profile-bound density currents if such currents occur.

3.7.3. Although Einsele found that the greater part of the ferric compounds produced in suspension by oxidation of ferrous iron at times of circulation were sedimented in a week or two, it was observed in the present investigation that appreciable amounts of ferric iron are normally present in suspension in Linsley Pond. It seemed possible, therefore, that part of the phosphate normally determined as soluble, might really be in a suspended but acid-soluble form. Accordingly a series of determinations was made of the soluble phosphate in both unfiltered water and in membrane filtrates, from a 35 second membrane filter which retains all solid particles of diameter over c. 0.5 μ. Differences of the order of 0.0002 mgrms. per litre, which are barely detectible, are probably due to differences in the color of filtered and unfiltered water, and are insignificant. The results, set out in Table 12, indicate that only at the time of the autumnal turn-over is any appreciable amount of suspended acid-soluble phosphate present. This may be regarded as ferric phosphate. In Linsley Pond, where the pH is normally between 7 and 8 and the calcium content is but moderately high, the occurrence of calcium triphosphate, recently discussed by

Gessner (1939), need not be considered. While the results confirm Einsele's conclusions as regards the fate of the phosphate at the time of turn-over, there is a strong possibility that some of this suspended phosphorus may be used by the phytoplankton, as Harvey (1937) has shown that suspended ferric compounds can be utilized by marine diatoms.

3.7.4. Although ferric phosphate is now known not to be a normal constituent of the seston ash, the supposed occurrence of free ferrous iron, presumably undergoing oxidation, in both the sea (Cooper, 1932) and in the surface waters of lakes (Hutchinson, Deevey and Wollack, 1939) suggested that possibly part of the control of the phosphorus content of the lake might be a hydrometabolic precipitation by iron which supposedly reached the free water from the mud in a reduced state, and when oxidized combined with phosphate present in solution. Although the results of a further study of the matter indicate that the presence of free ferrous iron in the epilimnion is probably an illusion, due to the dipyridyl method of determination, the details of this study, while not strictly relevant to the phosphorus cycle, are of sufficient interest to warrant mention. Aliquots from two samples, A, the untreated water, and B, a membrane filtrate from a 35 sec. filter, were analyzed in the following ways:

(1) Ferric iron, without oxidation or other treatment.

(2) Apparent ferrous iron, by dipyridyl (some reduction of ferric occurs).

(3) Reducible iron by dipyridyl and sodium sulphite.

(4) Total iron after evaporation, gentle ignition solution and reduction, by dipyridyl.

It is thus possible to determine

(a) Total iron in seston (A4-B4) and in solution (B4).

(b) Ferric iron in seston (A1-B1) and in solution (B1).

(c) Ferrous iron in seston (A3-B3)-(A1-B1) and in solution (B3-B1), which may be instructively compared with the total apparent ferrous iron (A2).

(d) Combined iron, presumably organic, in seston (A4-B4)-(A3-B3) and in solution (B4-B3).

The color of the water makes readings below 0.01 mgrms. per litre impossible; tripyridyl, which is more sensitive, could not be obtained. No dipyridyl being commercially available during the end of 1938 and beginning of 1939 it was not possible to make complete fractionations before March, 1939. During May, 1939, some substance, destroyed on ignition and presumably organic, developed in the seston which gave a brownish color in the presence of sodium sulphite and dipyridyl, at the same time decreasing the intensity of the red co-ordination compound formed with ferrous iron. Procedure three was therefore not possible on unfiltered water after the end of April, and the number of analyses is therefore very limited; the results are set out in Table 13. Some data on ferric iron were obtained in the fall of 1938 and are incorporated in Table 12. Ferric iron appears

TABLE 13

Date	Total iron		Ferric iron		Ferrous iron			Organic iron	
	sest.	sol.	sest.	sol.	sest.	sol.	app.	sest.	sol.
Mar. 1 (frozen)	0.15	0.03	0.03	0.00	0.01	0.00	.035	0.11	0.02
Mar. 22 (ice breaking)	0.25	0.05	0.04	0.00	0.04	tr. ≤ 0.01	0.07	0.18	0.05
Apr. 5 (open)	0.07	0.03	0.05	0.00	0.02	tr. ≤ 0.01	0.07	0.01	0.03
Apr. 28 (open)	0.04	0.03	0.03	0.00	0.01	0.00	0.03	0.00(4)	0.03
May 17 (open)	0.08	0.025	0.03	0.00	—	0.00	0.01	—	0.02
Mean March–April	0.13	0.035	0.04	0.00	0.02	<0.005	....	0.075	0.03

to be a normal constituent in suspension; in view of Harvey's (1937) observations, it may well be the main source of iron in the nutrition of the phytoplankton. Table 12 shows an abnormal quantity present after autumnal circulation, as found by Einsele. As is to be expected, no ferric iron appears in solution; doubtful traces were found in one incomplete analysis, May 29, 1939 (Table 12), when some suspended hydroxide may have passed the filter. Unequivocal evidence of ferrous iron in solution is also not forthcoming; on the two occasions on which traces are reported ferrous iron is high in the seston and traces may have passed the filter which does not retain colloidal material. The absence of detectible amounts of ferrous and ferric iron in ionic solution is to be expected since Cooper (1937) finds theoretically that at pH 7.0 on attainment of equilibrium not more than 4×10^{-8} mgrms. per litre of iron in true solution, mostly as ferrous and $FeOH^{++}$ ions, is present. The occurrence of ferrous iron in suspension at the time of the breaking of the ice and of the vernal circulation seems to be well established; on the other two dates the amounts recorded, being obtained by difference, are not significant. The high figures for apparent ferrous iron should be contrasted

TABLE 12. Phosphate phosphorus and iron in the surface waters of Linsley Pond, November 1938-June 1939.

Date	Temp.	Sol. P. Unfiltered	Sol. P. Filtered	Acid-sol. seston P.	Total seston P.	Total Fe.	Ferric seston Fe.
1938							
Nov. 2...	11.9	0.0037	—	—	—	0.27	—
Nov. 16...	10.3	0.0052	—	prob. present	—	0.43	—
Dec. 2...	1.2	0.0065	0.0039	0.0016	0.003	—	—
Dec. 5...	—	0.0091	0.0073	0.0018	—	0.55	0.14
1939							
Jan. 25...	2.91	0.0021	0.0021	0.0000	0.010	0.28	0.03
Feb. 15...	3.2	0.0008	0.0008	0.0000	0.032	0.16	—
March 22...	—	0.0022	0.0021	(0.0001)	0.010	0.30	0.04
April 5...	7.8	0.0023	0.0022	(0.0001)	—	0.10	0.05
April 13...	7.5	0.0020	0.0020	0.0000	0.019	—	—
April 27...	13.2	0.0009	0.0009	0.0000	0.014	0.07	0.03
May 10...	17.7	0.0015	0.0017	(—0.0002)	0.009	0.10	0.03
May 24...	16.2	0.0009	0.0008	(0.0001)	—	—	0.03
May 30...	—	0.0006	0.0006	0.0000	0.010	—	—
June 13...	24.1	0.0003	0.0003	0.0000	0.010	—	—
June 23...	24.0	0.0004	0.0004	0.0000	—	—	—
June 29...	24.3	0.0004	0.0004	0.0000	—	—	—

54   G. Evelyn Hutchinson   Ecological Monographs
Vol. 11, No. 1

with those obtained by difference, and clearly indicate the impossibility of using dipyridyl in the estimation of small amounts of ferrous iron in the presence of suspended ferric hydroxide, owing to the continual removal of ferrous ions from solution in the formation of the red co-ordination compound and the consequent destruction of the original equilibrium (Harvey, 1937). No significant difference between the apparent ferrous and the reducible iron of the membrane filtrates was ever detected. The high values for ferrous iron in surface waters given by Hutchinson, Deevey and Wollack are due to a failure to realize this limitation of the method. In the first two analyses the large amounts of suspended combined iron may well be present in the form of complexes derived from soils and brought into the lake by melt water running under the ice. The hardly detectible amounts present on later dates probably represent the normal iron content of the organisms of the plankton and their derivatives. Significant amounts of combined iron occur in solution on all dates.

PHOSPHORUS AND NITROGEN AS LIMITING FACTORS

3.8.1. In a later paper it is hoped to discuss the significance of the ratio of nitrate to phosphate, in its relation to the qualitative changes in the plankton of Linsley Pond. A few notes on other aspects of the interrelation of the cycles of these elements may not be out of place in the present contribution. The first aspect to be considered is the role of the two elements as limiting factors. Riley (1940), studying the matter statistically, found a significant correlation between the nitrate and the phosphate content of the surface water and the quantity of phytoplankton present a week later; he concluded that phosphate was the more important limiting factor, but the data were not extensive enough to permit final judgment. The matter has been also examined experimentally in a series of experiments in collaboration with Dr. Riley. In these experiments, bottles holding approximately 3.5 litres were filled with surface water and suspended to a frame or buoy, so that they hung

under water at the surface of the lake. Potassium phosphate or potassium nitrate was added in solution in quantities to raise the phosphorus or nitrogen by 1 mgrm. per litre to two bottles; both nutrients were added to a third bottle, a volume of distilled water equal to that of the solutions, to a fourth control bottle. The chlorophyll content of the water in the bottles after thorough mixing was determined after a week's exposure in the lake, the content in the nutrient bottles being compared with both the control and with the initial and final content in the lake. In discussing these experiments it is necessary to separate from those done during the spring and early summer, two performed when much Anabaena was present, the behavior of the latter organisms being (3.7.2) clearly atypical. The results of four experiments done before an Anabaena bloom appeared are presented in Table 14. It is obvious that little change occurred in the control bottle, while addition of nutrients singly, produced some increase in the phytoplankton. This increase, however, is small compared to that effected when both nutrients are added together. The results suggest that on the whole nitrate is somewhat more significant than phosphate in producing an increase, but that both elements must be present before any other possible limitation can come into play. The possibility that some potassium must be added to obtain the full result is unfortunately not excluded by these experiments, though the effect of this element is clearly small. As far as they go the observations recorded in the table suggest that the results obtained may well depend on the qualitative composition of the phytoplankton at the time under consideration. A considerable increase in the phytoplankton would undoubtedly occur if an accession of both elements were to occur simultaneously, but it is possible that the bloom so produced would shade the water below, and so would not greatly increase the productivity of the whole lake (Riley, 1940).

3.8.2. Unlike the phosphorus cycle, the nitrogen cycle is to some extent open to the atmosphere; the significance of nitrogen-fixing bacteria has long been

TABLE 14

Period	Initial Nitrate Nitrogen mgrms. per litre	Initial Phosphate Phosphorus mgrms. per litre	Initial Chloro- phyll mgrms. m³	Control		1 mgrm. of P per litre added			1 mgrm. of N per litre added			1 mgrm. of P and 1 mgrm. of N per litre added			Final Chlorophyll in Lake Surface	
Year 1937				Chlorophyll	Percent of Initial	Chlorophyll	Percent of Initial	Percent of Control	Chlorophyll	Percent of Initial	Percent of Control	Chlorophyll	Percent of Initial	Percent of Control	Chlorophyll	Percent of Initial
April 30–May 7	c. 0.02	0.001	14.2	13.2	93.2	15.3	108	116	18.7	132	142	99.4	702	754	12.7	89.5
May 17–May 24	0.022	0.001	7.5	5.5	71.8	11.1	148	207	7.7	102	143	49.7	665	927	8.0	107.0
June 15–June 22	0.019	0.0005	6.8	5.9	87.0	8.4	125	143	16.5	243	279	79.2	1163	1343	4.8	70.2
July 13–July 20	0.030	0.000	9.6	9.9	105.5	14.5	154	146	29.8	312	296	206	2165	2062	9.2	96.3
Mean of Percentages...	.....	.....	....	....	89.4	....	134	146	....	197	215	....	925	1272	....	90.8

considered in limnological literature; in small eutrophic lakes, the nitrogen-fixing capacity of the alga Anabaena, recently established by De (1939, Fritch and De, 1938) may be as important as an annually recurring source of nitrogen in the lake. In 1938 a few determinations of total (Kjeldahl) nitrogen in the surface water, before and after filtration through No. 42 Whatman paper, were made during the rise of the Anabaena bloom. The final determination was made on a bottle of water suspended for a week after the Anabaena had been killed in the lake by copper sulphate treatment. This bottle showed a huge increase in phytoplankton, practically all Anabaena, over the previous week, the saponifiable ether-soluble green pigments rising from 25.6 mgrms. per cubic meter of chlorophyll to 65.6 mgrms. per cubic meter. A very great increase in the combined nitrogen present is also to be observed. Although the Linsley water was presumably not bacteria-free, and the Anabaena present in the lake was *A. circinalis* (Francis Drouet det.), a species not among the three studied by De, there can be very little doubt that the intense nitrogen fixation observed was due to this alga.[13] The data obtained in this way are presented in Table 15.

TABLE 15

Date	Total Organic N. Mgrms.	Soluble Organic N. per litre	Chlorophyll	Remarks
19 June	0.54	—	8.0	Little Anabaena
25 June	0.79	0.51	20.3	Water-bloom
29 June	1.00	0.92	25.6	Bloom decaying
5 Aug.	1.98	1.11	65.6	In bottle, healthy bloom

A curious condition is exemplified by the Anabaena in the bottle exposed between June 29 and August 5, which has also been noticed on other occasions. The bottle when filled contained a great deal of decaying Anabaena filaments; the nitrogen figures suggest that much of the organic material of these filaments had been liberated in soluble form. During the week of exposure in the bottle a very great development of the bloom occurred, producing a chlorophyll content higher than has ever been observed in the lake. In an experiment done in 1937 with nitrogen and phosphorus, a very similar condition was noted, the chlorophyll in the control bottle rising from 34.5 on August 31 to 98.7 on September 7; in none of the experimental bottles was there a comparable rise. In some of Dr. Riley's experiments on oxygen production in clear bottles at the lake-surface in September, 1937, a similar rise in the number of Anabaena filaments was obvious, at a time when they had almost disappeared from the lake. These observations clearly throw little light on the nature of the phenomenon, but suggest that the growth of Anabaena is facilitated either by the non-turbulent state of the water in a

[13] Since the above was written, an entry in *Chemical Abstracts* (**33**: 5446, 1939) has brought to my attention the observations of B. S. Aleev and K. A. Mudretsova (*Microbiology* (*U. S. S. R.*) **6**: 329-338, 1937). — These investigators, whose original paper I have not seen, apparently observed similar phenomena in a pond during the period of water-bloom; unfortunately no indication is given, in the abstract, of the nature of the alga believed by them to be a fixer of nitrogen.

full stoppered bottle or by some substance produced, bacterially or otherwise, at the glass-water interface. The effect is so striking that it deserves further investigation; it clearly invalidates all conclusions in nutrient experiments at the time when a bloom is at its height or early decline.

3.8.3. Finally a few words may be added on the ratio of combined phosphorus and nitrogen in the seston and in solution. So few observations are available that if they were not concordant with the deduced experience of other investigators they would not be worth mentioning. An unknown source of ammonia, probably from a leaking gas outlet, vitiated the Kjeldahl determinations made at the end of the investigation to add to the evidence assembled. The following figures appear to be reliable.

Date	Seston N.	Seston P.	Sol. org. N.	Sol. org. P.
5 April	0.17	0.016	0.256	0.0021
27 April	0.14	0.014	0.359	0.0016
17 May	0.23	0.009	0.406	0.008

The ratio of the nitrogen to phosphorus in the three seston determinations lies between 9.4:1 and 25.5:1; in marine planktonic organisms, Cooper (1938a) considers a ratio of 15:1 in milligram atoms or 6.8:1 in milligrams as typical. The three values given above are all greater than this, the lower two are of the same order of magnitude. On the other hand, the soluble combined nitrogen bears a ratio to the soluble combined phosphorus of from 50:1 to 220:1. The ratio of total combined nitrogen, other than nitrate, to the total non-phosphate phosphorus lies between 23:1 and 37:1. In the Wisconsin lakes the mean total combined nitrogen is .457, the mean non-phosphate phosphorus (organic of Juday and Birge) is .0203, the ratio being 22.5:1. The soluble organic phosphorus appears on an average to be higher than in Linsley Pond; if the estimate given in 3.2.1. be correct, the ratio is 27.1:1. Similarly the ratio of sestonic nitrogen to sestonic phosphorus in Wisconsin appears to be about 14:1. These figures strongly suggest that organic nitrogen is liberated in inorganic form less easily during the decomposition of planktonic organisms than is organic phosphorus. Birge (in Birge and Juday, 1934) moreover has shown that a comparable situation exists with respect to nitrogen and carbon. It is therefore very probable that in general, phosphorus is liberated more easily than nitrogen, nitrogen more easily than carbon, in the regeneration of inorganic substances from organic in aquatic environments.

HORIZONTAL CURRENTS AS A GENERAL FACTOR IN LIMNETIC INTERMEDIARY METABOLISM

3.9. The two factors primarily responsible for the rise in the bicarbonate content of the hypolimnion are the pelometabolism of the mud, which causes the production of bicarbonate in the first instance, and the horizontal water movements by which the material is carried into the free water. There can be no possible path by which phosphorus is carried rapidly, at all levels of the hypolimnion, into the central part of the lake other than the same water movements. In the epilimnion, where the mode of indirect treatment adopted in the present paper is inapplicable, the relative importance of horizontal movements and turbulent exchange between water in immediate contact with mud in the littoral region, and the surface layers immediately above the contact water, cannot be estimated. All the energy required to produce the horizontal and turbulent movements induced in the first and second part of the paper must have originally been delivered at the surface. Much of this energy is certainly dissipated in the epilimnion, and it is safe to conclude that if a change involving transport from the mud can be demonstrated in the hypolimnion, *a fortiori*, it is mechanically, though not necessarily chemically, possible in the epilimnion. It has been adequately demonstrated that increases in total phosphorus can occur at all levels in phases I and II of stagnation, when chemical conditions are essentially epilimnetic so far as this substance is concerned. The dilemma resulting from the work of Juday and Birge therefore is resolved, for in all trophogenic layers a continual replacement is possible by the same mechanism that results in the rise in the bicarbonate content. The special conditions found at the bottom of the lake, producing the full development of phase III, may well be due not only to absence of illumination, but also to the supposed system of density currents which would have given the water finally appearing well away from the bottom, for example at 11 m. in the center of the lake, unusual opportunity to be influenced by the surface layer of the mud. The general theoretical scheme implied in the present paper is represented diagrammatically in Figure 34. The mechanisms of this scheme can be invoked in general, to explain all observed rises in concentration, in any of the substances dissolved in the lake. It is equally clear that the internal metabolic cycle of every different substance

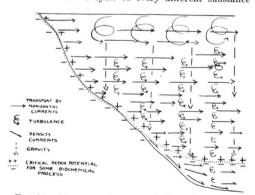

TRANSPORT BY
→ HORIZONTAL
CURRENTS
ε TURBULENCE
DENSITY
CURRENTS
GRAVITY
+ ‖ CRITICAL REDOX POTENTIAL
‒ FOR SOME BIOCHEMICAL
PROCESS

Fig. 34. Diagrammatic representation of the path of dissolved substances (not water-movements in general) passing from the mud into the free water according to the theory developed in the present paper (solid lines) and of the path (broken) of sedimenting substances whose concentration is non-conservative in the free water (i.e., phosphorus). The dotted line separating − from + signs indicates the boundary between redox potentials critical for the occurrence of some process that accelerates the loss of the solute from the mud.

will have specific characteristics, which though unintelligible without a knowledge of water movements, require specific biochemical treatment. The phosphorus cycle may be considered as one of the simplest of such specific cycles.

THE PHOSPHORUS CYCLE AS A SPECIFIC EXAMPLE OF INTERMEDIARY METABOLISM

3.10.1. The phosphorus cycle in Linsley Pond may be characterized by certain general properties, which are probably not exhibited together by the cycle of any other ordinary cyclical element.

(a) It is ideally *a closed cycle.* In an ideal abstraction no exchanges need occur between the lake and the paralimnion in phosphorus metabolism. In practice some small amount of phosphorus is brought into the lake by the inlets in excess of that lost through the outlet, and this excess presumably compensates for what is lost to the sediments of the lake. Over short periods of time *there is no need to consider such replacements as inevitable parts of the cycle,* which, in broad outline, can be understood without reference to anything but the events in the water and the mud with which it is in contact.

(b) The cycle is intimately bound up with the biocoenosis of the lake, and is of such a nature that at any one time *a very high proportion of the element present in the water is in particulate organic form.* This means that during a part of the cycle most of the quantity of the element present comes under the influence of gravity, so leading to greater removal by sedimentation than would be exhibited by an element such as nitrogen, in organic combination, where a large part of the atoms are present in soluble, or at least colloidal, forms which have a negligible tendency to sediment.

(c) *The rate of regeneration from the mud varies with the redox potential* of the layer into which the regenerated phosphorus is entering, or more probably is dependent on the redox potential at the mud-water interface of this layer. *This dependence, however, is not direct,* but *depends on the relationship of phosphorus and iron.*

3.10.2. The significance of these characteristics can best be made clear by reference to certain other elements. The nitrogen cycle in the biosphere has been described *ad nauseam,* but in any given biotope is frequently inadequately understood. Cooper's (1937a) theoretical treatment for the sea may be profitably consulted, as a more modern presentation of the most complex known elementary cycle than is usually given in textbooks. Such information as is available for lakes indicates very clearly that (a) the cycle is open to the atmosphere; (b) although the element is as important biologically as is phosphorus, its fate in the free water is probably quantitatively different, as a greater proportion of the nitrogen in lake waters appears to be present in soluble organic form, at least when comparison is made in a given lake or district; (c) the dependence on redox potential is clearly very marked and iron may be involved, but the effects of variations of the potential are largely mediated by bacterial action and are probably more qualitative (ammonification versus nitrification) than quantitative (Mortimer, 1939).

3.10.3. Riley has considered the copper cycle in Linsley Pond. Here there is a strong presumption in favor of an essentially open cycle, but one in which the inlets rather than the atmosphere are involved. The behavior of copper in relation to the proximate fractions of organic matter in the free water is curiously similar to that of phosphorus, and abundant opportunity for sedimentation occurs. The association, however, is probably a process of adsorbtion, rather than entry into the metabolic cycles of individual organisms. The vertical distribution of copper in full stagnation shows a far less marked maximum at the bottom than is found in the case of phosphorus (Riley, 1939a) so that the redox potential is probably not involved either directly or indirectly.

3.10.4. It is highly probable that the types of criteria used in 3.10.1. to characterize the phosphorus cycle could be used to classify elementary cycles in any aquatic biotope. A somewhat more general form of each criterion could be introduced. The importance of the direction in which the cycle is open, if at all, needs no emphasis. The second criterion, when generalized, relates to the effect of chemical and biological events on the direction of migration of the element. The effect of the biochemical properties of an element on its mobility must be considered with care; Yoshimura finds that sodium behaves in most lakes as a non-biological element, in spite of its great physiological importance. The third criterion, that of the conditions under which the simple inorganic compounds, capable of biological utilization, are liberated and accumulate need not be limited to a discussion of redox potential. It is, however, probable that the relative state of oxidation or reduction of a portion of an environment is of more general significance than many investigators have realized. In order to emphasize the great importance of this aspect of the elementary cycles, reference may be made to the recent work of Oparin (1938) for a discussion of the occurrence of free oxygen, carbon dioxide, and nitrogen in the earth's atmosphere. Evidence exists that points to the conclusion that the original condition of the earth's biosphere was one of general reduction. Some corollaries of this conclusion have been considered by Oparin with respect to the oxygen, carbon and nitrogen cycles. The significance of the hypothesis with respect to other biological elements, notably iron, phosphorus and calcium, remains to be elucidated. If speculations of the sort so ingeniously developed by Oparin have any validity, it is probable that biogeochemical studies of parts of the biosphere with low oxidation-reduction potentials may throw unexpected light on the early history of living beings.

SUMMARY

1. A study of the rate of change of temperature with respect to time, in Linsley Pond, North Branford, Conn., Lake Quassapaug, Middlebury, Conn., and in Lake Mendota from Birge's table of mean temperatures, indicates that thermally the hypolimnion can be divided into an upper *clinolimnion* in which the rate of heating falls exponentially with increasing depth, and a lower *bathylimnion* in which the rate of heating approaches a constant value independent of depth.

2. In the clinolimnion, the coefficient of turbulence can be calculated, using McEwen's method. For the mean temperatures and rates at the height of stagnation values of 0.86 m.² per month, for Linsley Pond, 6.55 m.² per month for Lake Mendota and about 1.1 m.² per month for Lake Quassapaug are obtained.

3. McEwen's detailed procedure, in which bathylimnetic values of the rate of change of temperature with respect to depth are used, is criticized, as not fulfilling his own condition for a constant value of the coefficient of turbulence. McEwen's values of the coefficient of turbulence in Lake Mendota are therefore too high, and the residuals obtained in the hypolimnion and explained by his statistical theory of convectional cooling appear to be unreal.

4. There is no evidence of a vertical variation of turbulence in inverse relationship to stability in the clinolimnia of any of the lakes studied.

5. In Lake Mendota, turbulence decreases with increasing stability throughout the heating season.

6. Application of the first and second order equations of Schmidt to determine the coefficient of turbulence in the bathylimnion leads to conflicting results. The conflict can only be resolved by supposing that the bathylimnion heats by some mechanism other than vertical turbulence. There is reason to believe that there is little change in the value of the coefficient of turbulence in passing from the clinolimnion to the bathylimnion.

7. If the total heat income at any level is calculated on a Birgean basis, the first and second order equations give reasonably consistent results in the clinolimnion. This means that some of the heat passing into the bathylimnion has previously passed through part of a horizontal plane in the clinolimnion lateral to the contour describing the top of the bathylimnion.

8. It is suggested that heating by profile-bound chemical density currents is largely responsible for the formation of the bathylimnion. Heating by biological oxidations is entirely inadequate.

9. The use of Grote's regional modulus of turbulence suggests that while the coefficients of turbulence of the clinolimnia of Lake Quassapaug and Linsley Pond are not significantly different, the epilimnion of the former lake is about four times as turbulent as that of the latter.

10. The morphometric relationship observed in the mean curve relating bicarbonate content to depth in Linsley Pond during summer stagnation, previously reported for 1937, is shown in its essential features by the 1938 data.

11. From the results of the study of temperature and turbulence, it is possible to determine what part of

58 G. EVELYN HUTCHINSON Ecological Monographs
Vol. 11, No. 1

the observed rate of increase of bicarbonate is due to vertical mixing and what part to other water movements. The latter are of much greater importance than turbulent vertical mixing. The relationship observed between morphometry and the bicarbonate curve shows that such other water movements are primarily horizontal currents, but a consistent scheme emerges if the supposed density current mechanism is taken into account.

12. The metabolism of the hypolimnion of a stratified lake may be regarded as composed of *hydrometabolism* or the series of chemical cycles taking place in the free water, and of *pelometabolism,* or the sum of chemical cycles occurring in the mud or at the mud-water interface, in so far as they affect the free water. In Linsley Pond the pelometabolism appears to constitute about 20% of the total metabolism estimated in terms of carbon dioxide production.

13. In the surface waters of Linsley Pond there is a slight excess of calcium and magnesium over bicarbonate. In the deep water of the hypolimnion, there is a considerable deficiency of calcium and magnesium, made up by ammonium and ferrous ions. Iron and magnesium appear to increase regularly with depth; calcium may show transitory maxima in the upper part of the hypolimnion; rapid and extensive variations in the manganese content of the hypolimnion and in the mode of stratification of this element were observed.

14. On an average, the phosphorus content of the surface waters of Linsley Pond is divided into soluble phosphate phosphorus, 8%, soluble organic phosphorus, 29%, and sestonic organic phosphorus, 63%. The proportion of sestonic phosphorus is probably unusually high.

15. The variations in sestonic phosphorus are correlated with both the mass of organic seston and the quantity of phytoplankton, as measured by its chlorophyll content. The phytoplankton probably contains more phosphorus per unit mass than does the detritus of the seston.

16. Acid soluble sestonic phosphate (ferric phosphate) is only present in the surface waters after the autumnal overturn.

17. Variations in the total phosphorus content of the whole lake are considerable and rapid. In the summer of 1937, the lake gained on an average at least 1.09 kilograms per week, and lost on an average at least 0.98 kilograms per week, independent of gain and loss through the inlets and outlet. The actual rate of replacement may be 4 to 9 times as great as this minimum estimate.

18. Variations can occur at all depths and can only be interpreted in terms of horizontal water movements that carry phosphorus from the mud-water interface into the free water, and of continual sedimentation of the considerable fraction present as sestonic phosphorus. It is suggested that faeces production by the zooplankton constitutes the chief sedimenting agency.

19. Comparable, but less rapid, variations in the total mass of phosphorus present in Lake Quassa-

paug are recorded; similar changes are deduced from data published by other investigators.

20. It is convenient to divide the processes of stagnation at any depth into three phases. In phase I oxygen falls rapidly, while bicarbonate begins to rise. In phase II there is a relatively constant low oxygen content, with rising alkalinity; iron and phosphorus are both low. In phase III the oxygen remains low while bicarbonate increases, though less rapidly; ferrous iron and soluble phosphate phosphorus increase rapidly.

21. Einsele's work on the resulting vertical distribution of iron and phosphorus is confirmed, and is of great significance. The limitation of the accumulation of phosphate in the presence of high oxygen concentrations is briefly discussed in relation to oxidation-reduction potentials and the ferrous-ferric poising system.

22. In the surface waters of Linsley Pond ferric iron is normally present in suspension but not in solution; ferrous iron may apparently sometimes be present in suspension, but not in solution; organic iron is present in variable amounts in suspension and in rather constant amounts in solution. It is suggested that suspended ferric hydroxide may be a source of iron available to phytoplankton.

23. Addition of a mixture of potassium phosphate (K_2HPO_4) and potassium nitrate to surface water suspended in a bottle in the lake leads to a great increase in the phytoplankton over that present in a control bottle. Addition of one or the other salt alone leads to a smaller increase, sometimes greater in the nitrate than in the phosphate bottle and sometimes the reverse. When an Anabaena bloom is present the results are irregular, mere isolation of the alga in a bottle of lake water appears to favor its multiplication.

24. Evidence is presented that the nitrogen-fixing capacity of Anabaena, recently established by De, is of importance in the metabolic cycle of a lake containing appreciable quantities of the alga.

25. The phosphorus cycle is characterized by being ideally closed, by being dependent on the gravitational sedimentation of organisms or their excreta, and by being indirectly dependent on the oxidation-reduction potential. Characterization of other elementary cycles by the use of similar criteria is possible.

26. The oxidation-reduction potential of environments is significant in general biological theory in view of the probable prevalence of highly reduced substances throughout the biosphere in the early history of the earth.

BIBLIOGRAPHY

Alsterberg, G. 1927. Die Sauerstoffschichtung der Seen. Bot. Notiser. Lund. 1927, pp. 255-274. Lund.

———— 1930. Die thermischen und chemischen Ausgleiche in den Seen zwischen Boden- und Wasserkontakt sowie ihre biologische Bedeutung. Int. Rev. Hydrobiol., **24**: 290-327. Leipzig.

———— 1931. Die Ausgleichströme in den Seen im Sommerhalbjahr bei Abwesenheit der Windwirkung. Int. Rev. Hydrobiol., **25**: 1-32. Leipzig.

American Public Health Association. 1936. Standard Methods for the Examination of Water and Sewage. 8th ed. New York.

Barnes, H. D. 1928. A Note on Two Sensitive Colour Reactions for Magnesium. J. S. Afr. Chem. Institute, **11**: 69. Johannesburg.

Behrens, H. 1937. Temperatur- und Sauerstoffuntersuchungen in Tümpeln und Brunnen. Arch. Hydrobiol., **31**: 145-162. Stuttgart.

Birge, E. A. 1916. The Heat Budgets of American and European Lakes. Trans. Wisc. Acad. Sci., Arts, Lett., **18**: 166-213. Madison.

—— 1916a. The Work of the Wind in Warming a Lake. Trans. Wisc. Acad. Sci., Arts, Lett., **18**: 341-391. Madison.

Birge, E. A. and C. Juday. 1911. The Inland Lakes of Wisconsin. The Dissolved Gases of the Water and Their Biological Significance. Wisconsin Nat. Hist. Surv. Bull., **22**: Sci. ser. **7.** Madison.

—— 1929. Transmission of Solar Radiation by the Waters of Inland Lakes. Trans. Wisc. Acad. Sci., Arts, Lett., **24**: 509-580. Madison.

—— 1934. Particulate and Dissolved Organic Matter in Inland Lakes. Ecol. Monog., **4**: 440-474. Durham.

Birge, E. A., C. Juday, and H. W. March. 1928. The Temperature of the Bottom Deposits of Lake Mendota; a Chapter in the Heat Exchanges of the Lake. Trans. Wisc. Acad. Sci., Arts, Lett., **23**: 187-231. Madison.

Buchanan, J. Y. 1879. On the Freezing of Lakes. Nature, **19**: 412-414. London.

Chandler, D. C. 1937. Fate of Typical Lake Plankton in Streams. Ecol. Monog., **7**: 445-479. Durham.

Cooper, L. H. N. 1932. Iron in the Sea and in Marine Plankton. Proc. Roy. Soc. London (B). **118**: 419-438.

—— 1935. The Rate of Liberation of Phosphate in Sea Water by the Breakdown of Plankton Organisms. J. Mar. Biol. Ass., **20**: 197-202. Plymouth.

—— 1937. Some Conditions Governing the Solubility of Iron. Proc. Roy. Soc. London (B). **124**: 299-307.

—— 1937a. The Nitrogen Cycle in the Sea. J. Mar. Biol. Ass., **22**: 183-204. Plymouth.

—— 1938. Salt Error in Determinations of Phosphate in Sea Water. J. Mar. Biol. Ass., **23**: 171-195. Plymouth.

—— 1938a. Redefinition of the Anomaly of the Nitrate-Phosphate Ratio. J. Mar. Biol. Ass., **23**: 196. Plymouth.

De, P. K. 1939. The Role of Blue-green Algae in Nitrogen Fixation in Rice-fields. Proc. Roy. Soc. London (B). **127**: 121-139.

Defant, A. 1932. Beiträge zur theoretischen Limnologie. Beitr. Physik der Freien Atmosphäre. **19**: (Bjerknes-Festband), 143-147.

Einsele, W. 1936. Ueber die Beziehungen des Eisenkreislaufs zum Phosphatkreislauf im Eutrophen See. Arch. Hydrobiol., **29**: 664-686. Leipzig.

—— 1938. Ueber chemische und kolloidchemische Vorgänge in Eisen-Phosphat-Systemen unter limnochemischen und limnogeologischen Gesichtspunkten. Arch. Hydrobiol., **33**: 361-387. Stuttgart.

Einsele, W. and H. Vetter. 1938. Untersuchungen über die Entwicklung der physikalischen und chemischen Verhältnisse im Jahreszyklus in einem mässig eutrophen See (Schleinsee bei Langenargen). Int. Rev. Hydrobiol., **36**: 285-324. Leipzig.

Elster, H. J. 1939. Beobachtungen über das Verhalten der Schichtgrenzen nebst einigen Bemerkungen über die Austauschverhältnisse im Bodensee (Obersee). Arch. Hydrobiol., **35**: 286-346. Stuttgart.

Fisher, R. A. 1934. Statistical Methods for Research Workers. 5th ed. Edinburgh and London (Oliver and Boyd).

Fritsch, F. E. and P. K. De. 1938. Nitrogen Fixation by Blue-green Algae. Nature, **142**: 878. London.

Gessner, F. 1939. Die Phosphorarmut der Gewässer und ihre Beziehung zum Kalkgehalt. Int. Rev. Hydrobiol., **38**: 202-211. Leipzig.

Grote, A. 1934. Der Sauerstoffhaushalt der Seen. Die Binnengewässer. **14.** Stuttgart.

—— 1936. Ist das absolute Defizit das Mass des biogenen Sauerstoffverbrauchs im See? Kritische Studien der diesbezüglichen Abhandlungen Alsterbergs. Arch. Hydrobiol., **29**: 410-544. Stuttgart.

Harvey, H. W. 1937. The Supply of Iron to Diatoms. J. Mar. Biol. Ass., **22**: 205-219. Plymouth.

Hayashi, K. 1921. Fünfstellige Tafeln der Kreis- und Hyperbelfunktionen sowie der Funktionen e^x und e^{-x} mit den natürlichen Zahlen als Argument. Berlin and Leipzig (Ver. wissensch. Verleger, Walter de Gruyter & Co.).

Hutchinson, G. E. 1938. Chemical Stratification and Lake Morphology. Proc. Nat. Acad. Sci., **24**: 63-69. Washington.

Hutchinson, G. E., E. S. Deevey, Jr., and Anne Wollack. 1939. The Oxidation-Reduction Potentials of Lake Waters and their Ecological Significance. Proc. Nat. Acad. Sci., **25**: 87-90. Washington.

Hutchinson, G. E. and Anne Wollack. 1940. Studies on Connecticut Lake Sediments. II. Chemical Analyses of a Core from Linsley Pond. Am. J. Sci. **238**: 493-517. New Haven.

Juday, C. and E. A. Birge. 1931. A Second Report on the Phosphorus Content of Wisconsin Lake Waters. Trans. Wisc. Acad. Sci., Arts, Lett., **26**: 353-382. Madison.

Juday, C., E. A. Birge, G. I. Kemmerer, and R. J. Robinson. 1928. Phosphorus Content of Lake Waters of Northeastern Wisconsin. Trans. Wisc. Acad. Sci., Arts, Lett., **23**: 233-248. Madison.

Juday, C., E. A. Birge and V. W. Meloche. 1938. Mineral Content of the Lake Waters of Northeastern Wisconsin. Trans. Wisc. Acad. Sci., Arts, Lett., **31**: 223-276. Madison.

McEwen, G. F. 1929. A Mathematical Theory of the Vertical Distribution of Temperature and Salinity in Water under the Action of Radiation, Conduction, Evaporation and Mixing due to the Resulting Convection. Bull. Scripps Institution of Oceanography. Technical Series, **2**: 197-306.

Maucha, R. 1932. Hydrochemische Methoden in der Limnologie. Die Binnengewässer. **12.** Stuttgart.

Meloche, V. M. and T. Setterquist. 1933. The Determination of Calcium in Lake Water and Lake Water Residues. Trans. Wisc. Acad. Sci., Arts, Lett., **28**: 291-296. Madison.

Möller, Lotte. 1928. Hydrographische Arbeiten am Sakrower See bei Potsdam. Z. Ges. Erdkunde. Sonderband 1828-1928. 535-551. Berlin.

—— 1933. Der Sakrower See bei Potsdam. Ein Beitrag zur Frage der Wirkung der Windzirkulation auf die physikalische und chemische Schichtung in einem See. Verh. int. Vereinig. Limnol. **6**: 201-216. Stuttgart.

Mortimer, C. H. 1939. The Work of the Freshwater Biological Association of Great Britain in Regard to Water Supplies: (a) The Nitrogen Balance of Large

Bodies of Water. Public Health Congress and Exhibition (1938). British Waterworks Association. Off. Circ. **21**: 2-10.

Ohle, W. 1934. Chemische und physikalische Untersuchungen norddeutscher Seen. Arch. Hydrobiol. **26**: 386-464 and 584-658. Stuttgart.

———— 1939. Zur Vervollkommnung der hydrochemischen Analyse III. Die Phosphorbestimmung. Angewandte Chemie **51**: 906. (Reprint paginated separately.)

Oparin, A. I. 1938. The Origin of Life. Trans. S. Morgulis. New York. (Macmillan Co.)

Pearsall, W. H. 1930. Phytoplankton in the English Lakes. I. The Proportions in the Waters of Some Dissolved Substances of Biological Importance. J. Ecology. **18**: 306-320. Cambridge.

Redfield, A. C., H. P. Smith and B. Ketchum. 1937. The Cycle of Organic Phosphorus in the Gulf of Maine. Biol. Bull., **73**: 421-443.

Richards, O. W. and P. M. Roope. 1930. A Tangent Meter. Science, **71**: 290. New York.

Riley, G. A. 1939. Limnological Studies in Connecticut. Ecol. Monographs, **9**: 53-94.

———— 1939a. Correlations in Aquatic Ecology. J. Marine Research (Sears Foundation). **2**: 56-73. New Haven.

————1940. Limnological Studies in Connecticut. III. The Plankton of Linsley Pond. Ecol. Monog. **10**: 279-306.

Robinson, R. J. and G. Kemmerer. 1930. Determination of Organic Phosphorus in Lake Waters. Trans. Wisc. Acad. Sci., Arts, Lett., **25**: 117-121. Madison.

Ruttner, F. 1931. Hydrographische und hydrochemische Beobachtungen auf Java, Sumatra und Bali. Arch. Hydrobiol. Suppl., **8**: 197-460. Stuttgart.

Saunders, J. T. 1927. The Hydrogen-Ion Concentration of Natural Waters; I. The Relation of pH to the Pressure of Carbon Dioxide. Brit. Journ. Exper. Biol., **4**: 46-72. Cambridge.

Schmidt, W. 1925. Der Massenaustausch in freier Luft und verwandte Erscheinungen. Probleme der Kosmischen Physik. **7**. Hamburg (Henri Grand).

Strøm, K. M. 1932. Tyrifjord. A Limnological Study. Skrift. d. Norske Videsk. Akad. Oslo. Math.-Nat. Kl. 1932. (3) 1-84.

Wedderburn, E. M. 1910. Current Observations in Loch Garry. Proc. Roy. Soc. Edinb., **30**: 312-323.

Wedderburn, E. M. and W. Watson. 1909. Observations with a Current Meter in Loch Ness. Proc. Roy. Soc. Edinb., **29**: 619-647.

Whipple, G. C. 1927. The Microscopy of Drinking Water. 4th ed. New York (John Wiley & Sons, Inc.)

Whitney, L. V. 1937. Microstratification of the Waters of Inland Lakes in Summer. Science, **85**: 224-225. New York.

———— 1938. Microstratification of Inland Lakes. Trans. Wisc. Acad. Sci., Arts, Lett., **31**: 155-173. Madison.

Williams, F. T. and E. McCoy. 1934. On the Role of Microorganisms in the Precipitation of Calcium in the Deposits of Fresh Water Lakes. J. Sedimentary Petrology. **4**: 113-126.

Yoshimura, S. 1933. Chloride as Indicator in Detecting the Inflowing into an Inland-water Lake of Underground Water Possessing Special Physico-chemical Properties. Proc. Imp. Acad., **9**: 156-158. Tokyo.

———— 1936. A Contribution to the Knowledge of Deep Water Temperatures of Japanese Lakes. Part 1. Summer Temperatures. Jap. J. Astron. Geophys., **13**: 61-120. Tokyo.

———— 1936a. A Contribution to the Knowledge of Deep Water Temperatures of Japanese Lakes. Part 2. Winter Temperatures. Jap. J. Astron. and Geophys., **14**: 57-83. Tokyo.

———— 1938. Dissolved Oxygen of the Lake Waters of Japan. Science Reports of the Tokyo Bunrika Daigaku (ser. C), **2**: 63-277.

———— 1939. Stratification of Dissolved Oxygen in a Lake During the Summer Stagnation Period. Int. Rev. Hydrobiol., **38**: 441-448. Leipzig.

The History of a Lake

Integrated Sciences Lead To New Fundamentals

G. EVELYN HUTCHINSON
Associate Professor of Zoology

THE studies discussed in the present article are directed towards gaining an understanding of the life history of a lake. An attempt is made to trace the history of the development of a newly formed body of water, probably almost devoid of living beings, through a stage of maturity, richly filled with plankton, pond weeds, insects and fish, to a senescent condition when the open water has disappeared, and the lake, filled with silt from the surrounding country and with organic sediments of its own making, has become a swamp and is at the moment of death. Such studies have been in progress over a period of about seven years. They developed out of biochemical investigations of certain Connecticut Lakes and have been prosecuted by a number of different workers both in this laboratory and elsewhere. In order to produce a fully integrated picture of the events involved, the investigator must be something of a geologist, botanist, chemist, zoologist, and climatologist, must have a great experience of modern lakes, and also will find that some acquaintance with prehistoric archaeology is useful. The greater part of the work to be reported has been accomplished by Dr. E. S. Deevey, Jr., now of the Rice Institute, Houston, Texas, who possesses most of these qualifications; the specifically chemical work has largely been done by Miss Anne Wollack and myself; Dr. Ruth Patrick of the Philadelphia Academy of Natural Sciences has practically finished a most intensive study of the diatoms; Dr. Minna E. Jewell of Thornton Junior College and Mr. T. S. Austin of this Laboratory have made special studies of sponge-spicules and certain Cladoceran remains of considerable importance. The full details of the investigation can be found in a series of reports now appearing in the American Journal of Science. In the later phases of the work we have been most fortu-

nate in having the opportunity of discussion with Dr. Raymond L. Lindeman, who has been working on a similar problem in a lake in Minnesota. Dr. Lindeman's study will probably be more complete even than our own and has already led to certain new and fundamental conclusions.

Methods

In ecological field work there are two ways of treating the dimension of time. In most of the botanical ecology of the past, a series of stands of vegetation have been selected, beginning with bare ground, and ending with the most permanent kind of plant cover known in the region under investigation. Relatively short term observation indicates a tendency for one vegetational type, with its associated animals, to pass over into another, so that the whole series of stands can be arranged in what is called a *sere*, leading from bare ground (either on land or under water) up to the vegetational climax of the region. This method has many advantages, but suffers from the

---

G. Evelyn Hutchinson. Born (30 January, 1903) in Cambridge, England. Educated at Gresham's School, Holt and Cambridge University (B.A., 1925, M.A., 1928). Senior Lecturer in Zoology, University of the Witwatersrand, Johannesburg, Union of South Africa (1926-1928). Instructor in Biology, Yale University, 1928. Ass't. Prof. Zool. 1931. Biologist, Yale North India Expedition, 1932. Has worked mainly on the physics, chemistry, and biology of lakes in South Africa, Indian Tibet and North America. More recently has also considered some of the broader aspects of chemical ecology.

---

defects that, in general the stages are mostly stages in the re-establishment of natural vegetation after human disturbance, and the sites are rarely *physiographically comparable*. The second method consists in a study of sediments containing the fossilized remains of the earlier associations, so that the dimension of time is here treated in a direct, albeit at present relative, way. This latter method is in practice only applicable in closed, concave, areas that retain undisturbed wet sediments; it further suffers from the fact that the stages discovered are often not *climatically comparable*. Both methods can be used in the study of the history of lakes, and have been so used with marked success.

Contemporary Comparison

We will first compare two lakes in Connecticut, though the results of our study will be merely introductory, as no results not generally known will emerge from our use of the first method, that of comparison of contemporary localities. One of these lakes to be examined is East Twin Lake, Salisbury, the other Linsley Pond, North Branford. The latter locality will then be considered historically, using the second method.

East Twin is a larger and much deeper lake (maximum depth 24 m., mean 9.9 m.) than Linsley Pond (maximum depth 14.8 m., mean 6.7 m.); the water is much more transparent, and the lake is well known to fishermen as the home of the indigenous Round White-fish (*Coregonus cylindraceum quadrilaterale*) and the introduced Sockeye Salmon (*Oncorhynchus nerka*) and Lake Trout (*Cristivomer namaycush*). In Lindsley Pond no salmonids or plankton-feeding White-fish occur. In general, lakes like East Twin are recognized as being in a less mature stage than are lakes like Linsley Pond. We may first inquire into the cause of the difference in fish fauna. We find that in any lake of adequate depth, only the surface waters become heated by the sun during the summer. The decrease in density that results from this heating is sufficient to prevent the wind mixing the warm water to the bottom; as a result, in such a lake, a freely mixed warm layer (epilimnion) is separated from a cold and relatively unmixed deeper layer (hypolimnion) by a discontinuity layer (thermocline). If a lake is deep, or poorly supplied with living organisms, the quantity of organic matter falling to the bottom will be smaller per unit volume of hypolimnion, than if the lake were either shallower with the same biological population, or richer in living organisms, but of the same depth. The constant supply of decomposable organic matter (dying organisms, faeces, cast skins, etc.) that rains down from the epilimnion is in part decomposed by aerobic bacteria in the hypolimnion or in the superficial layer of mud. Oxygen is thereby extracted from the water, so that in the shallower or the more productive lake, the deep stagnant layer becomes almost free of the gas and rich in carbon dioxide, ammonia, methane, ferrous and manganous salts and, in some cases, hydrogen sulphide. In the deeper or the less productive lake, however,

we find that all the water remains more or less oxygenated until the fall, when, with the cooling of the surface layers, the whole lake becomes of the same density and can freely mix whenever a slight breeze plays over its surface. Fishes that cannot long tolerate warm water must remain below the thermocline in the summer; in general the Salmonids and Coregonids require such cold water, but they are also sensitive to low oxygen concentrations, and therefore tend to occur only in the deeper, biologically less productive lakes. Beside the fishes, the whole fauna of the two types of lakes differs; this is particularly true of the midge larvae or the bloodworms and their allies that are the chief constituents of the fauna of the mud in freshwater lakes and ponds. Examination of a collection of such larvae from the bottom of a lake shows at once with which type of lake one is dealing. Fortunately the heads of such larvae are frequently found as fossils and most of the genera are highly characteristic.

Historical Method

Like most of the smaller lakes of Connecticut, Linsley Pond began its career as a block of ice, detached from the receding ice front of the last glaciation, and embedded in material washed out from the moraine. The investigation of its history is based on a series

of borings made through the mud at the bottom of the lake, so that samples can be studied, extending from the modern mud surface to the glacial sand which formed the original bottom of the lake. The borings are most easily made in winter when the lake is covered with ice. The first type of study to be made of such samples is to provide a chronology. This is done by counting the relative numbers of different kinds of pollen grains embedded in the mud from different depths. Since the climatic changes that occurred subsequent to the retreat of the ice were accompanied by a series of changes in the forest cover of the land, and since much pollen is blown into the water, a series of pollen periods can be established that enable comparison of profiles to be made in different lakes or in different parts of the same lake. In southern New England the first period recognized by Deevey and termed A is dominated by spruce, pine and fir. This coniferous period is divided into A_1 and A_2; during A_1 the relative amount of spruce and fir rose; during A_2 these trees declined. The division between A_1 and A_2 probably represents a secondary postglacial minimum in temperature. The whole of A was certainly cool and moist, becoming warmer and drier as A_2 gave place to B, characterized by pine being almost the only tree present. The pine forest gave place to a mixed

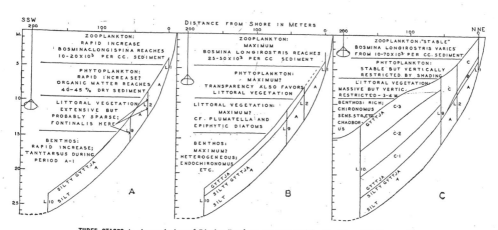

THREE STAGES in the evolution of Linsley Pond, as reconstructed by Dr. Edward S. Deevey, Jr.

(Courtesy of the American Journal of Science)

deciduous forest in period C. This deciduous period is divided into three phases. C_1 is characterized by an oak-hemlock forest; in C_2 in which oaks reached their maximum abundance, hickory became an important subdominant; in C_3 the conditions became more mesic, and chestnut accompanied the other species. C_2 probably represents the warmest postglacial time, corresponding to the climatic optimum of European climatologists and prehistorians.

The next studies are concerned with the chemistry of the sediments. It is obvious that the lowest part of the column consists of an inorganic silty clay which became richer in organic material as the lake matured. The rise in organic matter is unaffected by the incidence of low temperatures in the middle of period A. Later the proportion of organic matter settled down to a rather constant value of about 50% of the dry weight, of which about half consists of a lignin-protein complex; the proportion of easily hydrolysable material remains more or less constant in all the fossil sediments. The detailed form of the curve expressing organic matter (measured as ignition loss, or as lignin, or as total nitrogen) is initially sigmoid. Very similar curves have been found in the sediments of Windermere, in England, by Jenkin, Mortimer and Pennington, and in other Connecticut lakes. Hutchinson and Wollack came to the conclusion that such a curve might legitimately be compared to the growth curve of a single animal or of a homogeneous population. The proportion of organic matter gives a rough measure of the total production of living mat-

(Continued on page 22)

old man. The "Voder" can also be made to laugh or moo like a cow!

One of the most interesting accomplishments of the "Voder" is that by further development in the technique of operation, it can be made to sing. The steady character of the vocal sound source is given a musical quality by the introduction of a vibrato, which causes the sound to flutter or fluctuate over a small range about six times a second. The effect is like that obtained from a cello when the cellist moves his finger back and forth slightly in a periodic manner to get the vibrato sound. Under this condition the "Voder" operator can make the tones of the major triad, sing songs like "Home on the Range", "Loch Lomand", and many others. Another interesting effect in this connection is obtained by removing the vocal and breath sound sources from the "Voder" and supplying music in their place, to be moulded into speech. With the music of four violins playing in beautiful harmony the music of "Love's Old Sweet Song", the "Voder" operator forms the words of "Love's Old Sweet Song". The effect is wonderful; in fact it is indescribable!

Possible Uses of the "Voder"

I am certain that everyone who has witnessed a demonstration of the "Voder" has enjoyed it, but I also think the great majority of us went away from the "Voder" realizing that it is more than an interesting toy and an amazing novelty, but a great instrumentality or tool for accoustical research. Do you realize the purposes behind the intensive research that has gone into the production of the "Voder", or the practical use to which it may some day be put? In addition to an interesting study in voice operation, the "Voder" is also a significant step in serious telephone research. With the means of artificially creating speech at hand, one may speculate on the possibility of doing so at some distant point. Perhaps some day the words spoken into the telephone may be converted into narrow bands of frequency simulating telegraph signals, transferred as such to the distant end of the line, and then wired to control, by the impulses, electrical equipment to re-create the words of the speaker. Since telegraph signals may be transmitted over a much narrower band of

frequencies than is required for voice transmission, such sending as above described would make possible the multiplication of speech channels over telephone lines by methods other than the present highly satisfactory carrier systems.

Thus another talent of man, that is his ability to speak, has been synthesized with inanimate products under the control of a guiding intelligence, a result of intensive research and scientific advancement. After attempts to create the sounds of human speech, attempts as early as 1780, the Bell System has finally given us this truly remarkable apparatus, the "Voder", and the company and its men are to be congratulated on their work.

The History of a Lake

(*Continued from page* 15)

ter in the lake. At first this increases slowly, then more and more rapidly, then again slowly until an equilibrium value is achieved, which persists for hundreds, probably for thousands, of years. In default of an absolute time scale and any knowledge of the exact proportion of the organic production that is buried as sediment, we cannot obtain the constants of the growth curve, but the form is characteristic, and is unchanged, whatever assumptions are made as to the values of the rate of inorganic silting. Only at the extreme top of the profile where agricultural operations have caused a great increase of soil erosion, and consequently of inorganic sedimentation, does the relation we have described break down.

In his study of the animal microfossils Dr. Deevey has gone a step further in such analyses. He has found, first that the initial bottom fauna of the lake was entirely different from that now to be found in Linsley Pond, and resembled that of the deep, or less productive lakes beloved by lake-trout and trout fishermen. Now Linsley Pond was initially, of course, much deeper than it is today, indeed twice as deep, but by studying another lake, now in the dying stage, namely Lydhyt Pond, Totoket, the same initial fauna is found in sediments laid down in a lake

that was actually shallower than the modern Linsley. We may conclude, therefore, that the ancient lakes in the Linsley and Lydhyt basins were actually not merely deeper but also truly less productive than are the modern lakes. Moreover, by studying the relative numbers of shells of a small crustacean, *Bosmina longispina coregoni*, Deevey concludes that the increase in the population of this animal, throughout the period of growth in productivity, occurred much more rapidly than the increase in the total organic matter produced. The latter consists mostly of material derived from plants, so the animal evidently became more and more efficient in the utilization of the available food supply as the latter increased. The numbers of its shells in the sediments in fact vary with the fourth or fifth power of the organic matter of the mud. The final synthesis of our studies is so admirably expressed in Dr. Deevey's ideal sections of the lake (fig. 1) at different times in its history that further discussion is unnecessary, though the detailed reasons for his conclusions must be sought in the original papers.

Conclusion

The mechanism of the growth of the totality of living matter in a restricted environment, the dynamic relations of its component parts, and the changes in the efficiency of the utilization of sunlight, the sole source of biological energy, by the organisms of an isolated and well defined area of the earth's surface are very little understood. The lines on which such problems can be investigated have been briefly indicated in this article; a very remarkable set of further developments has been initiated by Dr. Lindeman in a paper shortly to appear in "Ecology". A complete knowledge of the laws involved in such a phenomena might be of the greatest practical importance; a rational feeling for the equilibria involved must be inculcated into the minds of future men of affairs if we are ever to achieve that harmonious existence which alone will justify the evolutionary ascendance of our species. We can only hope that the restricted studies here reported may do a little to add to the great body of knowledge of this aspect of ecology, which with patience and continual effort will slowly develop to ennoble the life of man.

in many races it would explain the oft-noted restriction of helmeted populations to the upper, turbulent waters of stratified lakes.   Those living in lower, less turbulent waters usually have shorter helmets.

Summary.—Preliminary experiments on a cyclomorphic race of *Daphnia longispina* indicate that turbulence of the water is one of the environmental factors controlling relative rate of helmet growth during postnatal life.

[1] Wesenberg-Lund, C., *Plankton Investigations of Danish Lakes*, General Part, Glyden-dalske Boghandel, Copenhagen (1908).

[2] Woltereck, R., "Variation und Artbildung," *Int. Rev. Hyd.*, **9**, 1–151 (1921).

[3] Coker, R. E., and Addlestone, H. H., "Influence of Temperature on Cyclomorphosis in *Daphnia longispina*," *Jour. Elisha Mitchell Sci. Soc.*, **54**, 45–75 (1938).

[4] Brooks, J. L., "Cyclomorphosis in Daphnia. I," *Ecol. Mon.*, **16**, 409–447 (1946).

[5] Huxley, J., *Problems of Relative Growth*, New York, MacVeigh (1932).

[6] Suggested in Coker, R. E., "The Problem of Cyclomorphosis in Daphnia," *Quart. Rev. Biol.*, **14**, 137–148 (1939).

[7] Butler, C. O., and Innes, J. M., "A Comparison of the Rate of Metabolic Activity in the Solitary and Migratory Phases of *Locusta migratoria*," *Proc. Roy. Soc. London*, *Ser. B*, **119**, 296–304 (1936).

A DIRECT DEMONSTRATION OF THE PHOSPHORUS CYCLE IN A SMALL LAKE*

By G. Evelyn Hutchinson and Vaughan T. Bowen

Osborn Zoological Laboratory, Yale University and American Museum of Natural History

Communicated March 24, 1947

It is well known that the total quantity of plankton present in the waters of a lake may undergo marked and rapid variation, so that in the course of a year a number of pulses or maximum populations may succeed each other. Juday and Birge[1] noted that such rises in the phytoplankton might occur without reducing the phosphate content of the water and that on occasions both phosphate and plankton might rise together.   It has also been noted for example by Pearsall,[2] that rises in·the population of blue-green algae may occur at the end of summer when it would seem that the phosphorus content of the water was totally inadequate to support an increased phytoplanktonic population.   In an earlier paper[3] much indirect evidence was assembled indicating that in Linsley Pond, a small inland lake which develops a very stable thermal statification in summer, there is a continual liberation of phosphorus from the mud into the free water.   Such of this phosphorus as enters the illuminated layers of the lake is believed to be taken up by the phytoplankton, later to be sedimented as a fine rain of particulate matter, partly no doubt dead phytoplankton, but also feces of zooplankton feeding on the plant cells. The resultant movement of phosphorus is thus believed to be a horizontal movement into the free water as phosphate, and a vertical downward movement as seston or particulate matter. The very low concentrations usually observed in summer in the surface waters of lakes are thus to be regarded as steady state concentrations, maintained at low levels by the activity of the phytoplankton, the rate of development of which depends rather on the rate of supply of phosphate ions from the mud than on their concentration in the free water. Such an hypotheses explains the rather paradoxical situations encountered by other workers and is in accord with the facts relating to chemical cycles in Linsley Pond and other lakes. If the hypothesis is correct, phosphorus present in the surface as phosphate on any given day should move to greater depths during the course of the succeeding few days or weeks, even in fully stratified lakes in which virtually no mixing is taking place. The possibility of obtaining relatively large amounts of the radioactive isotope of phosphorus P^{32} permits the hypothesis to be tested.

On June 21, 1946, a sample of radiophosphorus, of strength approximately ten millicuries, received as phosphoric acid and made up as sodium phosphate in hundredth normal sodium bicarbonate, was introduced into the surface water of Linsley Pond in twenty-four approximately equal portions. Twelve of the portions were placed at approximately equal distances along a line running across the middle of the lake from West to East. The other twelve samples were placed along a line between the first line and the outlet at the south end of the lake. As a light south wind was blowing at the time, and as the surface water of the lake could be seen to be drifting northward when the boat was anchored in the middle of the lake, it was believed that this distribution of the samples would secure a fairly uniform dispersion of radiophosphorus in the circulating surface waters.

Collections of water were made on June 28 in the deep central part of the lake normally used for limnological stations. For measurement of radioactivity, the vertical water column was divided into four layers, from each of which approximately 18 liters of water was collected, as is indicated in table 1. The depths of collection refer to the position of the top of the 1.25-liter Nansen reversing bottle used to collect the water, the figures after the depths indicate the number of times the bottle was filled at each depth. The deepest water to be included at each filling of the bottle came from approximately 50 cm. below the top of the bottle so that the composite samples may be regarded as representing I, 0–3 m.; II, 3–6 m.; III, 6–9 m. and IV, 9–14.5 m. layers. The composite sample from each layer was evaporated almost to dryness, the organic matter oxidized with nitric and perchloric acids, and after addition of a drop of phosphoric acid as a carrier, the phosphate was precipitated as ammonium phosphomolyb-

140   *Limnology*

150   *ZOOLOGY: HUTCHINSON AND BOWEN*   Proc. N. A. S.

date. The measurements of radioactivity were performed on the dry phosphomolybdate precipitates on filter paper.

Owing to the great dilution of the radioactive material, it was necessary to make a large number of counts, particularly on the sample from layer III, and to test the significance of the means obtained by Fisher's[4] table of t. The voltage stabilizer of the only Geiger counter circuit available was not sufficiently good to prevent alterations in the background count which completely obscured the increases, in single two-minute counts, due to the radioactivity of the samples. In the case of the samples from layers I and II, the mean count was found to be significantly different from the background when twenty-one 2-minute counts had been made; for quantitative determination nine more counts were included. For the lower activity of the sample from layer III, sixty-three runs of two minutes, alternated with background counts on a control sample of phosphomolybdate

TABLE 1

LAYER		COMPOSITION OF SAMPLE	COUNTS PER M.^{3} PER MIN.	PROBABILITY DUE TO CHANCE	VOL. LAYER M.^{3}	COUNTS PER LAYER PER MIN.
I	0–3 m.	0.0 m., 2; 0.5 m., 3 1.0 m., 3; 1.5 m., 2 2.0 m., 3; 2.5 m., 2	1100	0.01–0.001	241,800	266.10⁶
II	3.6 m.	3.0 m., 2; 3.5 m., 2 4.0 m., 3; 4.5 m., 2 5.0 m., 3; 5.5 m., 2	980	0.01–0.001	192,400	189.10⁶
III	6–9 m.	6.0 m., 2; 6.5 m., 2 7.0 m., 3; 7.5 m., 2 8.0 m., 3; 8.5 m., 2	620	0.02–0.01	120,400	75.10⁶
IV	9–14.8 m.	9.0 m., 4; 10.0 m., 3 11.0 m., 2; 12.0 m., 2 13.0 m., 2; 14.0 m., 1	(200	0.5–0.6)	Sum of significant values	530.10⁶

precipitated in concentrated oxidized surface water collected on June 13, 1946, were made. The counts for layer III only appear significant if the significance of the mean difference between each run and the immediately preceding control background run be considered. This is reasonable, since the voltage will vary little between two consecutive counting periods, but over the whole sixty-three pairs of observations the over-all changes in voltage will greatly increase the variance of both background and sample counts. The sample from layer IV gave no significant indication of radioactive phosphorus after forty-two observations had been made and must be regarded as negative. With improved counting facilities and a somewhat larger quantity of radiophosphorus it may be possible to measure the activity of phosphate phosphorus, organic soluble phosphorus and sestonic phosphorus spearately and to follow the cycle for longer periods on some future occasion.

Vol. 33, 1947   *ZOOLOGY: HUTCHINSON AND BOWEN*   151

The results of the study are set out in table 1, all counts being adjusted for decay and fluctuation of the counting rate, to correspond to the time and conditions of measurement of a phosphomolybdate preparation of an aliquot of 10^{-6} of the original sample, which gave a count above the background of 715 per minute on July 8. On this basis the total quantity of P^{32} put into the lake corresponds to 715.10^6 counts per minute. The total recovery therefore amounts to 74.2% of the P^{32} introduced.

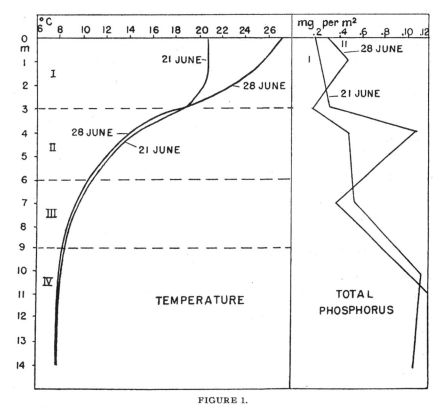

FIGURE 1.

Vertical distribution of temperature and total phosphorus in Linsley Pond 21 and 28 June, 1946, and limits of the four layers used in the present study.

The temperature curves (fig. 1) for the days at the beginning and end of the experiment are almost identical below three meters. The slightly lower temperature in the deep water on June 28 is certainly insignificant, and due to slight distortion of the isotherms by wind, as frequently occurs in all lakes. The chief change in the week of the experiment was a great heating of the surface water due to a sudden spell of hot weather.

Determinations of total phosphorus indicate a movement of phosphorus into the epilimnetic layer I during the period under consideration. The apparent decline in the total phosphorus at 4 m. in layer II is probably due to the descent of decomposing plankton which was clearly visible in the 4 m. sample on June 21, but which was probably at a greater depth and was consequently missed in sampling on June 28.

In spite of the fact that phosphorus has entered the epilimnion during a period when there can have been practically no vertical mixing, it is clear from the radioactivity measurements that 47% of the phosphorus present as phosphate in solution at the surface on the first day of the experiment had descended into the stratified part of the lake below 3 m. and about 10% had crossed the 6 m. level. These findings completely confirm the hypothesis put forward in 1941 and summarized in the first paragraph of the present contribution.

Of the 25.8% of the P^{32} put into the lake and not recovered, the greater part had probably entered the aquatic plants and sediments in contact with the 0–3 m. layer. Assuming uniform sedimentation, the bottom of the lake from the 0 m. to 3 m. contour, having an area of about 24.6% of the surface area of the lake, would have received a like fraction of the radio-phosphorus. Actually uniform vertical sedimentation is unlikely in the shallow water, but an appreciable amount of radiophosphorus entered the aquatic vegetation growing in the littoral zone. Littoral plants, mainly *Potamogeton* spp., were collected on June 28 and July 12 and showed the activities in counts per minute, given in table 2.

TABLE 2

	PER GRAM WET	PER GRAM DRY (80°c.)	PER GRAM P.
June 28	1.65	8.85	4700
July 12	0.68	4.47	2800

If the wet weight be taken as approximately equal to the volume it will be observed that on June 28, when the surface water gave a count of the order of 10^3 per m.³ per minute, the weed gave a count of the order of 10^6 per m.³ per minute, indicating a thousand-fold concentration of radio-phosphorus in the weed over that in the water. It is unfortunate that no quantitative data are available on the population of littoral plants in Linsley Pond, though, except at the south end, no extensive beds occur. The results just given suggest that in lakes in which there are wide expanses of littoral vegetation, this vegetation competes with the phytoplankton for phosphorus. Such competition is of course likely to be most severe during the growth period of the plants from April to June, and may conceivably account for phytoplankton minima often observed at the end of May or June.

* The authors wish to acknowledge their indebtedness to Dr. E. F. Pollard for his help and advice and for arranging for the manufacture of the sample of P^{32} employed. An unsuccessful attempt to perform an experiment of the kind described was made in 1941 by Pollard, Hutchinson and W. T. Edmondson.

[1] Juday, C., and Birge, E. A., "A Second Report on the Phosphorus Content of Wisconsin Lake Waters," *Trans. Wis. Acad. Sci.*, Arts Lett. **26,** 353–382, 1931.

[2] Pearsall, W. H., "Phytoplankton in the English Lakes. II. The Composition of the Phytoplankton in Relation to Dissolved Substances," *J. Ecol.*, **20,** 241–262, 1932.

[3] Hutchinson, G. E., "Limnological Studies in Connecticut. IV. The Mechanism of Intermediary Metabolism in Stratified Lakes," *Ecol. Monographs*, **11,** 21–60, 1941.

[4] Fisher, R. A., and Yates, F., Statistical Tables, London, 1938.

AMERICAN SCIENTIST *52*, 1964

THE LACUSTRINE MICROCOSM
RECONSIDERED

By G. E. HUTCHINSON

THE great intellectual fascination of limnology lies in the comparative study of a great number of systems, each having some resemblance to the others and also many differences.   Such a point of view presupposes that each lake can in fact be treated as at least a partly isolated system.

Today [1] I want to begin by considering two rather different approaches implicit in such treatment, partly in the work of Birge and Juday during the time when they were making Lake Mendota famous throughout the scientific world, and partly in the earlier work of S. A. Forbes from whom my title is of course derived.

Birge's mature point of view is expressed in his concept of the heat budget [2], which, though derived from ideas of Forel and others, represented a highly original and important contribution, because it first called attention to the lake as a natural system with an input and an output. This point of view has tended to underlie most of what has been done in lake chemistry and in the study of primary productivity during the past three or four decades. Such a way of thinking, in which the lake is considered, in the jargon of the moment, as a black box, has been called elsewhere [3] the *holological* approach. It has been extremely fertile, but, since water is transparent, the black box is too restrictive an analogy. The time has perhaps come for further development of the antithetical *merological* approach, in which we discourse on the parts of the system and try to build up the whole from them. This is what Forbes was trying to do in his classical lecture on "The Lake as a Microcosm" [4].

It is desirable to think for a moment about certain scale effects characterizing the lacustrine microcosm when viewed by a human observer. If we suppose that an organism reproduces about once every week for the warmer half of the year and on an average about once every month in the cooler half, it will have about thirty generations a year. This corresponds in time to about a millenium of human generations, and considerably longer for those of forest trees. In the case of the latter, we should expect in thirty generations some secular climatic change to be apparent. We should not expect in a tree the seeds or resting stages to remain viable while thirty generations passed, and in the larger animals no such stages exist. The year of a cladoceran or a chrysomonad, in both of which groups rapid reproduction may alternate with the formation of resting stages, is thus in some ways comparable to a large segment of postglacial time, though in other ways the comparison either to several millenia, or to a year in the life of a human being or tree, is definitely misleading. Another peculiar scale effect is that, in passing from the surface to the bottom of a stratified lake in summer, we can easily traverse in 10–20 m. a range of physical and chemical conditions as great or greater than would be encountered in climbing up a hundred times that vertical range on a mountain.

I would also emphasize how fantastically complicated the lacustrine microcosm is likely to be. There is probably no almost complete list of species of animals and plants available for any lake, but it would seem likely from the several hundred species of diatoms and of insects [5] known from certain lakes that a species list of the order of a thousand entries may be not unusual. This probably means that in the course of a season at least a thousand somewhat different ecological niches may for a time be recognizable. Most of this diversity is associated with the shallow marginal waters in which the bottom can form a solid substratum for attached aquatic plants.

Simpler situations in a lake are probably provided by the plankton, though it soon appears that they are not particularly simple and that we cannot regard the plankton, excluding the rest of the community, as an entirely satisfactory entity. We begin with the variously named and deductively respectable principle that two co-occurring organisms cannot form equilibrium populations in the same niche. In the phytoplankton we immediately meet the paradoxical situation of an enormously complicated association of phototrophic species all living together under conditions that do not seem to permit much niche specialization.

It is possible that the permanent and apparently almost monospecific *Anacystis* blooms recorded [6] under some conditions in tropical waters, notably temple tanks of South India, may represent a monospecific equilibrium of the kind to be expected from theory. Much more often, what I have elsewhere termed the paradox of the plankton intrudes itself. It is to be noticed that the paradox of a multispecific phototrophic phytoplankton only arises if we assume a closed system, providing a single niche, with enough time to permit the achievement of equilibrium. In general in a lake, we do not have a single niche system that is closed. The epilimnion if reasonably turbulent may approach a single niche system, but introductions from the littoral benthos are always possible. Moreover, there is no rule about the speed at which competitive exclusion excludes. As Hardin [7] has pointed out, in the theory all that is needed is an axiom which states that no two natural objects, or classes of objects, are ever exactly alike. What then happens is that under constant conditions one class, or population of reproducing objects, finally displaces the others. If the conditions are continually changing, the favored species might also change. This is what usually seems to be happening, but it must not be forgotten that a multispecific system never in equilibrium would be expected to suffer continual random extinctions and, if not quite closed, random reintroductions also, and should therefore drift in specific composition, probably more than is indicated by palaeolimnological data.

It is possible that, in a lake, random extinction is primarily a danger for the rarer species which are never likely to be observed. In a square basin 100 m. across and one meter deep we should have 10^6 organisms so common that one occurred per cubic meter, 10^9 of those occurring a thousand times more often at a rate of one per liter, and 10^{12} of those with one individual in the average cubic centimeter. If we are considering ordinary phytoplankton organisms, the first organism would be far too rare ever to find by ordinary techniques even though the population before us numbered a million.

I am now inclined to think that a large part of the diversity of the phytoplankton is in fact due to a failure ever to attain equilibrium so that the direction of competition is continually reversed by environmental changes, as suggested many years ago, moderated in two ways which insure that competitive exclusion does not continually and irreversibly remove bits of the association. The first moderating influence is the speed at which exclusion occurs. In spite of the ultimate validity of Hardin's axiom of inequality, Riley [8] feels that a sort of asymptotic approach to more or less equal adaptation is not unexpected in the phytoplankton. If we suppose two species S_1 and S_2, such that in niche N_1, S_1 displaces S_2, and in niche N_2, S_2 displaces S_1, if seasonal environmental changes occur, so that at first only N_1 and then N_2 are available. If competition went fast compared with the rate of environmental change, S_1 would be eliminated and would not be available for a new cycle, but if the two species were almost equally efficient over a wide range of environmental variables, competitive exclusion would be a slow process. Both species then might oscillate in varying numbers, but persist almost indefinitely.

The second way of moderating the tendency to random extinction is the provision of resting stages, so that if S_1 is eliminated completely as an active competitor in the plankton, when N_1 gives place to N_2, later next season when the reverse change occurs, resting stages of S_1 can recolonize the environment now again providing niche N_1. In practice, any plankter that really disappears and reappears rather than becomes alternatively rare or very common must have some such stages. Annual macrophytes and many small animals also have such stages as seeds, eggs, pupae, and the like. Perennials, moreover, may hibernate in ways that take the individual out of competition. In the diatoms in many of which resting zygospores are still unknown, it is possibly relatively unmodified littoral or shallow water benthic individuals that are involved in tiding the planktonic populations over competitively unfavorable conditions. In *Melosira*, Lund's beautiful work [9] shows how a relatively heavy diatom rests on the bottom for very long periods in a more or less unassimilative form; here there is doubtless some special physiological adaptation, so we are halfway between a species invading the plankton casually with a continuous littoral population and the condition in which morphologically specialized resting stages or cysts are produced. One of the most remarkable results of several recent palaeolimnological studies, notably Nygaard [10] on Store Gribsø and our own work on Lago di Monterosi [11], is the fantastic variety of Chrysophycean cysts recognizable in the sediments, at least, of rather soft water lakes. Resting stages of all sorts are of course particularly prone to occur in freshwater organisms, where they were doubtless developed primarily to promote survival under extreme physical conditions, notably desiccation and freezing. Once developed, they would however clearly be of great value in obviating extinction when conditions changed in favor of a competitor. It is therefore peculiar that Lund finds that planktonic desmids tend to lose such stages.

The diversity of the phytoplankton is clearly of primary importance in producing that of the zooplankton. Given the diversity of the phytoplankton, and some degree of food specificity in the animal forms, no striking paradoxical situation need arise. Moreover, it is clear from all the available work on the seasonal succession of closely allied forms, such as the species of *Daphnia*, from Birge's [12] early studies on Mendota up to the very beautiful and elaborate investigations of Dr. J. L. Brooks and of Dr. Donald W. Tappa at Yale, shortly to be published, that the same sorts of seasonal phenomena that damp competition in plants, also occur in animals.

MacArthur and Levins [13] have recently pointed out that two rather different extreme types of diversity between closely allied sympatric species (i.e., members of a genus or subfamily) are possible.

The two species may be specialized in such a way that they eat slightly different food, but hunt it over the same area. In this case morphological specializations, of which the simplest is a size difference, are to be expected. Probable examples, such as the hairy and downy woodpeckers, easily come to mind.

If the two species eat the same sorts of diversified food, they are likely to differ in the proportions in which they encounter it, and to specialize in habitat preferences without much morphological specialization becoming necessarily involved in feeding activity. MacArthur's own work on the American warblers provides a striking example. The existence of these two general situations has long been known, but MacArthur and Levins provide a good abstract theory of the phenomenon.

Over the whole vertical column in a stratified lake, even if only 10 m. deep, habitat differences are available in summer, at least as great as over the range from the bottom and to the top of a mountain several hundred times that vertical range. In the turbulent epilimnion it is in general hard to develop habitat preferences and, within any layer in which free movement is habitual, size differences may be expected as the simplest specialization increasing diversity, as is the case with copepoda. In view of the extreme vertical variation when we leave a turbulent freely mixed layer, the antithetic habitat difference type of specialization in the plankton is likely to be rather different from what is found terrestrially, involving fairly complete adaptation to very divergent physical factors rather than habitat preferences, though, in two species living together with vertical migration over partly overlapping ranges, we have the lacustrine analogue of birds feeding in different parts of the same tree. The rapid production of a number of generations per year permits a kind of seasonal succession in rotifers and cladocera, though to a less extent in copepoda, that is comparable to that in the phytoplankton. Considerable possibility of avoiding competitive exclusion is thus achieved by slow competition between species that have slightly different optima, and so succeed each other in time. Here the production of resting stages is of the greatest importance. That they are produced at the time of maximum population fits reasonably into this scheme quite independently of adaptation to unfavorable physical situations.

This succession in time may be coupled with size differences and habitat differences, probably producing, in for instance the genus *Polyarthra* [14] where five or six species can be sympatric but not strictly synchronic, a very pronounced niche specificity.

An interesting question arises, namely to what extent sympatric species of a given taxon, say genus or family, not merely have different ecologies, but also have ecologies that, though different, are closer than any would be likely to be to that of a sympatric nonmember of the taxon picked at random.

If we compare a desert assemblage with a limnoplanktonic one, it is obvious that, if the first organism captured in the desert is a beetle, the probability that the next one of another species is also another beetle is higher than that it is a rotifer, and vice versa. It is however rather surprising to find that in Carlin's data [14] four unallied perennial rotifers, including the microphagous sedimenters *Keratella*, *Notholca*, and *Conochilus* and the selective predator *Asplanchna*, all reacting similarly to an unidentified difference, possibly involving an earlier decline in the late summer bloom of *Oscillatoria*, that distinguished 1940 from the other years of his study.

The relatively small development that has been possible in the study of the interrelationships of the plankton since Forbes [4] wrote in 1887 and Birge [12] in 1898, and of which some examples have just been given, has been due to a very large amount of work both in field observations, in very meticulous taxonomy and in ecological theory. In the other parts of the lacustrine community the problems are more difficult though their solutions would have great fascination, as will be apparent from a single example. If we examine the lasion, "Aufwuchs" or fouling community of freshwaters, we find a variety of filamentous algae and diatoms with an associated fauna ordinarily of small motile forms. The biomass of the animals is doubtless ordinarily much less than that of plants, and on a surface near the bottom of the euphotic zone organisms will tend to be scarce. There may be a few sponges and bryozoa but they are not conspicuously important. In the sea, the parallel community, though largely algal in the tidal range, consists at most levels of an astonishing mass and variety of sessile animals, sponges, molluscs such as *Mytilus*, numerous hydroids, bryozoa, and tunicates. The difference is presumably due to the lack of pelagic larvae other than copepodan nauplii in freshwater. The only exceptions are a very few molluscs of far from worldwide distribution, notably *Driessena* and to a less extent *Corbicula*, the larval colonies of phylactolaematous bryozoa, hardly ever noted in open water, perhaps a few transitory planula larvae (*Cordylophora*, *Limnocnida*, *Craspedacusta*) and the free swimming larvae of Trematodes which show odd diversity in behavior when allied species are compared, but which presumably do not enter into competition with other plankters. This is a very poor showing compared to the dozen phyla likely to be found in any series of marine neritic plankton samples. This difference has received various explanations, the most reasonable, essentially due to Needham [15], probably being the difficulty that a small larval animal in freshwater, feeding on plant cells low in sodium and chloride, would have in acquiring enough salt before it could develop salt absorbing organs. In a certain sense the adult animals of the marine littoral benthos, not all sessile nor all microphagous, are the resting stages removed from competition at least in the open water plankton. We see in this type of relation the rather large scale example of the sort of interaction which fascinated Forbes. It is hard, in reading Forbes on The Lake as a Microcosm, to prevent the mind drifting back to the tangled bank of the last chapter of Darwin's *Origin of Species* [16], and from there it is permissible, at least in 1964, to look back still further to that other "bank where the wild thyme blows" [17]. Both Forbes and Darwin realize struggle but see that it has produced harmony. Today perhaps we can see just a little more. The harmony clearly involves great diversity, and we now know, in the entire range from subatomic particles to human artifacts, that every level is surprisingly diverse. We cannot say whether this is a significant property of the universe; without the model of a less diverse universe, a legitimate but fortunately unrealized alternative, we cannot understand the problem. We can, however, feel the possibility of something important here, appreciate the diversity, and learn to treat it properly [18].

REFERENCES

1. Text of address given after the transfer of keys of the Laboratory of Limnology, from the National Science Foundation to the Board of Regents of the University of Wisconsin; Wisconsin Center Auditorium, Madison, May 8, 1964. The author on this occasion represented the International Association for Pure and Applied Limnology.
2. E. A. BIRGE. The heat budgets of American and European Lakes. *Trans. Wis. Acad. Sci. Arts Lett.*, *18*, 166–213, 1915.
3. G. E. HUTCHINSON. Food, time and culture. *Trans. N. Y. Acad. Sci.*, Ser. II, *5*, 152–154, 1943.
4. S. A. FORBES. The Lake as a Microcosm. *Bull. Scientific Assoc. Peoria* 1887: 77–87, reprinted, *Illinois Nat. Hist. Surv. 15*, 537–550, 1925.
5. N. FOGED (On the diatom flora of some Funen lakes. *Folia Limnol. Scand.* no. 6. 75 pp. 1954) lists up to 260 different diatom taxa from a single lake. L. BRUNDIN (Chironomiden und andere Bodentiere der südschwedischen Urgebirgsseen. *Inst. Freshwater Res.*, Drottningholm. Rep. 30. 914 pp. 1949) finds up to 140 species of insects of the family Tendipedidae in a single lake.
6. S. V. GANAPATI. The ecology of a temple tank containing a permanent bloom of microcystis aeniginosa (Kutz) Henfr. I. *Bombay Nat. Hist. Soc.*, *42*, 65–77, 1940. G. E. HUTCHINSON. The paradox of the plankton. *Amer. Nat.*, *98*, 132–146, 1961.
7. G. HARDIN. The competitive exclusion principle. *Science*, *131*, 1292–1297, 1960.
8. G. A. RILEY *in* Marine Biology I. Proc. First Internat. Interdisciplinary Conference on Marine Biology. Amer. Inst. Biol. Sci., Washington, D.C. 286 pp. (see pp. 69–70), 1963.
9. J. W. G. LUND. The seasonal cycle of the plankton diatom, *Melosira italica* (Ehr.) Kütz. subsp. *subarctica* o. Müll. *J. Ecol.*, *42*, 151–179. 1954. Further observations on the seasonal cycle of *Melosira italica* (Ehr.) Kütz subsp. *subarctica* o. Müll. *J. Ecol.*, *43*, 90–102. 1955. See also his Baldi lecture. *Verh. int. Ver. Limnol.*, *15*, 37–56, 1964.
10. G. NYGAARD *in* K. BERG and I. C. PETERSEN. Studies on the humic acid Lake Gribsø. *Fol. Limnol. Scand.*, no. 8. 273 pp., 1956.
11. ELAINE LEVENTHAL (ms) to appear in a series of papers on this locality.
12. E. A. BIRGE. Plankton studies on Lake Mendota. II. The Crustacea of the plankton from July 1894, to December 1896. *Trans. Wis. Acad. Arts Sci. Lett.* 1898.
13. R. H. MACARTHUR and RICHARD LEVINS to appear in P.N.A.S. June 1964.
14. CARLIN. Die Planktonrotatorien des Motalaström: zur Taxonomie und Ökologie der Planktonrotatorien. *Meddel. Lunds Univ. Limnol. Inst.*, no. 5, 255 pp., 1943.
15. J. NEEDHAM. On the penetration of marine organisms into fresh water. *Biol. Zbl.*, *50*, 504–509, 1930.
16. The concluding paragraphs of Forbes' essay and of the *Origin of Species* may be profitably compared:

> Have these facts and ideas, derived from a study of our aquatic microcosm, any general application on a higher plane? We have here an example of the triumphant beneficence of the laws of life applied to conditions seemingly the most unfavorable possible for any mutually helpful adjustment. In this lake, where competitions are fierce and continuous beyond any parallel in the worst periods of human history; where they take hold, not on goods of life merely, but always upon life itself; where mercy and charity and sympathy and magnanimity and all the virtues are utterly unknown; where robbery and murder and the deadly tyranny of strength over weakness are the unvarying rule; where what we call wrong-doing is always triumphant, and what we call goodness would be immediately fatal to its possessor—even here, out of these hard conditions, an order has been evolved which is the best conceivable without a total change in the conditions themselves; an equilibrium has been reached and is steadily maintained that actually accomplishes for all the parties involved

the greatest good which the circumstances will at all permit. In a system where life is the universal good, but the destruction of life the well-nigh universal occupation, an order has spontaneously arisen which constantly tends to maintain life at the highest limit—a limit far higher, in fact, with respect to both quality and quantity, than would be possible in the absence of this destructive conflict. Is there not, in this reflection, solid ground for a belief in the final beneficence of the laws of organic nature? If the system of life is such that a harmonious balance of conflicting interests has been reached where every element is either hostile or indifferent to every other, may we not trust much to the outcome where, as in human affairs, the spontaneous adjustments of nature are aided by intelligent effort, by sympathy, and by self-sacrifice? (Forbes.)

It is interesting to contemplate a tangled bank, clothed with many plants of many kinds, with birds singing on the bushes, with various insects flitting about, and with worms crawling through the damp earth, and to reflect that these elaborately constructed forms, so different from each other, and dependent upon each other in so complex a manner, have all been produced by laws acting around us. These laws, taken in the largest sense, being Growth with Reproduction; Inheritance which is almost implied by reproduction; Variability from the indirect and direct action of the conditions of life, and from use and disuse: a Ratio of Increase so high as to lead to a Struggle for Life, and as a consequence to Natural Selection, entailing Divergence of Character and the Extinction of less-improved forms. Thus, from the war of nature, from famine and death, the most exalted object which we are capable of conceiving, namely, the production of the higher animals, directly follows. There is grandeur in this view of life, with its several powers, having been originally breathed by the Creator into a few forms or into one; and that, whilst this planet has gone cycling on according to the fixed law of gravity, from so simple a beginning endless forms most beautiful and most wonderful have been, and are being evolved. (Darwin.)

17. I know a bank where the wild thyme blows,
    Where oxlips and the nodding violet grows;
    Quite over-canopied with luscious woodbine
    With sweet muskroses, and with eglantine
    There sleeps Titania sometime of the night,
    Lull'd in these flowers with dances and delight;
    And there the snake throws her enamell'd skin,
    Weed wide enough to wrap a fairy in.
    —A Midsummer's Night's Dream
    II. i. 249–256

18. The evening program ended with a performance of Mozart's Quartet in G Minor for piano and strings, K. 478.

CHAPTER *1*

The Origin
of Lake Basins

Lakes seem, on the scale of years or of human life spans, permanent features of landscapes, but they are geologically transitory, usually born of catastrophes, to mature and to die quietly and imperceptibly. The catastrophic origin of lakes, in ice ages or periods of intense tectonic or volcanic activity, implies a localized distribution of lake basins over the land masses of the earth, for the events, however grandiose, that have produced the basins have never acted on all the lands simultaneously and equally. Lakes therefore tend to be grouped together in *lake districts*; within each district the various basins may resemble each other in certain general characters yet differ markedly in size and depth and so in their rate of maturation and senescence. It is this diversity in unity that gives the peculiar fascination to limnology. A group of lakes confronts the investigator as a series of very complex physicochemical and biological systems, each member of which has its own characteristics and yet also has much in common with the other members of the group. Moreover, the whole group of lakes of a given lake district may be compared with another group, often under widely different geographical and climatic conditions. In this way it is possible to identify the causal factors producing the differences between lake and lake or lake district and lake district. Much of the present work will be devoted to this type of analysis. Lakes, moreover, form more or less closed systems, so that they provide a

1

series of varying possible ecological worlds which permit a truly comparative approach to the mechanics of nature. It is in the study of the lake as a microcosm, as Forbes (1887) put it many years ago, that much of the intellectual significance of limnology lies, and this significance grows geometrically as a whole series of comparable yet different microcosms become available.

Throughout the succeeding pages, relations between the shape and size of a lake and the nature of the physicochemical and biological events taking place in its water will become apparent. The size and shape of a lake are, however, largely dependent on the forces that have produced the lake basin. Some discussion of the geomorphological aspects of lake basins and of the ways in which lakes have come into existence is therefore an appropriate point of departure.

THE ORIGIN OF LAKES

Some geomorphologists, notably W. M. Davis (1882) in his earlier work, have adopted a formal classification of the agencies which may produce basins, grouping them as *constructive, destructive,* or *obstructive.* Penck (1882) and Supan (1896) used similar primary divisions. This procedure is somewhat artificial and tends to obscure the regional grouping of lakes. Glacial action may form a basin by destruction, as in the case of the excavation of a cirque lake, or by obstruction, as when a valley is dammed by a moraine. Similarly, volcanic action may produce a basin by destruction when an explosion or volcanic subsidence occurs, by construction when a crater rim is built, or by obstruction when a valley is dammed by a lava flow. In general, glacial basins, whether formed destructively or by obstruction, will tend to occur in the same regions, and volcanic lakes, however formed, in other regions. The limnologist is usually more interested in keeping together the regionally grouped lakes than in separating them by formal intellectual barriers. Within the regional group, Davis's threefold classification of agencies producing basins is often most useful.

It is also possible to consider the origin of lakes and their geomorphological classification primarily in temporal terms, as Davis (1887) did in a later brief paper which emphasizes the contrast between the immature landscape rich in lakes and the relatively lakeless mature landscape. Such a temporal approach certainly calls attention to an important aspect of lake basins, namely their impermanence, but it is no more satisfactory than is the threefold formal classification of Davis's earlier paper in providing a classification of the lakes of the world.

In the present work, it will be more convenient to follow those authors who have considered the agencies producing lakes more empirically and the first.   The fact that water plants immediately take up radiophosphorus and more slowly lose it is well-established and shows that both the first and second processes are in fact going on all the time, though during the period of the experiment the first was certainly more rapid than the second.   The supposed greater rapidity of the second process when the littoral vegetation is growing rapidly in the spring presumably provides a mechanism by which the cyclical nature of the migration of phosphorus in the lake can be maintained.

All the processes listed except the second and seventh either have no effect on the direction of movement or they result in a concentration of phosphorus in the deep water.   Since in Linsley Pond and in many other small lakes the ferrous iron content of the deep water is at the end of summer stagnation certainly greatly in stoichiometric excess of the phosphate, it is probable that a good deal of the phosphorus derived from sedimenting seston, as well as that derived from the mud, will be precipitated at the time of circulation.   After winter stagnation, however, during the spring circulation period there is usually an appreciable amount of free phosphate in the water.   It is very reasonable to suppose that the uptake of some of this phosphorus, whatever its origin may be, by the littoral vegetation represents the laying up of a supply of the element at the margins of the epilimnion.   As soon as the main growth of the rooted plants has come to an end and the decay of some of the older leaves begins, the overall process is reversed.

Such a sequence of events is a very plausible explanation of the rather general low phytoplankton populations encountered in early summer, followed by immense blooms of blue-green algae when the superficial layers of the lake reach their highest temperature in July and August. On this hypothesis the presence of such blooms is primarily determined by the onset of rapid decomposition at high temperatures in the littoral, and a subsequent rapid output of phosphorus into the epilimnion.   As has been indicated, both Einsele (1941) and Hutchinson (1941) have noted superficial increases in total phosphorus at the time of great blue-green blooms.   Though utilization of phosphate would be so rapid that the intermediate ionic stage could not be detected, there can be no question but that the phosphorus in the algae must have come from the margins of the lake.

SUMMARY

Phosphorus is in many ways the element most important to the ecologist, since it is more likely to be deficient, and therefore to limit the biological productivity of any region of the earth's surface, than are the other major biological elements.

THE PHOSPHORUS CYCLE IN LAKES   *751*

Apart from the possible rare and transitory occurrence of phosphine under strongly reducing conditions, only orthophosphate and various undifferentiated organic phosphates are likely to be important in limnology. In ordinary cases, it is convenient to distinguish *soluble phosphate* phosphorus, sestonic *acid-soluble phosphate* phosphorus (mainly ferric and perhaps calcium phosphate), *organic soluble* (and colloidal) phosphorus, and *organic sestonic* phosphorus.

The total phosphate in natural waters varies from less than 1 mg. m.$^{-3}$ to immense quantities in a very few closed saline lakes. The total quantity depends largely on geochemical considerations, usually being greater in waters derived from sedimentary rock in lowland regions than in waters draining the crystalline rocks of mountain ranges. Most relatively uncontaminated lake districts have surface waters containing 10 to 30 mg. P m.$^{-3}$, but in some waters that are not obviously grossly polluted, higher values appear to be normal. The soluble phosphate usually is but a small fraction, of the order of 10 per cent of the total. The ratio of organic soluble to sestonic phosphorus is variable; the general orders of magnitude of the two fractions are the same. Acid-soluble phosphate may be found during autumnal circulation of small lakes of which the hypolimnia acquire much ferrous iron and phosphate during stratification.

Phosphorus is removed from water by all phytoplanktonic organisms. Some organisms, such as *Asterionella*, in natural lake water are extraordinarily efficient in the utilization of the element and can take it up from water containing less than 1 mg. m.$^{-3}$ In artificial media the efficiency of utilization appears to be less. In some cases quite moderate amounts of phosphate, suboptimal to one species, appear to be inhibitory to others.

The vertical distribution of phosphate during stratification is very variable and may be complicated. In unproductive lakes, there is usually little vertical variation in any fraction. In productive lakes with clinograde oxygen curves, there is an increase in soluble phosphate in the oxygen-deficient part of the hypolimnion. This is in part due to decomposition of sinking plankton, but in most cases is primarily caused by liberation of phosphate from sediments on reduction. An oxidized mud surface not merely holds phosphate but prevents diffusion of phosphate and ferrous ions from deeper layers in the mud, as the ferrous iron is always in excess and when oxidized precipitates all the phosphate. In lakes which proceed to phase III of stagnation, large amounts of ferric phosphate will be formed during autumnal circulation. This probably slowly hydrolyzes, and it may restore phosphate to the lake water and littoral vegetation. Where much H_2S is formed in phase IV of stagnation, the phosphate will be more mobile at circulation, and better regeneration is possible.

There is a good deal of evidence of rapid movement of phosphate from the sediments to the water, at all levels, in small lakes. At the height of summer there may be a great increase in total phosphorus in the surface waters at times of algal blooms, though soluble phosphate is undetectable. One condition for the maximum development of such blooms may well be rapid decomposition and consequent liberation of phosphate in the littoral sediments during spells of very warm weather. The phosphate would be taken up so fast by the growing algae that it never would be detectable.

Experiments with radiophosphorus have demonstrated rapid horizontal movement of the element in the hypolimnion, and cases of great increase in phosphorus at all levels of a stratified lake also imply such horizontal movement from the mud to the free water. The total phosphorus concentration in Linsley Pond does not show the steplike morphometrically determined distribution exhibited by bicarbonate. The phosphate must be engaged in too rapid a cycle to permit such a pattern to develop. When massive amounts of soluble phosphate are added to a lake, it is rapidly taken up by the phytoplankton and then sedimented. The productivity of the lake is increased, but only for a time. The chemical relations of phosphorus in mud and water evidently constitute a self-regulating system, at least within certain limits. Studies with radiophosphorus indicate rapid uptake by littoral plants and sediments, and by the plankton.

The rate of delivery of radiophosphorus into the hypolimnion permits an estimate of the fraction of epilimnetic phosphorus delivered in unit time to the hypolimnion. In the summer in Linsley Pond it appears that the phosphorus of the epilimnion is replaced from the littoral about once every three weeks. Other small lakes apparently exhibit equal or more rapid turnovers. During the period of observation, falling of seston into the hypolimnion accounted for rather less than half of the gain in phosphorus in the hypolimnion. The gain due to movement from the sediments was limited to the deep water, in which the mud surface may be regarded as reduced.

Part

4 *Theory*

Reflection Thereon: G. Evelyn Hutchinson and Ecological Theory

Melinda D. Smith and David K. Skelly

There is fundamentally only one way in which we learn anything,
namely by experience and reflection thereon.
—G. Evelyn Hutchinson (1962)

In 1940, Evelyn Hutchinson reviewed a new textbook.[1] Entitled *Bio-Ecology* and authored by Frederick Clements, the text lays out the state of ecological understanding according to one of the field's leading thinkers of the early twentieth century.[2] Clements had long described communities as "superorganisms" that undergo development across successive stages toward a climax state. This perspective focused researchers on delineating stages described with a set of terms that ultimately evolved into a baroque classification scheme. Hutchinson did not waste time in his assessment of what he believed was a defective approach: "The general principles of ecology therefore appear as a set of rules for the construction of language. . . . It is, however, uncertain that the language of Bio-ecology will ever become a universal ecological tongue." A fundamental problem as he saw it was the authors' neglect of the "alternative language of mathematics." This prescient review marks one of the first points in Hutchinson's career where he makes evident his views on ecological theory.

"INDUCTIVE GENERALIZATION"

Over the succeeding decades Hutchinson made foundational contributions to ecological theory. In this essay, we argue that Hutchinson's ability to create durable and transformative theory flowed directly from his unusual perspective on its inspiration and use. He was into middle age before he made theoretical contributions of his own. After many years of collecting and sifting natural history observations, Hutchinson turned to theory because he perceived disconnects between what he saw as critical patterns in the natural world and the conclusions emanating from the theory of the day. This attitude is evident throughout the selections included in this chapter. In each, Hutchinson begins with a natural history observation that forms the bedrock on which the need for new theory rests. Hutchinson termed this approach "inductive generalization." While recognizing that the vast body of natural history information "may bewilder the student and may appear useless and pedantic to the outside world" he believed that there was no other grist for useful insights of general importance.[3]

Hutchinson placed his perspective into action with the publication of "Ecological Aspects of Succession in Natural Populations."[4] The paper is the first of several in which Hutchinson considered theory focused on competition among species. Here, Hutchinson inaugurates the use of such models to describe and understand patterns of ecological succession. The paper begins with a simple yet critical observation: the predictive success of competition models differs between laboratory and natural settings. The fate of competing populations in laboratory culture experiments followed closely the predictions of models assuming a constant environment and a single limiting resource.[5] Populations in nature, however, rarely were observed to follow the dynamics consistent with competitive exclusion that were so often observed in the laboratory. Instead, individual populations waxed and waned, often in concert with seasonal change. Hutchinson recognized that equilibrium approaches adopted by ecological modelers yielded predictions that would never be achieved if the environment changed rapidly relative to the generation time of the competitors. Evidence available at the time suggested that seasonal changes were likely to alter the environment in ways that could affect competitive interactions. In essence, Hutchinson recognized decades before most ecologists that nonequilibrium dynamics may be important and that they may contribute to the patterns of successional change so evident in nature.

"FOREVER ON THE MOVE"

Hutchinson continued to think about competition during the 1940s. Subsequent papers focused on developments that better matched competition models with observations of natural populations. Key contributions included the consideration of social interactions and the introduction of time lags in density dependence that produced fluctuations in numbers mimicking those often observed in real populations.[6] Then in 1951 Hutchinson published a short paper entitled "Copepodology for the Ornithologist."[7] In signaling that even bird biologists had something to learn from the tiny crustaceans that captivated his interest, Hutchinson was making clear that he believed it was possible to develop empirically grounded theory that could transcend taxonomic boundaries. "Copepodology" continued to lay out the case for nonequilibrium mechanisms for species coexistence. Even if locally doomed to extinction by species with superior competitive abilities, a "fugitive species" may be able to persist if individuals are able dispersers capable of exploiting the times and places where their competitors are absent or rare. By remaining "forever on the move," regional-scale persistence is possible even if a species will eventually be extinguished within particular local environments. While commentators on Hutchinson's work typically mention other papers first, there is a strong case to be made for the fundamental importance of "Copepodology." Hutchinson's fugitive species ideas predate the development of metapopulation and metacommunity concepts.[8] Nevertheless, a clear line of thought extends from his 1951 paper to recent efforts to understand how movement among isolated environments interacts with local dynamics in determining the fate of species and the structure of communities.[9]

Hutchinson returned to the theme of nonequilibrium coexistence in "The Concept of Pattern in Ecology."[10] Hutchinson begins by arguing that ecologists will benefit from more rigorous approaches to the study of patterns and then puts the idea into practice by attempting to distinguish patterns produced by stochastic processes and those resulting from competition. Based on these different types of patterns, he recognizes the existence of both equilibrium and nonequilibrium communities, and he postulates that the nonequilibrium communities are likely to be more common than previously thought. Even while championing the importance of nonequilibrium perspectives, Hutchinson continued to use equilibrium ideas to make a fundamental contribution to ecological theory.

"FORMALIZATION OF THE NICHE"

No concept has been more central to the development of ecological theory than the niche. First used by Joseph Grinnell in a study of habitat use by a bird, the California thrasher, the concept was soon adopted and modified by Charles Elton to incorporate feeding relationships among species.[11] By considering who eats whom, Elton moved the concept to a place from which G. F. Gause could use it in thinking about competitive interactions and ultimately derive the competitive exclusion principle: no two species can simultaneously occupy the same niche in the same place at the same time.[12] In "Concluding Remarks" Hutchinson left an enduring stamp on the ecological niche by making it an explicitly quantitative concept.[13] Hutchinson used set theory to examine the conditions under which we might expect one species to eliminate another through shared use of "niche space." This space of many dimensions could now be measured, and for any group of species the strength of interaction might be estimated. These developments opened the door to closer integration between ecological theory and natural history. The influence of "Concluding Remarks" on the ecological community was transformational. Much of the research effort expended by ecologists during the 1960s and 1970s traced back to Hutchinson's insights. While some of this later research eventually came under attack, a reading of Hutchinson's original paper shows that he was under no illusions concerning the difficulties and uncertainties that lay ahead for niche-based ecological studies.

Hutchinson took some steps to further develop the use of niche theory with the publication of the next paper included in this chapter. "Homage to Santa Rosalia, or Why Are There So Many Kinds of Animals?" was published as Hutchinson's presidential address to the American Society of Naturalists.[14] Papers associated with presidential addresses are rarely memorable, but in this case Hutchinson produced one of the most widely read papers on ecological theory ever published. In it Hutchinson asks a basic question about the factors that promote and constrain the diversity of species. Based on his niche-founded consideration of coexistence he wonders whether body size differences may mediate coexistence because size is associated with food use. He finds that co-occurring water bugs found near a sanctuary in Sicily differ in body size according to a ratio of approximately 1.3:1. Looking in the literature he finds this ratio is roughly matched by those found among co-occurring birds and mammals. The body size differential he observed eventually became known as the Hutchinsonian ratio.[15] As with his previous consideration of the ecological niche, the paper prompted an enormous wave of research into body size ratios and the search for a degree of limiting similarity that could be used to predict coexistence.

Ultimately, the hunt for a universal ratio was not successful.[16] In hindsight, however, it remains clear that Hutchinson was on to something fundamental, and studies of body size have remained an important part of ecological research since "Homage" appeared. Hutchinson's influence is perhaps best seen in the study of character displacement. W. L. Brown and E. O. Wilson had already described patterns of body size variation showing that closely related species tend to show larger differences in body size when found living together than when found apart.[17] These authors suggested that character displacement was a relic of initial con-

Santuario di Santa Rosalia, Monte Pellegrino, Palermo, Sicily. Water bugs that Hutchinson collected near this sanctuary prompted him to consider the role of limiting similarity in allowing species coexistence. Photograph from *John L. Stoddard's Lectures,* supplementary volume no. 4 (Boston, 1905).

tact between newly evolved species. Hutchinson, on the other hand, postulated that the pattern resulted from ongoing pressure due to interspecific competition. Here, Hutchinson shows that ecologists can contribute to the understanding of geographic variation in characters as they relate to species interactions, while also providing inspiration for the development of evolutionary ecology. As for character displacement, a further half century of study into the underlying mechanisms of character displacement has supported Hutchinson's original hypothesis that body size differences can promote coexistence.[18]

"THE VERY PARADOXICAL SITUATION OF THE PLANKTON"

Twenty years after he first considered the possibility of nonequilibrium coexistence, Hutchinson returned to the topic in one of his most influential papers.[19] "The Paradox of the Plankton" is a bookend to the 1941 paper that leads off this chapter. Where the earlier paper is tentative and exploratory in tone, "Paradox" is direct and to the point. Hutchinson is now a confident and accomplished theorist at the peak of his powers. He revisits the same patterns that initially provoked his interest with a much fuller understanding of the factors limiting plankton and a more certain conviction that their dynamics provide evidence that the competitive exclusion principle is not acting. Interestingly, he does not view this conclusion as evidence that the principle needs to be discarded. Much of the niche-based research that he would continue to develop over ensuing years required an equilibrium perspective in which competitive exclusion was the rule. Hutchinson's great knowledge of natural history led him, in this and many other instances, to parse the natural world into sets of cases likely to fit one concept versus another. Writing here in 1961 Hutchinson appears thoroughly comfortable with the idea that a unified theory of ecology is unlikely to emerge. Ecologists would still be divided on this issue four decades later.[20]

"CASES OF A STRIKING KIND"

While Hutchinson continued to write and publish for another thirty years, the appearance of "Paradox" marked the close of a remarkable period of productivity. During the remainder of his career, Hutchinson published additional papers on niche theory, but some of his most notable writings showed his mind in the act of refining observations into insights. We conclude this chapter with two examples. While neither is strictly theoretical, both touch on themes that have since developed into important areas of research or, in a few cases, that remain to be explored. Hutchinson had achieved wide acclaim by this point, and his writing reflects efforts to reach broader audiences. "The Influence of the Environment" is illustrative.[21] In this compact but wide-ranging essay, Hutchinson aims to intrigue. Topics range from disparities in the elemental composition of living organisms relative to their abundance in the environment to a consideration of the rates at which processes occur in biological systems. Particularly interesting is his focus on the speed of ecological processes relative to evolutionary change. Subsequent study has shown that rates of evolution are commonly rapid enough to impinge on ecological interactions.[22]

The final paper in the chapter, "Thoughts on Aquatic Insects," considers a group of organisms Hutchinson had been thinking about for more than sixty years.[23] A series of topics are presented that reflect Hutchinson's extraordinary knowledge of natural history and his ability to find potentially important patterns from among enormous piles of data. Much of this material anticipates the development of adaptive theories of life history which came into

their own later in the 1980s.[24] The paper is a particularly clear example of his inductive generalization approach. Equally, it reveals his great interest in the way organisms work.

Around this time Hutchinson also published an ecology textbook of his own.[25] Writing in the *New York Review of Books,* Stephen Jay Gould found that "unlike most examples of the genre, it is accessible, intelligent, and enjoyable to read as literature."[26] Gould remarked particularly on the extensive footnoting (Hutchinson achieved the distinction of publishing an entire page that is all footnotes) and commented that Hutchinson clearly had "a love of detail for its own sake." While that is perhaps true, it is also clear that Hutchinson cherished observation of detail because he believed that it was the source of understanding. Hutchinson noted that "it is indeed extremely difficult to think of any class of phenomena which would not yield some generalizations of interest if properly studied."[27] Later in his life, as it became clear that interest in natural history among ecologists was on the decline, he declared that "an investigator may spend his whole life falsifying hypotheses and in the end discover nothing whatsoever about how the world actually works."[28] While he certainly believed in the utility of hypothesis testing, Hutchinson was concerned that ecologists had increasingly pursued their interests in ways that were disconnected from the natural world. As Gould put it, "An experimentalist, using technological machinery under a strict and repeated protocol, might get by in ignorance of the history and implications of his field. Those who work directly with nature's multifarious complexity cannot afford such narrowness."[29] The success of Hutchinson's theory, and particularly the durability of his ideas, offers its own ample testimony to the effectiveness of the approach. Few ecologists of the future can hope to have the same breadth of knowledge that Hutchinson enjoyed, but his legacy gives us a clear picture of the insight available to those who ground their work in an appreciation of the natural world.

ECOLOGICAL ASPECTS OF SUCCESSION IN NATURAL POPULATIONS

PROFESSOR G. EVELYN HUTCHINSON

YALE UNIVERSITY

THE striking changes that occur in both experimental mixed cultures and in natural populations have, as Dr. Woodruff has reminded us, been observed for a long time. In attempting a modern explanation of such changes it is first desirable to call attention to certain controlled laboratory experiments, that will provide a key to the more complex situation in nature. These experiments are due to Gause and are doubtless generally familiar; they are explicable by a relatively complete mathematical theory, provided by Lotka (1925), by Volterra (1926), and by Gause himself (1934, 1935). The particular experiments which are fundamental from the standpoint of the present contribution show that:

(1) If two species live in an identical niche, competing for the same food supply, maintained at a constant level, one species will entirely displace the other. This has been demonstrated with *Paramecium aurelia* and *Glaucoma scintillans;* the latter, having the higher coefficient of multiplication, alone remains.

(2) Dominance in competition is dependent on the environment conditions. This is strikingly shown in experiments with mixed cultures of *P. aurelia* and *P. caudatum,* in cultures that have, or have not, received biologically conditioned medium. If metabolic products of *Paramecium, Bacillus pyocyaneus,* etc., are not added, *caudatum* starts growing faster; if such products have been added, the reverse is true. In general, if the ecological conditions are such that utilization of food is the controlling factor, *caudatum* is dominant, if resistance to metabolic products, either hetero- or homotypical, is the controlling factor, *aurelia* is dominant (Gause, Nastukova and Alpatov, 1934).

The very elaborate development of Gause's work in which more than one niche is provided (*e.g.*, two foods of different sizes) need not concern us at present. The facts of fundamental importance are, firstly, the enormous difference that competition makes to organisms that singly exhibit superficially similar growth curves, and, secondly, that this dominance of one species over another is subject to environmental control. Comparable phenomena are known in other groups of animals, as for instance, the triclads (Beauchamp and Ullyott, 1932).

Successional changes have been studied primarily in systems of two kinds.

(1) Hay infusions, in which the free energy of the system is at a maximum at the moment of initiation of the experiment, and declines continually, though with presumably a decreasing rate.

(2) The open water of lakes and the ocean. The bioceonoses contained in such biotopes constitute machines causing a detour in the degradation of radiant energy from the sun. In the absence of seasonal changes the free energy might be expected to be constant, representing a steady state of production and decomposition of organic material, and a function of the supply of nutrient elements. In practice the free energy oscillates about a mean value. Such oscillations are largely seasonal, the mean value doubtless depending largely on the trophic standard, though the relations are very complex.

Any system in which external (seasonal) or internal changes modify properties which are involved in the struggle for existence will cause a successional series. In the hay infusion a number of workers (Fine, 1912, Bodine, 1921, Pruthi, 1927, Eddy, 1928, Darby, 1929) have shown marked changes in alkalinity, pH, CO_2 content, and by inference, O_2 content. These changes are obviously primarily determined by bacterial metabolism of the hay, the running down of the system, but differ in infusions with and without Protozoa. Both changes in oxygen content and in pH are probably involved in succession. The hay infusion, as commonly prepared, is emphatically not a single niche system (*cf.* Woodruff, 1912, Eddy, 1928) and the competition phenomena are liable to be complex. Metabolic products undoubtedly play an important part, either directly, or indirectly, in altering pH, but apart from Woodruff's pioneer work, their significance for infusion succession is largely inferential.

Somewhat simpler conditions are provided by the open waters of lakes and the ocean. Here we avoid the complexities of a solid substratum. A marked light gradient, a temperature gradient, and some chemical gradients occur, but by confining our attention to the epilimnion or epithalassa which is very thick compared with the length of a protozoan or, in most cases, the vertical component of its mean path, we can avoid consideration of spatial changes in chemical or thermal properties, so we have, at any moment, what is as close as possible to a single niche. The properties of this niche, however, vary seasonally.

In considering the populations of lakes and of the ocean, the most interesting forms are certainly autotrophic, particularly the Dinoflagellates and certain Chrysomonads, such as *Dinobryon.* Emphasis may be placed on these groups as they are not to be considered in great detail by later contributors to this symposium. Since such organisms must be supposed to enter into competition with other autotrophic members of the plankton, particularly diatoms and blue-green algae, it is impossible to discuss their periodicity without reference to that of the non-flagellate groups. The primary requirements of all these forms relate to *light, temperature,* the supply of *nutrient elements,* of which N, P, Fe, Mn and in the case of diatoms, Si, are the most important, and sometimes the presence of *organic accessory substances,* which may be regarded as equivalent to, and in certain cases are identical with, the vitamins of importance in vertebrate nutrition.

The annual cycles of light and temperature, being both dependent on the annual variations in the position of the sun, are in general similar, but since heat can be stored as molecular motion, whereas light can not, the rate of change of temperature and of light intensity is somewhat different in the spring, and minor fluctuations in the weather are more clearly reflected in the variation of light than of temperature. This makes it possible to separate by statistical methods their effects in nature. It would appear that wherever this has been done in considering the general spring increase of autotrophic plankton, the effect of light is at least as, if not more, significant than that of temperature. Almost fifty years ago Calkins (1893) noted the increase of diatoms during the "period of lengthening days, in spite of the almost freezing temperature of the water." An example of a more recent statistical study is provided by Riley's (1941) investigation of Long Island Sound. Richards (1929), moreover, believes that there is a statistically recognizable light factor influencing the division rate of pure cultures of ciliates (*P. aurelia, Blepharisma undulans* and *Histrio complanatus*). There is further evidence that even within a single genus the vertical distribution of populations is controlled by the illumination. This is most strikingly demonstrated by the *Ceratia* of the open ocean (Steemann Neilsen, 1934), where the depth distribution depends on the total amount of plankton shading the upper water. Many species of the genus are therefore shade species in just the same sense as are the plants composing the ground cover in a tract of woodland. In view of these various observations it seems not improbable that seasonal variation in light intensity may directly or indirectly play a rôle in determining succession of single species. Unequivocal separation of the effect of light and temperature in regulating seasonal replacement of one species by another does not yet seem to have been demonstrated, in spite of its inherent probability.

Turning to the variations in chemistry in the surface

layers of lakes and the ocean, we find complex but well-marked seasonal changes. These are largely determined by the organisms themselves. In general the increasing light and temperature in the spring lead to an outburst of photosynthetic plankton, which, in the epilimnion causes an increase in O_2 concentration and pH, a fall in the concentration of nitrate and phosphate. These changes are most marked in small lakes, but here subsequent events are irregular, though there is usually a marked fall in total plankton at midsummer followed by a rise of Cyanophyceae about the time of maximum water temperature. There is evidence that this latter event accompanies in some cases intense fixation of atmospheric nitrogen. In large lakes and in the sea there is generally an autumnal rather than an aestival secondary bloom, correlated with the breakdown of thermal stratification. In surface waters changes in oxygen concentration are probably of little importance; changes in oxygen tension in the water of lakes are of greater protozoological interest in the hypolimnion, where the reduction of O_2 during the summer, permits the development of a remarkable ciliate fauna.

Owing to the supposed ease with which the concentration of hydrogen ions can be determined by dropping solutions of dyes into water samples, pH became a fashionable symbol. It is, however, exceedingly doubtful if more than a single case has been brought forward demonstrating unequivocally that the natural variation in numbers of any species of animal is due to variation in pH, though it is not improbable that a number of ciliates in relatively saprobiotic habitats are affected by this factor. The work of Saunders (1924) and of Darby (1929) does indicate that *Spirostomum ambiguum* is really limited in nature by rises in pH over 7.8 and that comparable phenomena are probably exhibited by other species. Saunders's experiments are particularly important in that the pH was controlled both by Sørensen buffers and by calcium bicarbonate and CO_2, so that the effects are shown to be independent of all ions save H· and OH′. Numerous further examples of apparently acid or alkali water animals are known, but it has yet to be demonstrated that these are not due to variations in Ca or CO_2 content rather than H·. The apparent ecological insignificance of hydrogen ion concentration, in the numerous euryionic species that are known, contrasts so strongly with the immense and universal importance of this variable within the organism, that one is tempted to conclude that some independence of external variations in pH was one of the first prerequisites in the evolution of higher organisms. However this may be, the seasonal changes in pH, CO_2 or Ca in the open waters of lakes and the ocean are in general too small to be of importance in regulation of natural succession.

Combined nitrogen and phosphate are of paramount importance to all autotrophic forms. It is first important to remember the very great dilutions at which these substances occur. Concentrations of 100 γ per litre N or 20 γ P per litre are to be regarded as high; frequently even at times of increase in the autotrophic plankton, not more than 20–40 γ $N.NO_3$ and 0.5–2 γ P can be detected.

Secondly, it is clear that the space provided by the illuminated waters of lakes and the ocean is rarely filled to capacity. Only in the production of intense water blooms is any spatial crowding probable and in almost every case addition of appropriate substances to water under natural conditions of illumination will increase the crop. The autotrophic forms therefore live under intense conditions of competition for nutrient substances. It should also be pointed out that a relatively few divisions are needed to raise a species to a dominant position. One cell per cc. is usually below the limit of accurate counting, even though it may correspond to a population of 10^{10} cells in the top meter layer of a small lake. Yet after ten divisions a species at a concentration of one cell per cc. will be present, if all descendants survive, in a concentration of over 1000 per cc. and may well

have become a dominant, if the other species present are
dividing more slowly or are surviving poorly.

Little information exists as to the optimal nutrient concentrations required by autotrophic plankters. The most
recent and probably the most accurate work is that of
Ketchum (1939), who found that reduction of nitrate
nitrogen to as low a concentration as 50 γ per litre did
not retard the growth of *Nitzschia closterium* populations. Reduction of phosphate phosphorus, however,
below 17 γ per litre, did cause a marked fall in division
rate of this marine species. Such a quantity is much
above the maximal amount of free phosphate phosphorus
(often less than 10 γ per litre) present in many lake
waters, so that it is highly probable that differences in
the minimal concentration required for maximal division
would be found if different species were compared. If
this is so, it is obvious that a mechanism exists for the
production of succession during the period of nutrient
exhaustion after a maximum of any species, for as the
concentration falls, the species responsible for the depletion, which dominates at an initial high nutrient level,
will suffer from competition with forms dominant at low
nutrient levels. Pearsall's (1932) field data strongly
suggest that the replacement of the diatom *Asterionella*
by *Tabellaria* in the English lakes represents such a shift
in population due to change in nutrient level. In Linsley
Pond, North Branford, Connecticut, where *Tabellaria* is
not of importance in the plankton, the appearance of
Asterionella formosa is not regulated in the way indicated by Pearsall. In considering combined nitrogen as
a primary nutrient it is important to bear in mind that
it may occur in many inorganic forms; ammonia and
nitrate are the most important, though nitrite and perhaps hyponitrite may also occur (Cooper, 1937). In
some lakes (*e.g.*, Lake Mendota) the balance between
bacterial production of ammonia, nitrification, and denitrification produces a stationary concentration of ammonia greater than that of nitrate; in other lakes

(English lake district and northeast Wisconsin) the reverse is true. Other cases occur in which the ratio of nitrogen as ammonia to nitrogen as nitrate shifts with the seasons (Yoshimura, 1932). Zobell (1935) and Harvey (1940) have shown that marine diatoms use ammonia in preference to nitrate when both forms of nitrogen are supplied. It is therefore probable that in waters with a low $N.NH_4 : N.NO_3$ ratio, variations in ability to use nitrate effectively may be of importance in regulating competition.

It is worthwhile noting that in any case (*i.e.*, the ocean or a large lake) in which a certain concentration of nutrients is established at the spring over-turn or full circulation, which concentration can not be maintained by additions during the summer period of utilization, the maximum population would inevitably be one developed largely of high nutrient forms. If, however, a slow production of nutrients is possible from the mud of shallower parts of the lake, as is certainly the case in small lakes (Hutchinson, 1941), a steady state can theoretically be established, leading to a peak in the production of species adapted to any nutrient level.

In relation to the thesis that competition for nutrients is of paramount importance in regulating succession, we may examine the seasonal incidence of the colonial Chrysomonad *Dinobryon*. In a large number of lakes, several species of this genus show a well-marked maximum after the spring phytoplankton bloom. Pearsall suggested that this was determined by a fall in the SiO_2 content of the water below 0.5 mgrs. per litre and by a rise in the N/P ratio to over 40 or so, the exact value being dependent on the silicate and calcium content. In Linsley Pond, where the Ca is always high and relatively constant, we have found *Dinobryon* (mainly *D. divergens*) to develop a maximum with a SiO_2 content of over 7 mgrs. per litre, and although the maximum is usually preceded or accompanied by a rise in the N/P ratio, this may be but little over 40 even with such a high SiO_2 concentration. Moreover, a study of the literature of the occurrence of this organism indicates that, while its incidence at periods of low SiO_2 is not confined to Pearsall's localities, remarkable exceptions to all of Pearsall's determining factors are known. It is noteworthy that *Dinobryon* maxima may intervene after a phase composed of either diatoms (*Asterionella* and *Tabellaria*) or of Cyanophyceae, groups which are probably very different biochemically; the production of some specific substance favoring *Dinobryon* by the antecedent species appears therefore unlikely. *A priori* it may be expected that in any water in which the N/P ratio is above the mean biological ratio of about 7, increasing utilization increases the ratio; this is confirmed by Ketchum's experiments, though in the inverse situation, an abnormally low N/P ratio is brought nearer to normal by absorption of nutrients. There are, moreover, usually great quantities of diatoms in the spring bloom. It is obvious that an increasing N/P ratio and a declining silicate content may both be symptoms of depletion. Where such symptoms are often, but not invariably, correlated with a change in the population, it seems natural to suppose that competitive relationships are involved. Owing to the complexity of such relationships, particularly when the species of competitors and the values of other factors regulating competition are not constant, we should hardly expect any hard and fast limiting value in the nutrient level to determine the incidence of *Dinobryon*.

Apart from combined nitrogen and phosphorus, iron and manganese are probably the most important inorganic substances to be considered. Iron raises special problems. In simple ionic form it is practically insoluble at the pH normally encountered in both fresh and salt waters. Suspended and colloidal ferric hydroxide are probably always available, and at least in fresh waters soluble or colloidal organic compounds also. The ferric hydroxide of fresh waters may well be normally present partly in the form of bacterial sheaths. Harvey has shown that at least some diatoms can utilize suspended and colloidal iron; practically no information exists as to a possible seasonal influence. In the case of Mn, however, Harvey finds that not infrequently summer sea water which will not support the winter diatom *Ditylum brightwellei,* even after enrichment with N, P, and Fe, can produce a good growth if Mn be added. Mn depletion, therefore, is probably involved in the seasonal cycle of the sea.

More and more attention is now being given to those accessory organic substances that would be called vitamins in vertebrate biochemistry. Though primarily known by their effects in laboratory experiments, it seems desirable to consider briefly their possible rôle in natural environments. Harvey, again working with *Ditylum,* finds two groups of substances which he calls A (adsorbable on charcoal) and N (not adsorbable on charcoal) which are necessary for growth of this diatom. These may be absent from enriched summer water. Both can be derived from soils and from decaying sea weed. Substance ''A'' can be replaced by cystine, and by smaller amounts of thiamin (= vitamin B_1). A crude biotin (= vitamin H = coenzyme R) preparation was also effective. The natural ''A'' can not be thiamin, and *a fortiore* not cystine, as the amount (.25 mgr. per litre) required is disproportionate to the organic content (5 mgrs. per litre) of sea water. It is, moreover, apparently not an amino acid. Substance ''N'' can be replaced by α-alanine, lactic acid, gluconic acid, and dextrose. Harvey suspects its function is to form a soluble iron or manganese complex.

That these substances are apparently sometimes present, sometimes absent, in the sea, indicates that they may have an effect in the production of seasonal succession. A matter of particular ecological significance is raised by them because they may explain the otherwise puzzling fact that the depletion of combined nitrogen and of phosphate often goes much farther in small eutrophic lakes than in oligotrophic lakes or in the sea. It is very usual to find, therefore, large crops of plankton continuing to develop at a nutrient level apparently less satisfactory than that provided by water which never produces much planktonic growth. Moreover, great differences in vitamin requirement are already known in protists, fungi, and bacteria, and not improbably many puzzling facts in distribution may find their explanation here. Steemann Nielsen suggests on the contrary that organic matter in neritic waters excludes many species of oceanic *Ceratia*. It is, however, possible that competition with otherwise more successful forms which require a higher amount of accessory substances is involved.

Finally a word must be said about relationships between prey and predators, as this concerns holozoic forms. As is well known, Lotka and Volterra have shown that cyclical variation in the numbers of two forms, one feeding on the other, can theoretically be set up and perpetuated. In any case, there are definite particular numbers of the two species which can coexist constantly at a singular point; however, if either number is altered no return to the singular point is possible, but a cycle is generated, so that the number of one species plotted against the other goes round and round the singular point. Gause, who studied this matter experimentally with *Didinium* and *Paramecium,* found that in a uniform microcosm no agreement with theory occurred. In general this is due to the fact that if predation is very efficient, the singular point lies so near the origin that chance statistical variations will soon bring one or other population to zero. If the predator is kept artificially rarified in proportion to its abundance, *i.e.,* if the predation is reduced, a well-developed cycle can be introduced into the system, as has been shown by Gause for *P. caudatum* feeding on *Schizosaccharomyces*. In most cases in nature it is probable that rhythmic fluctuations in numbers are not due to simple prey-predator relationships, though other types of internal rhythm may develop. Gause, however, suggests that one remarkable example involving the protozoa is dependent on the prey-predator relationship, namely the striking variation in numbers of bacteria and of protozoa in soils, recorded by Cutler, Crump, and Sandon (1923).

In conclusion, it is apparent that a variety of environmental factors may control succession in protistan populations. In laboratory experiments clear-cut results can often be obtained; in nature a bewildering number of possibilities present themselves, and the further analysis is carried the more difficult is it to isolate controlling physico-chemical variables. This difficulty should be admitted as one of the relevant facts to be considered; it is indeed one of the most important of the data, for it strongly suggests that in many cases modification of the dominance exhibited among numerous competing species is the major rôle of the environmental factors, rather than the direct transgression of limits of tolerance, so easily studied in laboratory experiments with pure cultures.

LITERATURE CITED

Beauchamp, R. S. A., and P. Ullyott
    1932. *Jour. Ecol.*, 20: 200.
Bodine, J. H.
    1921. *Biol. Bull.*, 41: 73.
Calkins, G. N.
    1893. *Massachusetts, State Board of Health, 24th annual report (for 1892)*: 381.
Cooper, L. H. N.
    1937. *Jour. Mar. Biol. Asso.* (n.s.), 22: 183.
Cutler, D. W., L. M. Crump and H. Sandon
    1923. *Phil. Trans. Roy. Soc. London*, Ser. B., 211: 317.
Darby, H. H.
    1929. *Arch. Protistenk.*, 65: 1.
Eddy, S.
    1928. *Trans. Amer. Micro. Soc.*, 47: 283.
Fine, M. S.
    1912. *Jour. Exp. Zool.*, 12: 265.
Gause, G. F.
    1934. "The Struggle for Existence." Baltimore.
    1935. *Actualités Scientifiques et Industrielle*, No. 277. Paris.
Gause, G. F., O. K. Nastukova, and W. W. Alpatov
    Jour. Animal Ecol., 3: 222.

Harvey, H. W.
    1939. *Jour. Mar. Biol. Asso.* (n.s.), 23: 499.
    1940. *Jour. Mar. Biol. Asso.* (n.s.), 24: 115.
Hutchinson, G. E.
    1941. *Ecol. Monogr.*, 11: 21.
Ketchum, B.
    1939. *Amer. Jour. Bot.*, 26: 399–407.
Lotka, A. J.
    1925. ''Elements of Physical Biology,'' Baltimore.
Pearsall, W. H.
    1932. *Jour. Ecol.*, 20: 241.
Pruthi, H. S.
    1927. *Brit. Jour. Exper. Biol.*, 4: 292.
Richards, O. W.
    1929. *Biol. Bull.*, 56: 298.
Riley, G. A.
    1941. *Bull. Bingham Oceanogr. Coll.*, Yale University, 7: no. 3.
Saunders, J. T.
    1924. *Proc. Cambridge Philos. Soc. Biol. Sci.*, 1: 189.
Steemann Nielsen, E.
    1934. *The Carlsberg Foundation's Oceanographical Expedition round the World*, 1928–30. *Dana Report*, No. 4.
Volterra, V.
    1926. *Mem. R. Accad. Lincei*, ser. 6, II, fasc. 36.
Woodruff, L. L.
    1912. *Jour. Exp. Zool.*, 12: 205–264.
Yoshimura, S.
    1932. *Arch. Hydrobiol.*, 24: 155.

clusion that biting, scratching and similiar violent forms of love-making are habitual only in cultures, here termed permissive, that allow a good deal of adolescent sexual freedom, though the inverse proposition that such freedom implies the development of violent forms of sexual play is not true. Whatever the significance of this correlation may be among the preliterate cultures studied by Ford and Beach, "Phoebus amorous pinches black," belong to a culture that would probably not have been regarded as permissive in the Human Relations Area Files. The only other statistical conclusion that appears, seems to be that the occurrence of sexual interest in the breasts, which is not universal in human cultures, is not correlated with whether the breasts are habitually covered or left bare. The atomistic inadequacy of the approach is strikingly apparent when the covering or uncovering of the breasts is considered. As that learned Jesuit, Father Alberti, pointed out long ago in his unpublished tract on this question, everything depends on the prevailing circumstances. A Bavenda woman takes off her shapeless cotton blouse and appears in the full dignity of a sepia torso and a carefully beaded skirt because she is home in her village and has left the white man's town where certain regulations are enforced. The Virgin, in some mediaeval representations of the Last Judgment, opens her robe and cries in agony for all mankind *Fili, aspice ubera quae te lactaverunt.* An Italian peasant woman bares her breast in a railroad coach full of men for the excellent reason that her baby is hungry. A chorus girl at the Casino de Paris does the same to lend erotic charm to a theatrical spectacle. Exposure to public view of the same special part of the body, covered a short time before, is involved in all these instances, yet functionally the acts are extraordinarily different. It is extremely curious to see that a generation of functionalist anthropologists, who criticized Sir James Frazer for wrenching divine kings out of context, though when he wrote around the Mediterranean he really knew what he was writing about, should give place to a statistical generation whose members are now taking tiny bits of sexuality out of context without the slightest shame simply because they have an IBM machine rather than a classical education to help them.

The authors of *Patterns of Sexual Behavior* claim to have "consistently eschewed any discussion of rightness or wrongness of a particular type of sexual behavior." Actually, it is probably impossible to free any statements about the matters they consider from all normative overtones. It is extremely unlikely that most of the material in the book would be taken, by any large group of readers, with the same emotional calm as would result from a discussion of the evolution of suture pattern in Jurassic ammonites or a statement about the decrease in ionic radius with increasing atomic number in the rare earth elements.

It is impossible to avoid implying value judgments, and the only value judgment that is validly implied is one that is based on a consideration of at least a carefully weighted sample of all the evidence. It may, in fact, be doubted whether it is really of value to present comparative material based mainly on the statistical and exclusively non-functional study of preliterate societies to a reading public who are believed to need the dimensions of the human genitalia in inches and fractions thereof (which to this reviewer is definitely pornographic) rather than in millimeters. Readers accustomed to the metric system will, however, find ways of using the book profitably.

G. E. HUTCHINSON

YALE UNIVERSITY,
NEW HAVEN, CONNECTICUT

COPEPODOLOGY FOR THE ORNITHOLOGIST

Recently in the course of preparation of chapters on the zooplankton for a forthcoming comprehensive work on all aspects of limnology, it has been necessary to study in detail a considerable number of papers on the freshwater copepoda, particularily the Calanoida and planktonic Cyclopoida. During the course of this study it became apparent that the literature contained a number of disconnected facts of considerable evolutionary significance. Some of these facts provide striking analogues to phenomena recorded in other groups, particularly the vertebrata. Since it is probable that the somewhat more detailed account prepared for my book will not come to the attention of ornithologists and other students of terrestrial ecology, and since early publication is desirable to stimulate further research on these matters, the preparation of the following summary seems justified.

Selective and non-selective feeding. The calanoid copepods feed largely by filtration. In this process the antenna is used as a screw to produce a water current, from which food particles are filtered, largely by the maxillae. The details of the process have been elucidated in detail by Storch and Pfisterer (1925) and more recently by Lowndes (1935). It has commonly been supposed that the process of feeding must under these circumstances, be entirely automatic. Quite apart from some evidence of diversified feeding habits to be derived from a study of the structure of the mouth parts, Lowndes demonstrated directly selective feeding by supposedly non-selective species. Large particles can be rejected by the simple expedient of separating the maxillae, so letting the particle pass the filter. Certain substances added to the water can be avoided and field evidence indicated that even *Eudiaptomus gracilis* (Sars), a species commonly regarded as quite unselective, could seek out benthic desmids when living in a pond soupy with *Kirchneriella*, which was entirely rejected. Burckhardt (1935) found that the males of *Eurytemora affinis* (Poppe) rejected a greater proportion of *Coscinodiscus* in favour of smaller diatoms than did the females though there is no significant difference in size between the sexes. There is some morphological evidence that the maxilla is less well adapted as a filter in some *Diaptomidae* than in others and Lowndes found that even in the obvious filter feeder *Eurytemora velox*, there is often evidence in the faeces of predation on other Crustacea. We may expect therefore that, since what is large for a small species may be small for a large species, the food of two species of different sizes will not be identical and that, at least in some cases, the most confirmed filter feeders will exhibit some selectivity in the kind of neighbourhood they select for their filtering operations.

Size differences in associations of closely allied species. It has been observed (Hutchinson, Pickford and Schuurman 1932; Hutchinson 1937) that in the playa lakes or pans of semi-arid regions, the plankton often consists of a single species of *Daphnia* and two Diaptomid copepods of widely different sizes. In the Transvaal, *Lovenula falcifera* (Lóven) of length 3.5–5.0 mm in temporary less alkaline, and *L. excellens* Kiefer of length 3.0–3.5 mm, usually in perennial and more alkaline, localities are associated with *Metadiaptomus transvaalensis* Methuen of length about 1.9 mm. In Big Washoe Lake, a playa in Nevada very similiar to the Transvaal pans, the large *Hesperodiaptomus franciscanus* (Lillj.) and small *Leptodiaptomus tenuicaudatus* (Forbes) similiarly co-occur. More complex associations are found in some of the temporary waters of the southeastern Palaearctic. In the pluvial zone of North Africa, Gauthier (1928) found the immense *Hemidiaptomus ingens* (Gurney), about 5.2 mm long, associated with *D. cyaneus* Gurney about 3.2 mm long and with both *Eudiaptomus numidicus* (Gurney) and *Myxodiaptomus lilljeborgi* (Guerne and Richard) about 1.5 mm long. Similiarly in Turkey, where two other species of *Hemidiaptomus* of great size occur, Mann (1940) found *H. brehmi* Mann of length 4.2–5.0 mm associated with *Arctodiaptomus belgrati*

Mann of 0.98–1.24 mm, and *H. kummerlöwei* Mann of length 3.5–4.45 mm with *A. baccillifer* (Kölbel) of length 0.87–1.19 mm. It is probable that in Central Asia also the large species of *Hemidiaptomus* always have small associates (Sars 1903). Though the existence of marked size differences is often very conspicuous in temporary waters it is by no means confined to them. Carl (1940) concluded that in the lakes of British Columbia when two Calanoids co-occurred, they are generally separated by a size difference. The most frequent case that he encountered was the association of *Epischura nevadensis* Lillj. of length 1.8–2.1 mm and *Leptodiaptomus tyrelli* (Poppe) of length 1.2–1.3 mm. Gurney (1931) gives two particularly interesting cases of very closely allied species living together being separated by small size differences.

The first is that of the copepods regarded below as two separate species, living together in Loch Ness in Scotland:

	♂	♀
Eudiaptomus gracilis	1.26–1.35 mm	1.4 mm
E. pusillus (Brady)	0.98–1.17	0.99–1.2

The second case is that of the very closely allied *Arctodiaptomus wierzejskii* (Richard) and *A. laticeps* (Sars) which differ only in the most minute characters of certain segments of the antennule. Considering these as a whole, the rather wide range of size variation overlap completely. On the rare occasions when they co-occur they are separable by small size differences. Gurney gives the following mean values for the populations in Loch Moracha, North Uist:

	♂	♀
A. wierzejskii	1.4 mm	1.53 mm
A. laticeps	1.3 mm	1.43 mm

Cases are certainly known in which calanoids of the same size inhabit the same water. Sufficient data however exist to suggest that a diversity of size, probably implying, in view of what has been said about feeding habits, some difference in food, frequently makes possible the co-occurrence of two or more species of Calanoida.

The evidence in the Cyclopoida is less striking because the Cyclopidae include a great many benthic forms. It is therefore possible to have a number of allied species in a lake without their habitats overlapping appreciably. When two very closely allied species occur in the plankton, they tend to be of different sizes. Thus in Lake Toba, Sumatra, Kiefer (1933) records a population of large specimens of the very widespread *Mesocyclops leuckarti* (Claus), the females being about 1.12 mm long, in company with the endemic *M. tobae* Kiefer of which the females were only 0.7 to 0.8 mm long.

Similarly in Lake Lanao, Mindanao, Philippine Islands, the widespread *Thermocyclops hyalinus* (Rehberg) is accompanied by a smaller endemic species *T. wolterecki* Kiefer (1938).

A comparable case among supposedly benthic species is given by Harding (1942) who obtained four species of *Eucyclops* from Lake Young in Northern Rhodesia. Two of these were new and possibly endemic species, probably derived from *E. serrulatus* (Fischer). They resembled each other and differed from their nearest allies in the much shorter antennae. The most conspicuous difference between the two species of the pair is in size, *E. parvicornis* (Harding) having adult females 1.02–1.06 mm long, *E. spathanum* 0.64–0.71 mm long. The latter species has flattened spines on the pleopods which presumably may also indicate some functional difference.

On account of the less automatic feeding mechanisms of the Cyclopoids, some selection of food seems likely but, though many zoologists have reared Cyclopidae in the laboratory, little is really known of their food habits. What little information does exist in no way conflicts with the idea that larger individuals would not eat exactly the same kind of food as smaller.

The role of size differences in permitting the coexistence of closely allied species of birds and mammals has been stressed by Huxley (1942, p. 280–281) and by Lack (1944). Brooks (1950a) has employed the same idea in his interpretation of the distribution of *Odontogammarus* in Lake Baikal. It is obviously possible that in invertebrates with little metamorphosis and prolonged life histories, competition of adults of the smaller species with sub-adults of the larger species may produce complications. In spite of this and perhaps some other difficulties, the existence of size differences seems to provide one way by which sympatric species of restricted groups of planktonic animals can exist under conditions in which, in general, competition might be supposed inevitably to eliminate one form.

The variation in clutch size in the freshwater Calanoida. When, as in most freshwater Calanoida, the female carries the eggs in an egg-sac, it is easy to observe the variation in the size of the clutch within a given species. Such variation was noted by Schacht (1897) in *Diaptomus siciloides* Lillj. described from Lake Tulare in California as carrying but four eggs, but in the Illinois River having egg-sacs containing up to eighteen eggs. Schacht supposed that since the type locality is a mountain lake, the larger number of eggs carried in the Illinois River is due to higher temperature and better food supply. In some cases, the variation can be quite extraordinary; Mann (1940) records a minimum of 2 and a maximum of 64 in specimens of *Arctodiaptomus bacillifer* from Turkey.

Mean values for whole samples varied from 6 to 60, but the data are inadequate to demonstrate any seasonal trend.

Wesenberg-Lund (1904) found that *Eudiaptomus gracilis*, in most of the Danish lakes in which the species occurred, carried from 25 to 30 eggs in the spring, but only from 6 to 8 during the summer and autumn. Though there was some irregularity in the second half of the year with a possible hint of a second or autumnal maximum, the main and very striking feature of the variation is the rise to and fall from the well-marked spring maximum. Rzóska (1925) found a similiar variation in Poland, where *E. gracilis* carries 22 to 28 eggs in the spring and but 4 to 6 in July. Within a given region, there is, however, also some variation from lake to lake, for Wesenberg-Lund found that although the variation in the clutch-size in the population in Fureso [1] followed the same seasonal pattern as in the other lakes, the absolute number of eggs per clutch is smaller, falling from 14 to 20 at the time of the spring maximum to 4 to 5 at the time of the minimum. Von Klein (1938) in an admirable study of the Crustacea of the Schleinsee found that *E. gracilis* carried a mean clutch of 11.0 to 11.4 in April. This sank to a minimum of 3.0 on August 2 and rose to an autumnal maximum of 8.8 on November 5. Later, but five or six eggs were carried through the winter. It is evident that the variation follows the variation in abundance and reproductive capacity of *Daphnia cucullata* in the same lake. Though it is tempting to regard variations of this sort as controlled by food, von Klein points out that *Diaphanosoma* and *Ceriodaphnia*, which are filter feeders like *Daphnia*, rise to a maximum between the *Daphnia* peaks and the periods of maximum egg production by *E. gracilis*. Gurney, also writing of *E. gracilis* considers that in Britain, fewer eggs are carried by females in large lakes than in ponds; these remarks are in part due to a confusion about *E. pusillus* Brady but are no doubt partly correct. Again it is tempting to suppose a greater food supply in the smaller shallower waters.

In *Eudiaptomus graciloides* Wesenberg-Lund noted a variation comparable to that exhibited by *E. gracilis*, save that in the case of the former species so little reproduction occurred after the late spring that individuals carrying any eggs in the summer were very rare. In Esromsö, at the beginning of the breeding season in April, at a temperature of 3° C. the females carried

---

[1] The Fureso population is, by an obvious misprint, referred to *E. graciloides* in that part of the English summary that refers to clutch size; reference to the Danish text or to the earlier paragraphs of the English summary make clear that *E. gracilis* is meant.

from 7 to 9 eggs, in the first week in May at 8° C., 10 to 12 eggs, at the end of May at 13° C., 4 to 5 eggs. The very small number of females reproducing later in the year generally carried 4, 5 or 6 eggs. In other less well studied lakes, the same sort of variation occurred; occasionally up to 16 eggs may be carried in spring.

A similiar variation in egg number during the reproductive season is exhibited under entirely different circumstances by *Phyllodiaptomus annae* (Apstein) in the lake at Colombo, Ceylon (Apstein 1907). Here, when reproduction begins in June the females carry about 30 eggs, but this number sinks progressively as the population rises to a maximum and falls again, till in September when little reproduction is occurring and the species is becoming scarce, the egg-sacs contain only seven to nine eggs.

Cases such as that of *Eudiaptomus graciloides* in which the clutch size increases and decreases while a marked rise in temperature is occurring and cases such as that of *Phyllodiaptomus annae* where marked changes occur at a time when little variation in temperature is to be expected, suggest that a variation in food-supply is more likely to underlie the phenomenon than is a change in the physical environment. It is to be noted that such variations in clutch size are not limited to the Calanoida but occur in the Cyclopoida and in certain pelagic Cladocera also. Wesenberg-Lund noted a decline in the number of eggs carried by a species of *Cyclops* that he refers to *C. strenuus* from 40 to 50 in the spring to 20 to 30 in the summer. He also noted a large decrease in the number of the brood carried by *Daphnia* as the season progressed. Later work by Brooks (1946) strongly suggests that in the lakes of Connecticut, the summer individuals with small broods are indeed somewhat starved. Slobodkin (verbal communication) finds that food-supply controls fecundity in *Daphnia obtusa* Kurz.

Wesenberg-Lund claims that in the three copepods that he studied, not only does the clutch size decrease, but the egg diameter increases as spring gives place to summer. He gives measurements only in the case of *Eudiaptomus graciloides* where the spring eggs have a diameter of about 120 μ and the summer eggs of from 160 to 200 μ. Assuming a spherical form, the volume of the spring eggs must be about $9 \times 10^5 \mu^3$ and of the summer eggs 21–$42 \times 10^5 \mu^3$. Thus a summer clutch of 4 will contain as great a volume of eggs as will a spring clutch of from 9 to 18. These considerations at least suggest that the variation in clutch size throughout the breeding season is not necessarily merely a direct effect of starvation, and, at least in species in which reproduction occurs on a large scale throughout spring and summer, part of the variation may prove to

have some adaptive meaning. Wesenberg-Lund suspects that the large summer eggs may hatch at a naupliar stage later than the first. Several workers have noticed that the limited number of nauplii produced in spring survive better than do the summer nauplii which are responsible for the small rise in population in late August and September. It is therefore not unlikely that early in the season when the population is expanding into a lake rich in food, individuals producing many small eggs will leave most descendants, while later in the year when the population is larger and the food supply less, it is desirable to increase as far as possible the individual life expectancy of the nauplii produced. This case appears to be comparable to some studied by Lack (1947, 1950) in birds and provide a loose temporal analogy to the geographical variation in egg number and size noted by Svärdson (1949) and Määr (1949) in *Salmo alpinus* (cf. Brooks 1950b.)

Specific differentiation in clutch size. Brehm and Zederbauer (1902) recorded up to seven eggs in the egg-sacs of *Eudiaptomus gracilis* in the lakes of the Austrian Alps in summer, but usually only four in winter. They later (1906) observed that the winter population differed from that present in summer in having the spinous process at the tip of the twenty-third segment of the geniculate antennule of the male much reduced. Haempel (1918) noted a comparable or greater reduction in Austrian populations referred to *gracilis,* but elsewhere this supposed cyclomorphosis has not been observed (Brehm 1927). Gurney (1931) specifically states that he has looked for such temporal variation without success in the perennial populations of *E. gracilis* in the Norfolk Broads. However, in Loch Ness, Gurney describes a small slender form which he refers to *gracilis,* though noting that Brady has described just such a form as *pusillus.* It is distinguished from typical *gracilis* which has also been found in Loch Ness not only in its slenderer build and slightly smaller absolute dimensions but also in carrying from one to four eggs, the usual clutch consisting of two. It also differs in the male in having the apical process of the twenty-third segment of the geniculate antennule reduced, just as in Brehm and Zederbauer's supposed winter form of *gracilis.* It is most reasonable to suppose that two species [2] are involved, dif-

---

[2] There is a possibility that dimorphism within a single species is really involved. Several South American diaptomids are supposed by Wright (1938) to be dimorphic; a comparable case in the genus *Neodiaptomus* is treated by Kiefer (1939) by giving both forms specific status. Most cases of supposed dimorphism are doubtless due to the adult moulting (Gurney

Copepodology for the Ornithologist 179

July, 1951   REVIEWS   575

fering in minute morphological characters, in the average number of eggs per clutch in any given locality and, at least in the Austrian Alps, in the breeding season. It appears probable from Gurney's figures that the eggs of *E. pusillus* are slightly larger than those of *E. gracilis*, though this requires further investigation. The co-occurrence of two hardly distinguishable species differing in reproductive physiology is reminiscent of similiar phenomena in the genus *Coregonus*, for instance, in the lakes of Sweden where sympatric species, differing mainly in breeding season have been described by Svärdson (1949).

Fugitive species. In ecological succession of any sort, a series of species becomes established early in the process; almost any member of this series is destined to become extinct locally by the time the climax is reached. This is, in general, due to the fact that the earlier species have certain properties, often related to dispersal and ecesis, which the later species lack, while the later species are always able to displace the earlier in any competitive process. In an undisturbed landscape in which successional stages are present over a relatively small area, the species limited to the earlier stages of succession are obviously extremely local.

It is conceivable that species, not obviously associated with successional stages on a large scale, may actually be able to exist primarily by having good dispersal mechanisms, even though they inevitably succumb to competition with *any other species* capable of entering the same niche. Such species will be termed *fugitive* species. They are forever on the move, always becoming extinct in one locality as they succumb to competition and always surviving by reestablishing themselves in some other locality as a new niche opens. The temporary opening of a niche need not involve a full formal successional process. Very small, seemingly random changes in the physical environment might produce locally a new niche suitable for the fugitive, or some unfavorable circumstance might cause the temporary disappearance of a competitor from such a suitable niche over a small region. Any such change would give the fugitive species its chance, for a time, until some more slowly moving competitor caught up with it. Later it would inevitably become extinct, but some of its descendants could occupy transitorily a temporarily available niche in some other locality. A species of this sort will enjoy freedom from competition so long as small statistical fluctuations in the environment give it a refuge into which it can run from competitors. Although it is usual in large scale ecological succession for the species of one stage to differ considerably

1931), but this is not a likely explanation in the present case.

from those of another, there is nothing in the concept of fugitive species to suggest that this will be so. It is in fact likely that many fugitive species will be extremely closely allied to their principal competitor, since the two occupy identical niches. One of a pair of sibling species could well be fugitive.

While it is possible that the deductive situation described above is a limiting case, comparable to the bread-and-butterfly of *Alice Through the Looking Glass*, the probable existence of fugitive species is indicated by the appearance of otherwise rare species in artificially constructed habitats such as dams. A case of this sort described by Kiefer (1933, 1938) gave rise to the speculations just set out.

In Indonesia two widespread, and several endemic, species of the copepod genus *Thermocyclops* have been recorded, together with certain species of the very closely allied genus *Mesocyclops*. One of the widespread species, *T. hyalinus*, is distributed through the Oriental region northwestward into the western Palaearctic. The other widespread species *T. decipiens* Kiefer is found throughout the Oriental Region and tropical Africa. The two species are so closely allied that Gurney (1933) placed *decipiens* in the synonymy of *hyalinus*, but Kiefer's (1938) later paper dispels any doubt as to their distinctness.

There is no clear interpretation of the ecological differentiation between *Mesocyclops leuckarti* and the two widespread species of *Thermocyclops*, which permits, for example, *M. leuckarti* to co-occur with either *T. decipiens* or *T. hyalinus* in the plankton of ten out of the eleven lakes in which *M. leuckarti* was found by the German Limnological Expedition to the Sunda Islands. Whatever the cause, it is quite evident that *M. leuckarti* does not compete with the two species of *Thermocyclops* in Indonesia. It would seem equally clear that these two species do compete with each other, for out of thirty-five localities (Heberer and Kiefer 1932; Kiefer 1933, 1938) for *hyalinus* and eighteen for *decipiens* only one is common to the two species. Under most circumstances it is evident that *hyalinus* and *decipiens* exclude each other or have ecological requirements that do not overlap. None of the published data suggest any difference in the range of immediate environmental variables tolerated by the two species. Kiefer, however, points out that while the localities for *hyalinus* appear to be natural, those for *decipiens* include several dams and comparable artificial bodies of water. This indicates that *decipiens* though less common than *hyalinus* must have good powers of dispersal. The whole picture of the distribution would be explained if *decipiens* were a fugitive species in the sense defined above.

This idea has indeed been applied to the freshwater Copepoda before, by Elton (1927, 1929) who concluded that *Eurytemora velox* Lillj. and *Eudiaptomus gracilis* (Sars) behaved in the way postulated above, *E. velox* being the fugitive species which ultimately and inevitably succumbs to competition with *E. gracilis* when the latter gains an entry to the habitat in which *E. velox* is living. Lowndes (1929, 1930) has disputed Elton's conclusions, and Gurney (1931) notes that the two species can co-occur though he admits the rather frequent association of *E. velox* with artificial waters.

Although statistical presentation makes the very large number of rare species normally encountered in a flora or fauna seem reasonable, it is intuitively not obvious, in view of the fact that some organisms are extraordinarily eurytopic, why most species have to be content with such a limited degree of success. The problem as to why rarity is so common a phenomenon needs more detailed merological analysis than it has been accorded; a too great emphasis on the overall holological or statistical picture may obscure important issues. The above considerations on *fugitive species* are put forward, not on account of their originality, which is inconsiderable, but rather to emphasize in as formal a way as possible, one rather neglected mechanism by which rare species can maintain themselves.

Concluding remarks. By far the most important contributions to ecology and evolutionary theory recently based on the study of the Copepoda are the beautiful series of researches carried out by the Italian school of limnologists on the existence of slight racial differences when both planktonic (Baldi, Buzzati-Traverso, Cavalli and Pirocchi 1945; Tonolli 1949; Baldi, Cavalli, Pirocchi and Tonolli 1949) and benthic (Pirocchi 1947) copepods from different parts of the same lake are studied. Since an admirable review of the matter has recently been published by Baldi (1950) and since the conclusions arrived at are already becoming well-known, it has seemed best in the present note to concentrate on those aspects of copepological research for which some vertebrate analogies exist and which have been inadequately emphasized before.

References

Apstein, E. 1907. Das Plankton im Colombo-See auf Ceylon. Zool. Jahrb., **25** (Syst.): 201–244.

Baldi, E. 1950. Phénomènes de microévolution dans les populations planktiques d'eau douce. Viertel. Naturf. Ges., Zurich, **95**: 89–114.

Baldi, E., A. Buzzati-Traverso, L. L. Cavalli, and L. Pirocchi. 1945. Frammentamento di una popolazione specifica (*Mixodiaptomus laciniatus* Lill.) in un grande lago in sottopopolazioni geneticamente differenziate. Mem. Ist. It. Idrobiol., **2**: 167–216.

Baldi, E., L. L. Cavalli, L. Pirocchi, and V. Tonolli. 1949. L'isolamento delle popolazioni di *Mixodiaptomus laciniatus* Lill. del Lago Maggiore e i suoi nuovi problemi. Mem. Ist. It. Idrobiol., **5**: 295–305.

Brehm, V. 1937. 3. Ordnung der Crustacea Entomostraca: Copepoda. *In:* W. Kükenthal und T. Krumbach, "Handbuch der Zoologie," Bd. 3, Hälfte 1, pp. 435–496.

Brehm, V., and E. Zederbauer. 1902. Untersuchungen über das Plankton des Erlaufsees. Verh. d. zool.-botan. Ges., Wien, **52**: 388–402.

———. 1906. Beiträge zur Planktonuntersuchung alpiner Seen. Vehr. zool.-bot. Ges., Wien, **56**: 19–32.

Brooks, J. L. 1946. Cyclomorphosis in Daphnia. I. An analysis of *D. retrocurva* and *D. galeata.* Ecol. Monogr., **16**: 409–447.

———. 1950a. Speciation in ancient lakes. Quart. Rev. Biol., **25**: 30–60.

———. 1950b. Recent advances in limnology. Ecology, **31**: 659–660.

Burckhardt, A. 1935. Die Ernährungsgrundlagen der Copepodenschwärme der Niederelbe. Int. Rev. ges. Hydrobiol. Hydrogr., **32**: 432–500.

Carl, G. C. 1940. The distribution of some Cladocera and free-living Copepoda in British Columbia. Ecol. Monogr., **10**: 55–110.

Elton, C. 1927. Animal ecology. London: Sidgwick and Jackson. xxi + 207 pp., 8 pls., text illustrated.

———. 1929. The ecological relationships of certain freshwater Copepods. J. Ecol., London, **17**: 383–391.

Gauthier, H. 1928. Recherches sur la faune des eaux continentales de l'Algérie et de la Tunisie. Alger: Imp. Minerva. 419 pp., 3 pls., 6 maps.

Gurney, R. 1931. British freshwater Copepoda, vol. 1 (Calanoida). London: Ray Society. lii + 238 pp.

———. 1933. British freshwater Copepoda, vol. 3 (Cyclopoida). London: Ray Society. xxix + 384 pp.

Haempel, O. 1918. Zur Kenntnis einiger Alpenseen, mit besonderer Berücksichtigung ihrer biologischen und Fischerei-Verhältnisse. I. Der Hallstädter See. Int. Rev. ges. Hydrobiol. Hydrogr., **8**: 225–306. 2 Tb., 4 fig.

Harding, J. P. 1942. Cladocera and Copepoda collected from East African lakes by Miss C. K. Ricardo and Miss R. J. Owen. Ann. Mag. N. Hist. (11 ser.), **9**: 174–191.

Heberer, G., and F. Kiefer. 1932. Zur Kenntnis der Copepodenfauna der Sunda-Inseln. Arch. naturgesch., N. F., **1**: 225–274.

Hutchinson, G. E. 1937. A contribution to the limnology of arid regions primarily founded on observations made in the Lahontan Basin. Trans. Conn. Acad. Arts Sci., **33**: 47–132.

Hutchinson, G. E., G. E. Pickford, and J. F. M. Schuurman. 1932. A contribution to the hydrobiology of pans and other inland waters of South-Africa. Arch. Hydrobiol., **24**: 1–136.

Huxley, J. S. 1942. Evolution. The modern synthesis. New York and London: Harper and Bros. 645 pp.

Kiefer, F. 1933. Die freilebenden Copepoden der Binnengewässer von Insulinde. Arch. Hydrobiol., Suppl., **12**: 519–621.

——. 1938. Die von Wallacea-Expedition gesammelten Arten der Gattung *Thermocyclops* Kiefer. Int. Rev. ges. Hydrobiol. Hydrogr., **38**: 54–74.

——. 1939. Scientific results of the Yale North India Expedition: Biological Report No. 19. Freilebende Ruderfusskrebse (Crustacea Copepoda) aus Nordwest und Südindien (Pandschab, Kaschmir, Ladak, Nilgirigebirge). Mem. Ind. Mus., **13**: 83–203.

Klein, H. von. 1938. Limnologische Untersuchungen über das Crustaceanplankton des Schleinsees und zweier Kleingewässer. Int. Rev. ges. Hydrobiol. Hydrogr., **37**: 176–233.

Lack, D. 1944. Ecological aspects of species-formation in passerine birds. Ibis, **86**: 260–286.

——. 1947. The significance of clutch-size. Ibis, **89**: 302–352.

——. 1950. Family size in titmice of the genus *Parus*. Evolution, **4**: 279–290.

Lowndes, A. G. 1929. The occurrence of *Eurytemora lacinulata* and *Diaptomus gracilis*. J. Ecol., **17**: 380–382.

——. 1930. Some freshwater Calanoids. Direct observation *v*. indirect deduction. J. Ecol., **18**: 151–155.

——. 1935. The swimming and feeding of certain calanoid Copepods. Proc. Zool. Soc. London **1935** (3): 687–715.

Määr, A. 1949. Fertility of char (*Salmo alpinus* L.) in the Faxälven water system, Sweden. Rep. Inst. Freshwater Res., Drottningholm, No. 29: 57–70.

Mann, A. K. 1940. Uberpelagische Copepoden türkischer Seen. Int. Rev. ges. Hydrobiol. Hydrogr., **40**: 1–87.

Pirocchi, L. 1947. Isolamento ecologico e differenziazione di popolazioni di *Megacyclops viridis* Jur. nel Lago Maggiore. Mem. Ist. It. Idrobiol., **3/4**: 307–322.

Rzóska, J. 1925. Contribution à l'étude des Copepodes de la Grande Pologne. Bull. Acad. Amis des Sci. de Poznan, Ser. B, No. 1: 34–43, 1 pl.

Sars, G. O. 1903. On the Crustacean fauna of Central Asia. Part 3. Copepoda and Ostracoda. Annu. Mus. Zool., St. Petersb., **8**: 195–232. 8 pls. Appendix. Local faunae of Central Asia, pp. 253–264.

Schacht, F. W. 1897. The North American species of *Diaptomus*. Bull. Illinois State Lab., Natural History, **5**: 97–203.

Storch, O., and O. Pfisterer. 1925. Der Fangapparat von *Diaptomus*. Z. vergl. Physiol., **3**: 330–376.

Svärdson, G. 1949. Natural selection and egg number in fish. Rep. Inst. Freshwater Res., Drottningholm, No. 29: 115–122.

Tonolli, V. 1949. Distribuzione in quota e tempo di entità fenotipiche biometricamente differenziabili, entro la popolazione di *Mixodiaptomus laciniatus* Lill. del Lago Maggiore. Mem. Ist. It. Idrobiol., **5**: 317–325.

Wesenberg-Lund, C. 1904. Plankton investigations of the Danish lakes, Special part, Cophenhagen, 223 pp., 8 maps, 10 pls., 9 tbs.

Wright, S. 1938. A review of the *Diaptomus bergi* group with descriptions of two new species. Trans. Amer. Micro. Soc., **57**: 297–315.

G. E. Hutchinson

Osborn Zoological Laboratory,
Yale University,
New Haven, Connecticut

PROCEEDINGS

OF THE

ACADEMY OF NATURAL SCIENCES

OF

PHILADELPHIA

1953

THE CONCEPT OF PATTERN IN ECOLOGY *

BY G. EVELYN HUTCHINSON

Director of Graduate Studies in Zoology, Yale University

In any general discussion of structure, relating to an isolated part of the universe, we are faced with an initial difficulty in having no a priori criteria as to the amount of structure it is reasonable to expect. We do not, therefore, always know, until we have had a great deal of empirical experience, whether a given example of structure is very extraordinary, or a mere trivial expression of something which we may learn to expect all the time.

If, with the surrealists, we imagine ourselves encountering in the middle of a desert a rock crystal carving of a sewing machine associated with a dead fish to which postage stamps are stuck, we may suspect that we have entered a region of the imagination in which ordinary concepts have become completely disordered. Macroscopically, we are in the realm of what Elizabeth Sewell (1951), in her remarkable book *The Structure of Poetry*, defines as nightmare. On a smaller scale, since we can recognize the individual objects and give them names, we are still in the familiar world. The fish may be expected to have the various skull bones which have been enumerated by vertebrate morphologists; if it departed too radically from the accepted structure, we should see at once that it was not a fish. The rock crystal would have the ordinary physical properties of quartz; if it did not, we should not recognise and name it as such.

---

* An address given upon presentation of the Leidy Medal to the author on December 4, 1952, at the Academy of Natural Sciences of Philadelphia. (See notice of the award in *Proceedings* of the Academy, vol. 104, p. 249, 1952.)

(1)

When we push our analysis as far as we can, we end up with a series of statements of relations between entities, which at the present state of development of science are apparently unanalysable. What we call knowledge appears to consist of a series of known relationships between unknown elements. The latter may become known as new techniques permit their study, but it is reasonable to suppose that they too will become in the process of investigation relationships between new unknown entities of a higher degree of abstraction. Actually, the degree of abstraction which has been reached in modern theoretical physics is already so great that it is practically impossible to say anything intelligible in words about what the universe is made of. Our preliminary exploration thus suggests that the completely disordered is unimaginable and that the known consists of a collection of relationships between temporarily unknown entities. If we are going to say anything at all, some structure is certain to be involved, but, as has already been indicated, the amount of structure per unit volume cannot be guessed in advance.

Very roughly, in an empirical and qualitative way, we may distinguish a number of kinds of structure. The ordinary small-scale structure of the inorganic world, as exemplified in crystals, we may call, as is usually done, *order*. *Disorder* in physical science usually means random as opposed to placed in a particular order, such as that of a crystal lattice.

There is another important kind of structure in purely physical systems, which is in a sense a sort of converse of order, and which may be called *arrangement*. By this is meant the kind of structure exhibited by having the sun in one place, radiating energy, the earth in another receiving some of it. *Arrangement* in this limited sense decreases as entropy increases. Measured as negative entropy it is essentially what organisms eat. It is obviously a very different concept from order, thought the two are often confused by biologists.

The order of a system increases as we lower its temperature and is maximal at absolute zero. Order is an equilibrium phenomenon. Arrangement in the energetic sense in which it has been used, decreases as the *whole* system exhibiting it approaches absolute zero. It is essentially a non-equilibrium phenomenon, and most of modern cosmology is devoted ultimately to trying to find out how it came or comes into existence.

The characteristic structure of the living world will be called *organization*. Much order is also present, and, as organisms lay up an energy supply, arrangement is there also. The really characteristic structure of organisms, however, only exists near transition points. The art of living consists fundamentally of just crystallizing or just going into solution at the right time and place. Living matter is poised precariously between the solid and liquid states.   Organization is never an equilibrium phenomenon in the physical sense.

The structure which results from the distributions of organisms in, or from, their interactions with, their environments, will be called *pattern*. As is organization, pattern is essentially a steady state rather than an equilibrium phenomenon, though it will be convenient to speak of equilibrium and non-equilibrium communities in a later paragraph when the phenomena are completely abstracted from physico-chemical categories. Pattern is obviously closely related to the arrangement of the inanimate world in which it developed.

The structure which organisms may impose on material systems to convey information, or in the construction of tools, may be called *design*. Human artifacts of all sorts, including works of art, come into this category. Design may be an equilibrium phenomenon and may be unchanged by cooling.   A sentence written with appropriate materials is still the same sentence in the neighborhood of absolute zero, though the organization of the man who wrote it and the pattern of the community in which he lived, could not survive such extensive cooling. There are, however, remarkable formal mathematical analogies between arrangement and design.

These categories are to be considered as qualitative and suggestive; they are set up mainly to indicate how complex the problem of structure becomes even when an effort is made to keep the matter as simple as possible.   The justification for the use of such categories is that confusion may often be avoided by asking which are appropriate to any structure under discussion.

Pattern, in the sense used above, appears to be of five kinds.   The distribution of organisms and of their effects on their environment may be determined by external forces, such as light, temperature, humidity or density gradients, changes of state in certain directions, currents, winds, etc.   Patterns produced in this way will be termed *vectorial*.   The distribution may be determined by genetic continuity, offspring remaining near the parent, giving a *reproductive* pattern.   The distribution may be determined by signalling of various kinds, leading either to spacing or aggregation, producing *social pattern*.   The distribution may be determined by interaction between species in competition leading to *coactive* pattern. The distribution may depend on random forces producing a *stochastic* pattern.

The main theme of the present address is the coactive and stochastic types of pattern and their interaction.

Stochastic patterns.—The distribution of a plankton organism by night is commonly conceived as becoming increasingly random.   Moreover, when horizontal distribution across a vertical light gradient is considered, many investigators have unconsciously assumed the same random distribution to hold. It is a mistake to consider such a random distribution to be structureless or lacking in pattern. The probabilities involved may be regarded as quite definite. Only if such probabilities were completely indeterminate, could we get an irrational series of surprises and enter the world of nightmare above a certain size level.

If a number of samples of a habitat is considered, and the probability of the occurrence of any particular species is low, the incidence of that species will appear at first very irregular. Examination of the number of samples containing no specimen, one specimen, two ... n specimens, etc., however, will indicate the existence of stochastic patterns. The simplest of such patterns is that in which the number of samples of each rank from 0 to n approximates to a Poisson series. It is the property of such a distribution that the variance in the statistical sense is equal to the mean. Where the variance is much greater than the mean (*superdispersion*) the organisms are grouped together more than would be expected on the simple random hypothesis; where the variance is much less than the mean (*infradispersion*) they are much more evenly spaced than in a randomly distributed population (fig. 1).

In the plankton of lakes, the distribution of different species of animals has been studied from this point of view by Ricker (1937), by Langford (1938) and perhaps most beautifully by Tonolli (1949). The last-named investigator made series of horizontal tows with the Clarke-Bumpas plankton sampler in Lago Maggiore in November, three series of tows being made at each depth. The results of some of his investigations are shown in figures 1 and 2. Out of a large number of series of comparisons, ninety-three in all, infradispersion appeared to be significantly demonstrated in only two. Since the significance limit was set as the degree of infradispersion which would occur by chance only in 1% of the cases studied, we are probably at liberty to regard infradispersion as a chance phenomenon. Superdispersion was observed far more often and is certainly significant. In three common species, *Eudiaptomus vulgaris*, *Daphnia longispina* and *Asplanchna priodonta* there is a very obvious tendency for the superdispersion to be most marked in the 5-10 m. layer. Presumably in these species, the tendency of the individuals to disperse at random in a horizontal plane, to move like the molecules of a gas, is modified by some hydrographic factor, probably turbulent movement, itself a random phenomenon. The organisms may be supposed to react to such random turbulent movements, so that they collect in certain regions and not in others. In *Cyclops strenuus* the superdispersion is differently distributed and must be due to different factors. Bliss (1953) in a very important recent paper

EUDIAPTOMUS VULGARIS ♂

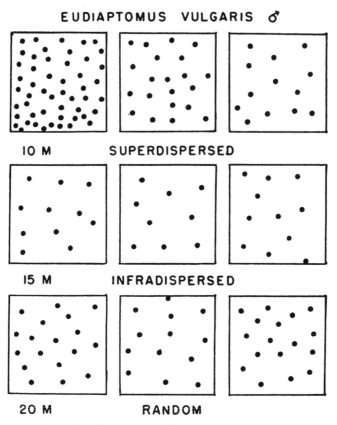

10 M   SUPERDISPERSED

15 M   INFRADISPERSED

20 M   RANDOM

Fig. 1.—Approximate distribution of males of *Eudiaptomus vulgaris* in three successive plankton samples at three different depths in Lago Maggiore, showing *superdispersion, infradispersion* and *random distribution.* (From the data of Tonolli 1949.)

has shown that in one marine copepod the superdispersed distribution approximates to the so-called negative binomial, which is to be expected when one stochastic process is superimposed on another. Tonolli's data are not appropriate for testing this particular distribution, but it is very likely that stochastic patterns dependent on the superimposed operation of random events involving different size dimensions, of the microorganisms themselves and of much larger convention cells for instance, may ultimately be found to produce a number of different kinds of stochastic pattern.

Coactive Pattern. The fundamental regularity underlying the distribution of all organisms in a community is Gause's principle, or, as it is more properly termed, the Volterra-Gause principle, that in an equilibrium

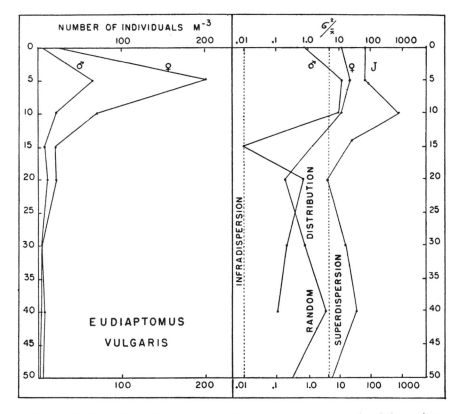

Fig. 2.—Vertical distribution of males and females, and of the ratio of the variance to the mean for males, females and immature individuals of *Eudiaptomus vulgaris* in Lago Maggiore. The absolute numbers of immature specimens are much greater than of mature individuals but are similarly distributed. The dotted lines indicate superdispersion and infradispersion significant to the 1% level. (From the data of Tonolli 1949.)

community no two species occupy the same ecological niches. A formal statement of the principle emerged early in Volterra's mathematical studies of biological associations (Volterra 1926), and Gause (1934, 1935) showed by an elegant series of experiments that in cultures of protozoa under conditions in which two species were forced to occupy a space of such simple structure that no niche diversification was possible, only one species could survive indefinitely.

In natural communities of a kind in which equilibrium may be expected, all subsequent studies have indicated that the generalization is true. Recently a number of studies largely by ornithologists have indicated that allied species which apparently live together under equilibrium conditions, may actually be occupying niches which are largely distinct. A particularly beautiful case is provided by three species of African weaver birds, *Ploceus intermedius cabanisi*, *P. collaris nigriceps* and *P. melanocephalus duboisii*, which all live together near Lake Mweru, the last two species even sharing the same communal nests. All feed on different foods; in the case of the two species which share a communal nest, one is a seed-eater and one is insectivorous (White 1951). A similar situation has recently been described among the owls, the Saw-whet, *Aegolius a. acadicus* and the Long-eared, *Asio otus wilsonianus*, in coniferous plantations in Ohio. Both occupy the same sleeping territory by day, but by night the Saw-whet hunts mainly in wooded areas catching large numbers of *Peromyscus*; the Long-eared owl mainly in open land catching many more *Microtus* and *Cryptotis* (Randle and Austing 1952).

Instances of this sort can be multiplied indefinitely and, where apparent exceptions to the Volterra-Gause principle of niche-specificity occur, we may legitimately suspect that a true equilibrium between the species is not established. We can in fact speak of *equilibrium* and *non-equilibrium* communities which may be distinguished by observing whether the principle holds.

Whenever two species are competing, the direction of competition is largely dependent on environmental factors. This is critically shown in Gause's (1935) experiments in which *Paramecium candatum* tended to replace *P. aurelia* in frequently renewed media, and *P. aurelia* to replace *P. candatum* when metabolic products were allowed to accumulate.

Precisely similar results have been long recognised in plant ecology. Many species which appear to be calciphil or calciphobe in nature can actually grow quite well when isolated in cultivation in soils of a wide range of calcium contents. The apparent restriction shows up only when the plants have to enter into competion with the rest of the flora to which they belong.

Ecological zonation is largely dependent on the competitive relations of species as controlled by the environment, and so is as much a coactive as a vectorial type of pattern. The production of a discontinuous discrete type of biological zonation in a continuous gradient of salinity, soil moisture or other physical variable is easily understood in terms of competition theory, since the direction of competition will change at a definite point on the gradient, below which one species, above which another species, will be successful (Gause and Witt 1935). It is probable that this process plays a considerable part in regulating the invasion of fresh waters from the sea. It is often apparent that species exhibit far greater salinity tolerances in the laboratory than in nature. In Joseph Leidy's day, the Schuylkill near Philadelphia contained a serpulid worm *Manayunkia speciosa* Leidy, one of the very few species of fresh-water polychaets. Allied species occur in the Great Lakes drainage, in Lake Baikal and in salt water in the Arctic. The nearest local marine ally of *M. speciosa* is *Fabricia sabella*, a marine species found on the Atlantic seaboard. J. P. Moore (in Johnson 1903) long ago showed that at least the adults of these worms could be acclimated to water of the normal salinity of each other's environments. Similar situations are found among the amphipoda of western Europe (Sexton 1939, Reid 1939, Beadle and Cragg 1940). It is probable that, when two species of slightly different tolerances compete in a salinity gradient, selection will cause the optima of values of the salinity of the two species to diverge. The operation of selection on a zonal pattern has doubtless played an immense part in evolution (cf. also Brooks 1950).

The very definite types of pattern which we have just considered are characteristic of equilibrium communities in biotops which contain physico-chemical gradients. Much of the diversity of the living world is due to this sort of pattern, but much is also probably due to the existence of non-equilibrium communities. The first type of non-equilibrium community characterizes those regions in which certain more or less catastrophic events are continually creating new empty biotops. If such biotops are colonized by more than one species, and if the species occupy the same niche, competition will begin and one species will tend to exterminate its weaker competitors. If before this happens, a new adjacent habitat is opened, a new mixed population may be set up. If the original habitat is now destroyed by a catastrophic event, and the process is repeated indefinitely, the mixed population will appear to persist. It is probable that, in order for this to happen, the tendency for the weaker species to disappear by competition must be balanced by a tendency for it to spread a little more easily than the stronger; it must in fact be a *fugitive species* (Hutchinson 1951). Wynne-Edwards (1952) has concluded that the co-existence of very closely allied species of birds in the Arctic, where local populations are easily exterminated by adverse climatic conditions, actually provides a case of this sort.[1]

A more widespread type of non-equilibrium population is dependent on the relation of the life-span or generation time to the seasonal cycle. If we consider two competing species with an annual or longer life cycle, such as is found in mammals, birds, many insects, the larger marine invertebrates, and some quite small aquatic animals such as many copepods, it is obvious that the species must be adapted throughout their life histories to a great variety of conditions. In these circumstances, we can properly

---

[1] I am indebted to James Bond for calling my attention to this case.

consider the two species to compete under some long term mean condition; transitory fluctuations may alter temporarily the direction of competition, but the final result will be the elimination of one of the competitors.

If we now consider two species with exceedingly short life histories, such as those of bacteria dividing rapidly in a favorable medium, it is quite possible that competition leading to extermination might occur so rapidly that no environmental change sufficient to reverse the direction of competition, would have time to occur before one species had been exterminated by the other.

The Volterra-Gause principle of one species per niche should, therefore, hold for very rapidly reproducing and very slowly reproducing organisms. In the intermediate region, in which a number of generations may occur in a year, but the generation time of several days or weeks is sufficient to permit considerable variation in environment in the course of a few generations, there is no reason to suppose that the law would hold. If one species displaced the other at low temperatures and the reverse at high, it is easy to see that, if both appeared from some resting stage in Spring, competition would first favor the first, then the second, and then again the first species.

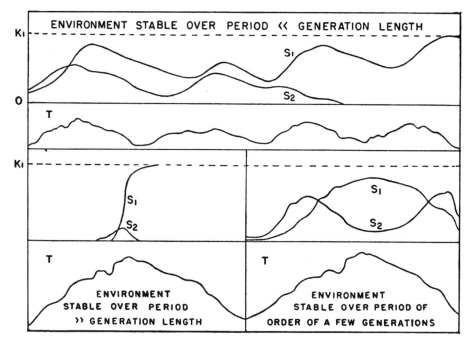

Fig. 3.—Ideal course of competition between two species as regulated by the relation between generation length and the period over which the environment may be taken as stable.

There can be little doubt that the great diversification of the phytoplankton of lakes, the turbulent epilimnia of which can hardly provide to autotrophic euplanktonic organisms greater niche diversity than existed in Gause's culture tubes, is due to the fact that phytoplanktonic organisms in nature probably divide at a mean rate of once every few days or every few weeks (see particularly Grim 1950) and so fall in the intermediate or potentially non-equilibrium category of the three which we have been considering.

It is also probable that some species of limnetic zooplankton form a non-equilibrium community, but here the situation is complicated in some cases by a tendency for the reproductive rate to vary, but with a lag-period, with the feeding rate. Very marked non-equilibrium populations of single species must often be set up during the Spring; the oscillations which follow and which are slowly damped out, may be extensive enough to prevent the species, in the time available for active feeding and reproduction, from ever achieving equilibrium, even with a constant food supply. This phenomenon, which has been most ingeniously studied by Slobodkin (1951) emphasizes the non-equilibrium nature of the plankton, though in its details it differs somewhat from the coactive non-equilibrium of the phytoplankton.

The Dividing of the Biotophy Coactive Processes

If we consider, not a single species but a whole series of species of a certain taxonomic group, and examine a large collection made at random in some specified habitat, we can enumerate the species which occur once, twice, three times and so on. Recently a good deal of attention has been paid to the regularities exhibited by such an enumeration. Designating the number of specimens per species as r, the rank of the species and the number of species which have that rank as n, Fisher, Corbet and Williams (1943) have concluded that for many kinds of organisms

$$n = \frac{R}{r} x^r$$

where x is a number slightly less than unity, and R a number which characterizes the diversity of the population under examination. The existence of the relationship, which is related to the negative binomial already mentioned, is attributed to a combination of random processes determining the incidence of a species and of individuals of that species in the collection.

Preston (1948), however, has shown that, when the rank coordinate is graduated logarithmically (he used $\log_2 r$) the resulting curves do not take the form implied by the expression given by Fisher, Corbet and Williams. What Preston finds is that there is a definite mode in the number of species for one of the logarithmic rank categories, the precise rank depending on the size of the collection. He believes that

$$n = n_0 e^{-(aR)^2}$$

when R is the logarithmic rank measured from the mode. This is, of course, the well-known log normal distribution. It is not entirely clear intuitively what such a distribution means biologically. In discussing the matter with Dr. E. S. Deevey, he suggested that in such cases what we are dealing with is not primarily a distribution of specimens, but rather a distribution of fractions of environments. Actually, nearly thirty years ago Dr. C. F. A. Pantin expressed the same idea (he has now probably forgotten it) when we were discussing the same general type of problem, which was beginning to interest biologists owing to the work of J. C. Willis. Every specimen, whether of diatom, moth, bird or elephant, will have required a certain amount of space for its development. The number of specimens, provided we stick to a single taxonomic group, gives in a certain sense, a measure of the space needed by the successful members of that species. It is probably reasonable that, in dividing up a space by coactive processes, a log normal type of distribution should result. What is really extraordinary is that the constant a, which is actually the reciprocal of $\sqrt{2}$, should have practically the same numerical value wherever it is encountered. In collections of moths from North America and Europe, Preston found values from 0.152 to 0.227; in a local bird census he obtained 0.194. Dr. Ruth Patrick, who is making very important studies along these lines, using the statistics of diatoms settling on slides submerged in streams, tells me that the constants found by her group of investigators are always close to 0.2. The value, therefore, appears to be independent of the size and reproductive rate of the organisms under investigation and probably applies to both equilibrium and non-equilibrium communities. It is likely that something very important is involved here, but for the present what it may be is a mystery, a very good thing with which to end a discourse.

REFERENCES

→ BEADLE, L. C. and J. B. CRAGG. 1940. The intertidal zone of two streams and the occurrence of Gammarus spp. on South Rona and Raasay (Inner Hebrides). *Jr. Animal Ecol. 9:* 289-295.

BLISS, C. I. 1953. Fitting the negative binomial distribution to biological data. To appear in *Biometrics.*

BROOKS, J. L. 1950. Speciation in ancient lakes. *Quart. Rev. Biol. 25:* 30-60, 131-176.

→ FISHER, R. A., A. S. CORBET, and C. B. WILLIAMS. 1943. The relation between the number of species and the number of individuals in a random sample of an animal population. *Jr. Animal Ecol. 12:* 42-58.

GAUSE, G. F. 1934. *The Struggle for Existence.* Baltimore. ix + 163 pp.

——. 1935. Verification expérimentales de la théorie mathématique de la lutte pour la vie. *Actualités scientifiques et industrielles,* no. 277. Paris. 63 pp.

GAUSE, G. F. and A. A. WITT. 1935. Behavior of mixed populations and the problem of natural selection. *Amer. Nat. 69:* 596-609.

GRIM, J. 1950. Versuche zur Ermittlung der Produktionskoeffizienten einige Planktophyten in einem flachen See. *Biol. Zentralbl. 69:* 147-174.

→ HUTCHINSON, G. E. 1951. Copepodology for the ornithologist. *Ecology 32:* 571-577.

JOHNSON, H. P. 1903. Fresh-water nereids from the Pacific coast and Hawaii, with remarks on fresh-water Polychaeta in general. *Mark Anniversary Volume.* New York. pp. 205-224.

LANGFORD, R. R. 1938. Diurnal and seasonal changes in the distribution of the limnetic crustacea of Lake Nipissing, Ontario. *Univ. Toronto Studies, Biol. Ser. 45:* 1-142.

PRESTON, F. W. 1948. The commonness, and rarity, of species. *Ecology 29:* 254-283.

RANDLE, W. and R. AUSTING. 1952. Ecological notes on the long-eared and saw-whet owls in southwestern Ohio. *Ecology 33:* 422-426.

REID, D. M. 1939. On the occurrence of *Gammarus duebeni* Lillj. (Crustacea, Amphipoda) in Ireland. *Proc. R. Irish Acad. 45 B:* 207-214.

RICKER, W. E. 1937. Statistical treatment of sampling processes useful in the enumeration of plankton. *Arch. f. Hydrobiol. 31:* 68-84.

SEXTON, E. W. 1939. On a new species of *Gammarus* (*G. tigrinus*) from Droitwich District. *Jr. Mar. Biol. Ass. 23:* 543-551.

SEWELL, ELIZABETH. 1951. *The Structure of Poetry.* London. Routledge and Kegan Paul Ltd. x + 196 pp.

SLOBODKIN, L. B. 1953. Population dynamics in *Daphnia obtusa* Kurz. (Yale thesis 1951.) To appear in 1953.

TONOLLI, V. 1949. Stuttura spaziale del popolamento mesoplanctico, eterogeneità delle densità dei popolamenti orizzontale e sua variazion in funzione della quota. *Mem. Ist. ital. Idrobiol.* " Dott. Marco de March " *5:* 189-208.

VOLTERRA, V. 1926. Variazioni e fluttuazioni del numero d' individui in specie animali conviventi. *Mem. Accad. Lincei* (6) 2: 31-113.

WHITE, C. M. N. 1951. Weaver birds at Lake Mweru. *Ibis 93:* 626-627.

WYNNE-EDWARDS, V. C. 1952. Zoology of the Baird Expedition (1950). I: The birds observed in central and south-east Baffin Island. *Auk 69:* 352-391.

Concluding Remarks

G. Evelyn Hutchinson

Yale University, New Haven, Connecticut

This concluding survey[1] of the problems considered in the Symposium naturally falls into three sections. In the first brief section certain of the areas in which there is considerable difference in outlook are discussed with a view to ascertaining the nature of the differences in the points of view of workers in different parts of the field; no aspect of the Symposium has been more important than the reduction of areas of dispute. In the second section a rather detailed analysis of one particular problem is given, partly because the question, namely, the nature of the ecological niche and the validity of the principle of niche specificity has raised and continues to raise difficulties, and partly because discussion of this problem gives an opportunity to refer to new work of potential importance not otherwise considered in the Symposium. The third section deals with possible directions for future research.

The Demographic Symposium as a Heterogeneous Unstable Population

In the majority of cases the time taken to establish the general form of the curve of growth of a population from initial small numbers to a period of stability or of decline is equivalent to a number of generations. If, as in the case of man, the demographer is himself a member of one such generation, his attitude regarding the nature of the growth is certain to be different from that of an investigator studying, for instance, bacteria, where the whole process may unfold in a few days, or insects, where a few months are required for several cycles of growth and decline. This difference is apparent when Hajnal's remarks about the uselessness of the logistic are compared with the almost universal practice of animal demographers to start thinking by making some suitable, if almost unconscious, modification of this much abused function.

[1] I wish to thank all the participants for their kindness in sending in advance manuscripts or information relative to their contributions. All this material has been of great value in preparing the following remarks, though not all authors are mentioned individually. Where a contributor's name is given without a date, the reference is to the contribution printed earlier in this volume. I am also very much indebted to the members (Dr. Jane Brower, Dr. Lincoln Brower, Dr. J. C. Foothills, Mr. Joseph Frankel, Dr. Alan Kohn, Dr. Peter Klopfer, Dr. Robert MacArthur, Dr. Gordon A. Riley, Mr. Peter Wangersky, and Miss Sally Wheatland) of the Seminar in Advanced Ecology, held in this department during the past year. Anything that is new in the present paper emerged from this seminar and is not to be regarded specifically as an original contribution of the writer.

The human demographer by virtue of his position as a slow breeding participant observer, and also because he is usually called on to predict for practical purposes what will happen in the immediate future, is inevitably interested in what may be called the microdemography of man. The significant quantities are mainly second and third derivatives, rates of change of natality and mortality and the rates of change of such rates. These latter to the animal demographer might appear as random fluctuations which he can hardly hope to analyse in his experiments. What the animal demographer is mainly concerned with is the macrodemographic problem of the integral curve and its first derivative. He is accustomed to dealing with innumerable cases where the latter is negative, a situation that is so rare in human populations that it seems to be definitely pathological to the human demographer. Only when anthropology and archaeology enter the field of human demography does something comparable to animal demography, with its broad, if sometimes insufficiently supported generalisations and its fascinating problems of purely intellectual interest, emerge. From this point of view the papers of Birdsell and Braidwood are likely to appeal most strongly to the zoologist, who may want to compare the rate of spread of man with that considered by Kurtén (1957) for the hyena.

It is quite likely that the difference that has just been pointed out is by no means trivial. The environmental variables that affect fast growing and slow growing populations are likely to be much the same, but their effect is qualitatively different. Famine and pestilence may reduce human populations greatly but they rarely decimate them in the strict sense of the word. Variations, due to climatic factors, of insect populations are no doubt often proportionately vastly greater. A long life and a long generation period confer a certain homeostatic property on the organisms that possess them, though they prove disadvantageous when a new and powerful predator appears. The elephant and the rhinoceros no longer provide models of human populations, but in the early Pleistocene both may have done so. The rapid evolution of all three groups in the face of a long generation time is at least suggestive.

It is evident that a difference in interest may underlie some of the arguments which have enlivened, or at times disgraced, discussions of this subject. Some of the most significant modern work has arisen from an interest in extending the concepts of the struggle for existence put forward as an evolutionary mechanism by Darwin practically a century ago. Such work, of which Lack's recent contributions provide a distinguished example, tends to concentrate on relatively stable interacting populations in as undisturbed communion as possible. Another fertile field of research has been provided by the sudden increases in numbers of destructive animals, often after introduction or disturbance of natural environments. Here more than one point of view has been apparent. Where emphasis has been on biological control, that is, a conscious rebuilding of a complex biological association, a view point not unlike that of the evolutionist has emerged—where emphasis has been placed on the actual events leading to a very striking increase or decrease in abundance, given the immediate ecological conditions, the latter have appeared to be the most significant variables. Laboratory workers have moreover tended to keep all but a few factors constant, and to vary these few systematically. Field workers have tended to emphasize the ever changing nature of the environment. It is abundantly clear that all these points of view are necessary to obtain a complete picture. It is also very likely that the differences in initial point of view are often responsible for the differences in the interpretation of the data.

The initial differences of point of view are not the only difficulty. In the following section an analysis of a rather formal kind of one of the concepts frequently used in animal ecology, namely that of the *niche*, is attempted. This analysis will appear to some as compounded of equal parts of the obvious and the obscure. Some people however may find when they have worked through it, provided that it is correct, that some removal of irrelevant difficulties has been achieved. It is not necessary in any empirical science to keep an elaborate logicomathematical system always apparent, any more than it is necessary to keep a vacuum cleaner conspicuously in the middle of a room at all times. When a lot of irrelevant litter has accumulated the machine must be brought out, used, and then put away. It might be useful for those who argue that the word environment should refer to the environment of a population, and those who consider it should been the environment of an organism, to use the word both ways for a couple of months, writing "environment" when a single individual is involved, "Environment" when reference is to a population. In what follows the term will as far as possible not be used, except in the non-committal adjectival form environmental, meaning any property outside the organisms under consideration.

THE FORMALISATION OF THE NICHE AND THE VOLTERRA-GAUSE PRINCIPLE

Niche space and biotop space

Consider two independent environmental variables x_1 and x_2 which can be measured along ordinary rectangular coordinates. Let the limiting values permitting a species S_1 to survive and reproduce be respectively x'_1, x''_1 for x_1 and x'_2, x''_2 for x_2. An area is thus defined, each point of which corresponds to a possible environmental state permitting the species to exist indefinitely. If the variables are independent in their action on the species we may regard this area as the rectangle $(x_1 = x'_1, x_1 = x''_1, x_2 = x'_2, x_2 = x''_2)$, but failing such independence the area will exist whatever the shape of its sides.

We may now introduce another variable x_3 and obtain a volume, and then further variables $x_4 \ldots x_n$ until all of the ecological factors relative to S_1 have been considered. In this way an n-dimensional hypervolume is defined, every point in which corresponds to a state of the environment which would permit the species S_1 to exist indefinitely. For any species S_1, this hypervolume $\mathbf{N}_1$ will be called the *fundamental niche*[2] of S_1. Similarly for a second species S_2 the fundamental niche will be a similarly defined hypervolume $\mathbf{N}_2$.

It will be apparent that if this procedure could be carried out, all X_n variables, both physical and biological, being considered, the fundamental niche of any species will completely define its ecological properties. The fundamental niche defined in this way is merely an abstract formalisation of what is usually meant by an ecological niche.

As so defined the fundamental niche may be regarded as a set of points in an abstract n-dimensional $\mathbf{N}$ space. If the ordinary physical space $\mathbf{B}$ of a given biotop be considered, it will be apparent that any point $p(\mathbf{N})$ in $\mathbf{N}$ can correspond to a number of points p_i ($\mathbf{B}$) in $\mathbf{B}$, at each one of which the conditions specified by $p(\mathbf{N})$ are realised in $\mathbf{B}$. Since the values of the environmental variables $x_1 x_2 \ldots x_n$ are likely to vary continuously, any subset of points in a small elementary volume $\Delta\mathbf{N}$ is likely to correspond to a number of small elementary volumes scattered about in $\mathbf{B}$. Any volume $\mathbf{B}'$ of the order of the dimensions of the mean free paths of any animals under consideration is likely to contain points corresponding to points in various fundamental niches in $\mathbf{N}$.

Since $\mathbf{B}$ is a limited volume of physical space comprising the biotope of a definite collection of species S_1, $S_2 \cdots S_n$, there is no reason why a given point in $\mathbf{N}$ should correspond to any points in $\mathbf{B}$. If, for any species S_1, there are no points in

[2] This term is due to MacArthur. The general concept here developed was first put forward very briefly in a footnote (Hutchinson, 1944).

B corresponding to any of the points in $\mathbf{N}_1$, then **B** will be said to be *incomplete* relative to S_1. If some of the points in $\mathbf{N}_1$ are represented in **B** then the latter is *partially incomplete* relative to S_1, if all the points in $\mathbf{N}_1$ are represented in **B** the latter is *complete* relative to S_1.

Limitations of the set-theoretic mode of expression. The following restrictions are imposed by this mode of description of the niche.

1. It is supposed that all points in each fundamental niche imply equal probability of persistance of the species, all points outside each niche, zero probability of survival of the relevant species. Ordinarily there will however be an optimal part of the niche with markedly suboptimal conditions near the boundaries.

2. It is assumed that all environmental variables can be linearly ordered. In the present state of knowledge this is obviously not possible. The difficulty presented by linear ordering is analogous to the difficulty presented by the ordering of degrees of belief in non-frequency theories of probability.

3. The model refers to a single instant of time. A nocturnal and a diurnal species will appear in quite separate niches, even if they feed on the same food, have the same temperature ranges etc. Similarly, motile species moving from one part of the biotop to another in performance of different functions may appear to compete, for example, for food, while their overall fundamental niches are separated by strikingly different reproductive requirements. In such cases the niche of a species may perhaps consist of two or more discrete hypervolumes in **N**. MacArthur proposed to consider a more restricted niche describing only variables in relation to which competition actually occurs. This however does not abolish the difficulty. A formal method of avoiding the difficulty might be derived, involving projection onto a hyperspace of less than n-dimensions. For the purposes for which the model is devised, namely a clarification of niche-specificity, this objection is less serious than might at first be supposed.

4. Only a few species are to be considered at once, so that abstraction of these makes little difference to the whole community. Interaction of any of the considered species is regarded as competitive in sense 2 of Birch (1957), negative competition being permissible, though not considered here. All species other than those under consideration are regarded as part of the coordinate system.

Terminology of subsets. If $\mathbf{N}_1$ and $\mathbf{N}_2$ be two fundamental niches they may either have no points in common in which case they are said to be *separate*, or they have points in common and are said to *intersect*.

In the latter case:

$(\mathbf{N}_1 - \mathbf{N}_2)$ is the subset of $\mathbf{N}_1$ of points not in $\mathbf{N}_2$
$(\mathbf{N}_2 - \mathbf{N}_1)$ is the subset of $\mathbf{N}_2$ of points not in $\mathbf{N}_1$

$\mathbf{N}_1 \cdot \mathbf{N}_2$ is the subset of points common to $\mathbf{N}_1$ and $\mathbf{N}_2$, and is also referred to as the *intersection subset*.

Definition of niche specificity. Volterra (1926, see also Lotka 1932) demonstrated by elementary analytic methods that under constant conditions two species utilizing, and limited by, a common resource cannot coexist in a limited system.[3] Winsor (1934) by a simple but elegant formulation showed that such a conclusion is independent of any kind of finite variations in the limiting resource. Gause (1934, 1935) confirmed this general conclusion experimentally in the sense that if the two species are forced to compete in an undiversified environment one inevitably becomes extinct. If there is a diversification in the system so that some parts favor one species, other parts the other, the two species can coexist. These findings have been extended and generalised to the conclusion that two species, when they co-occur, must in some sense be occupying different niches. The present writer believes that properly stated as an empirical generalisation, which is true except in cases where there are good reasons not to expect it to be true,[4] the principle is of fundamental importance and may be properly called the Volterra-Gause Principle. Some of the confusion surrounding the principle has arisen from the concept of two species not being able to co-occur when they occupy identical niches. According to the formulation given above, identity of fundamental niche would imply $\mathbf{N}_1 = \mathbf{N}_2$, that is, every point of $\mathbf{N}_1$ is a member of $\mathbf{N}_2$ and every point of $\mathbf{N}_2$ a member of $\mathbf{N}_1$. If the two species S_1 and S_2 are indeed valid species distinguishable by a systematist and not freely interbreeding, this is so unlikely that the case is of no empirical interest. In terms of the set-theoretic presentation, what the Volterra-Gause principle meaningfully states is that for any small element of the intersection subset $\mathbf{N}_1 \cdot \mathbf{N}_2$, there do not exist in **B** corresponding small parts, some inhabited by S_1, others by S_2.

Omitting the quasi-tautotogical case of $\mathbf{N}_1 = \mathbf{N}_2$, the following cases can be distinguished.

(1) $\mathbf{N}_2$ is a proper subset of $\mathbf{N}_1$ ($\mathbf{N}_2$ is "inside" $\mathbf{N}_1$)

    (a) competition proceeds in favor of S_1 in all the elements of **B** corresponding to $\mathbf{N}_1 \cdot \mathbf{N}_2$; given adequate time only S_1 survives.

    (b) competition proceeds in favor of S_2 in all elements of **B** corresponding to some part of the intersection subset and both species survive.

(2) $\mathbf{N}_1 \cdot \mathbf{N}_2$ is a proper subset of both $\mathbf{N}_1$ and $\mathbf{N}_2$; S_1 survives in the parts of **B** space

---

[3] I regret that I am unable to appreciate Brian's contention (1956) that the Volterra model refers only to interference, and the Winsor model to exploitation.

[4] *cf.* Schrödinger's famous restatement of Newton's First Law of Motion, that a body perseveres at rest or in uniform motion in a right line, except when it doesn't.

corresponding to ($\mathbf{N}_1 = \mathbf{N}_2$), S_2 in the parts corresponding to ($\mathbf{N}_2 = \mathbf{N}_1$), the events in $\mathbf{N}_1 = \mathbf{N}_2$ being as under I, with the proviso that no point in $\mathbf{N}_1 \cdot \mathbf{N}_2$ can correspond to the survival of both species.

In this case the two difference subsets ($\mathbf{N}_1 - \mathbf{N}_2$) and ($\mathbf{N}_2 - \mathbf{N}_1$) are, in Gause's terminology, refuges for S_1 and S_2 respectively.

If we define the realised niche $\mathbf{N}'_1$ of S_1 in the presence of S_2 as ($\mathbf{N}_1 - \mathbf{N}_2$), if it exists, plus that part of $\mathbf{N}_1 \cdot \mathbf{N}_2$ as implies survival of S_1, and similarly the realised niche $\mathbf{N}'_2$ of S_2 as ($\mathbf{N}_2 - \mathbf{N}_1$), if it exists, plus that part of $\mathbf{N}_1 \cdot \mathbf{N}_2$ corresponding to survival of S_2, then the Volterra-Gause principle is a statement of an empirical generalisation, which may be verified or falsified, that realised niches do not intersect. If the generalisation proved to be universally false, the falsification would presumably imply that in nature resources are never limiting.

Validity of the Gause-Volterra Principle. The set-theoretic approach outlined above permits certain refinements which, however obvious they may seem, apparently require to be stated formally in an unambiguous way to prevent further confusion. This approach however tells us nothing about the validity of the principle, but merely where we should look for its verification or falsification.

Two major ways of approaching the problem have been used, one experimental, the other observational. In the experimental approach, the method (*e.g.* Gause, 1934, 1935; Crombie, 1945, 1946, 1947) has been essentially to use animal populations as elements in analogue computers to solve competition equations. As analogue computers, competing populations leave much to be desired when compared with the more conventional electronic machines used for instance by Wangersky and Cunningham. At best the results of laboratory population experiments are qualitatively in line with theory when all the environmental variables are well controlled. In general such experiments indicate that where animals are forced by the partial incompleteness of the **B** space to live in competition under conditions corresponding to a small part of the intersection subset, only one species survives. They also demonstrate that the identity of the survivor is dependent on the environmental conditions, or in other words on which part of the intersection subset is considered, and that when deliberate niche diversification is brought about so that at least one non-intersection subset is represented in **B**, two species may co-occur indefinitely. It would of course be most disturbing if confirmatory models could not be made from actual populations when considerable trouble is taken to conform to the postulates of the deductive theory.

The second way in which confirmation has been sought, namely by field studies of communities consisting of a number of allied species also lead to a confirmation of the theory, but one which may need some degree of qualification. Most work has dealt with pairs of species, but the detailed studies on *Drosophila* of Cooper and Dobzhansky (1956) and of Da Cunha, El-Tabey Shekata and de Olivera (1957), to name only two groups of investigators, the investigation of about 18 species of *Conus* on Hawaiian reef and littoral benches (Kohn, in press) and the detailed studies of the food of six co-occurring species of *Parus* (Betts, 1955) indicate remarkable cases among many co-occurring species of insects, mollusks and birds respectively. However much data is accumulated there will almost always be unresolved questions relating to particular species, though the presumption from this sort of work is that, in any large group of sympatric species belonging to a single genus or subfamily, careful work will always reveal ecological differences. The sceptic may reply in two ways, firstly pointing out that the quasi-tautological case of $\mathbf{N}_1 = \mathbf{N}_2$ has already been dismissed as too improbable to be of interest, and that when a great deal of work has to be done to establish the difference, we are getting as near to niche identity as is likely in a probabilistic world. Occasionally it may be possible to use indirect arguments to show that the differences are at least evolutionarily significant. Lack (1947b) for instance points out that in the Galapagos Islands, among the heavy billed species of *Geospiza*, where both *G. fortis* Gould, and *G. fuliginosa* Gould co-occur on an island, there is a significant separation in bill size, but where either species exists alone, as on Crossman Island and Daphne Island the bills are intermediate and presumably adapted to eating modal sized food. This is hard to explain unless the small average difference in food size believed to exist between sympatric *G. fortis* and *G. fuliginosa* is actually of profound ecological significance. The case is particularly interesting as most earlier authors have dismissed the significance of the small alleged differences in the size of food taken by the species. Few cases of specific ecological difference encountered outside *Geospiza* would appear at first sight so tenuous as this.

A more important objection to the Volterra-Gause principle may be derived from the extreme difficulty of identifying competition as a process actually occurring in nature. Large numbers of cases can of course be given in which there is very strong indirect evidence of competitive relationships between species actually determining their distribution. A few examples may be mentioned. In the British Isles (Hynes, 1954, 1955) the two most widespread species of *Gammarus* in freshwater are *Gammarus deubeni* Lillj: and *G. pulex* (L.). The latter is the common species in England and most of the mainland of Scotland, the former is found exclusively in Ireland, the Shetlands, Orkneys and most of the other Scottish Islands and in Cornwall. On northern mainland Scotland only

G. lacustris Sars is found. Both *deubeni* and *pulex* occur on the Isle of Man and in western Cornwall. Only in the Isle of Man have the two species been taken together. It is extremely probable that *pulex* is a recent introduction to that island. *G. deubeni* is well known in brackish water around the whole of northern Europe. It is reasonable to suppose that the fundamental niches of the two species overlap, but that within the overlap *pulex* is successful, while *deubeni* with a greater tolerance of salinity has a refuge in brackish water. Hynes moreover shows that *G. pulex* has a biotic (reproductive) potential two or three times that of *deubeni* so that in a limited system inhabitable by both species, under constant conditions *deubeni* is bound to be replaced by *pulex*. This case is as clear as one could want except that Hynes is unable to explain the absence of *G. deubeni* from various uninhabited favorable localities in the Isle of Man and elsewhere. Hynes also notes that Steusloff (1943) had similar experiences with the absence of *Gammarus pulex* in various apparently favorable German localities. Ueno (1934) moreover pointed out that *Gammarus pulex* (*sens. lat.*) occurs abundantly in Kashmir up to 1600 meters, and is an important element in the aquatic fauna of the Tibetan highlands to the east above 3800 miles, but is quite absent in the most favorable localities at intermediate altitudes. These disconcerting empty spaces in the distribution of *Gammarus* may raise doubts as to the completeness of the picture presented in Hynes' excellent investigations.

Another very well analysed case (Dumas, 1956) has been recently given for two sympatric species of *Plethodon*, *P. dunni* Bishop, and *P. vehiculum* (Cooper), in the Coastal Ranges of Oregon. Here experiments and field observations both indicate that *P. dunni* is slightly less tolerant of low humidity and high temperature than is *P. vehiculum*, but when both co-occur *dunni* can exclude *vehiculum* from the best sites. However under ordinary conditions in nature the number of unoccupied sites which appear entirely suitable is considerable, so that competition can not be limiting except in abnormally dry years.

In both these cases, which are two of the best analysed in the literature, the extreme proponent of the Volterra-Gause principle could argue that if the investigator was equipped with the sensory apparatus of *Gammarus* or *Plethodon* he would know that the supposedly suitable unoccupied sites were really quite unsuitable for any self respecting member of the genus in question. This however is pure supposition.

Even in the rather conspicuous case of the introduction of *Sciurus caroliniensis* Gmelin and its spread in Britain, the popular view that the bad bold invader has displaced the charming native *S. vulgaris leucourus* Kerr, is apparently mythological. Both species are persecuted by man; *S. caroliniensis* seems to stand this persecution bet-

ter than does the native red squirrel and therefore tends to spread into unoccupied area from which *S. vulgaris leucourus* has earlier retreated (Shorten, 1953, 1954).

Andrewartha (see also Andrewartha and Birch, 1954) has stressed the apparent fact that while most proponents of the competitive organisation of communities have emphasised competition for food, there is in fact normally more than enough food present. This appears, incidentally, most strikingly in some of Kohn's unpublished data on the genus *Conus*.

The only conclusion that one can draw at present from the observations is that although animal communities appear qualitatively to. be constructed as if competition were regulating their structure, even in the best studied cases there are nearly always difficulties and unexplored possibilities. These difficulties suggest that if competition is determinative it either acts intermittently, as in abnormally dry seasons for *Plethodon*, or it is a more subtle process than has been supposed. Thus Lincoln Brower (*in press*) investigating a group of species of North American *Papilio* in which one eastern polyphagous species is replaced by three western oligophagous species, has been impressed by the lack of field evidence for any inadequacy in food resources. He points out however, that specific separation of food might lower the probability of local high density on a given plant, and so the risk of predation by a bird that only stopped to feed when food was abundant (*cf.* de Ruiter, 1952).

Unfortunately there is no end to the possible erection of hypothesis fitted to particular cases that will bring them within the rubric of increasingly subtle forms of competition. Some other method of investigation would clearly be desirable. Before drawing attention to one such possible method, the expected limitations of the Volterra-Gause principle must be examined.

Cases where the Volterra-Gause principle is unlikely to apply. (a) Skellam (1951; see also Brian, 1956b) has considered a model in which two species occur one of (S_1) much lower reproductive potential than the other (S_2). It is assumed that if S_1 and S_2 both arrive in an element of the biotops S_1 always displaces S_2, but that excess elements are always available at the time of breeding and dispersal so that some are never occupied S_1. In view of the higher reproductive potential, S_2 will reach some of these and survive. The model is primarily applicable to annual plants with a definite breeding season, random dispersal of seeds and complete seasonal mortality so all sites are cleared before the new generation starts growing, S_2 is in fact a limiting case of what Hutchinson (1951, 1953) called a fugitive species which could only be established in randomly vacated elements of a biotop. Skellam's model requires clearing of sites by high death rate, Hutchinson's qualitative statement a formation of transient sites by random small catastrophes in the biotop. Otherwise the two concepts developed independently are identical.

(b) When competition for resources becomes a contest rather than a scramble in Nicholson's admirable terminology, there is a theoretical possibility that the principle might not apply. If the breeding population be limited by the number of territories that can be set up in an area, and if a number of unmated individuals without breeding territory are present, food being in excess of the overall requirements, it is possible that territories could be set up by any species entirely independent of the other species, the territorial contests being completely intraspecific. Here a resource, namely area, is limiting but since it does not matter to one species if another is using the area, no interspecific competition need result. No case appears yet to be known, though less extreme modifications of the idea just put forward have apparently been held by several naturalists. Dr. Robert MacArthur has been studying a number of sympatric species of American warblers of the genus *Dendroica* which might be expected to be as likely as any organism to show the phenomenon. He finds however very striking niche specificity among species inhabiting the same trees.

(c) The various cases where circumstances change in the biotop reversing the direction of competition before the latter has run its course. Ideally we may consider two extreme cases with regard to the effect of changing weather and season on competition. In natural populations living for a time under conditions simulating those obtaining in laboratory cultures in a thermostat, if the competition time, that is, the time needed to permit replacement of one species by another, is very short compared with the periods of the significant environmental variables, then complete replacement will occur. This can only happen in very rapidly breeding organisms. Proctor (1957) has found that various green algae always replace *Haematococcus* is small bodies of water which never dry up, though if desiccation and refilling occur frequently enough the *Haematococcus* which is more drought resistant than its competitors will persist indefinitely. If on the contrary the competition time is long compared with the environmental periods, then the relevant environmental determinants of competition will tend to be mean climatic parameters, showing but secular trends in most cases, and competition will inevitably proceed to its end unless some quite exceptional event intervenes.[5]

[5] If there were really three species of giant tortoise (Rothschild, 1915) on Rodriguez, and even more on Mauritius, and if these were sympatric and due to multiple invasion (unlike the races on Albemarle in the Galapagos Islands) it is just conceivable that the population growth was so slow that mixed populations persisted for centuries and that the completion of competition had not occurred before man exterminated all the species involved.

Between the two extreme cases it is reasonable to suppose that there will exist numerous cases in which the direction of competition is never constant enough to allow elimination of one competitor. This seems likely to be the case in the autotrophic plankton of lakes, which inhabits a region in which the supply of nutrients is almost always markedly suboptimal, is subject to continual small changes in temperature and light intensity and in which a large number of species may (Hutchinson, 1941, 1944) coexist.

There is interesting evidence derived from the important work of Brian (1956a) on ants that the completion of competitive exclusion is less likely to occur in seral than in climax stages, which may provide comparable evidence of the effect of environmental changes in competition. Moreover whenever we find the type of situation described so persuasively by Andrewartha and Birch (1954) in which the major limitation on numbers is the length of time that meteorological and other conditions are operating favorably on a species, it is reasonable to suppose that interspecific competition is no more important than intraspecific competition. Much of the apparent extreme difference between the outlook of, for instance, these investigators, or for that matter Milne on the one hand, and a writer such as Lack (1954) on the other, is clearly due to the relationship of generation time to seasonal cycle which differs in the insects and in the birds. The future of animal ecology rests in a realisation not only that different animals have different autecologies, but also that different major groups tend to have fundamental similarities and differences particularly in their broad temporal relationships. The existence of the resemblances moreover may be quite unsuspected and must be determined empirically. In another place (Hutchinson, 1951) I have assembled such evidence as exists on the freshwater copepoda, which seem to be reminiscent of birds rather than of phytoplankton or of terrestrial insects in their competitive relationships.

It is also important to realize, as Cole has indicated in the introductory contribution to this Symposium, that the mere fact that the same species are usually common or rare over long periods of time and that where changes have been observed in well studied faunas such as the British birds or butterflies they can usually be attributed to definite environmental causes in itself indicates that the random action of weather on generation is almost never the whole story. Skellam's demonstration that such action must lead to final extinction must be born in mind. It is quite possible that the change in the phytoplankton of some of the least culturally influenced of the English Lakes, such as the disappearance of *Rhizosolenia* from Wastwater (Pearsall, 1932), may provide a case of random extinction under continually reversing competition. The general evidence of considerable stability under most conditions would suggest that competitive action of some sort is nearly always of significance.

Rarity and commonness of species and the non-intersection of realised niches. Several ways of approaching the problem of the rarity and commonness of species have been suggested (Fisher, Corbet and Williams, 1943; Preston, 1948; Brian, 1953; Shinozaki and Urata, 1953). In all these approaches relatively simple statistical distributions have been fitted to the data, without any attempt being made to elucidate the biological meaning of such distribution. Recently however MacArthur (1957) has advanced the subject by deducing the consequences of certain alternative hypotheses which can be developed in terms of a formal theory of niches.

It has been pointed out in a previous paragraph that the Volterra-Gause principle is equivalent to a statement that the realised niches of co-occurring species are non-interesting. Consider a **B** space containing an equilibrium community of n species $S_1 S_2 \cdots S_n$, represented by numbers of individuals $N_1 N_2 \cdots N_2$. For any species S_K it will be possible to identify in **B** a number of elements, each of which corresponds to a whole or part of $\mathbf{N}'_K$ and to no other part of **N**. Suppose that at any given moment each of these elements is occupied by a single individual of S_K, the total volume of B which may be regarded as the specific biotop of S_K will be $N_K \Delta \mathbf{B} (S_K)$, $\Delta \mathbf{B}(S_K)$ being the mean volume of **B** occupied by one individual of S_K. Since the biotop is in equilibrium with respect to the n species present, all possible spaces will be filled so that

$$\mathbf{B} = \sum_{K=1}^{n} N_K \Delta \mathbf{B} (S_K)$$

We do not know anything *a priori* about the distribution of $N_1 \Delta \mathbf{B}(S_1)$, $N_2 \Delta \mathbf{B}(S_2) \cdots N_n \Delta \mathbf{B}(S_n)$,

except that these different specific biotops are taken as volumes proportional to $N_1 N_2 \cdots N_n$, which is a justifiable first approximation if the species are of comparable size and physiology. In general some of the species will be rare and some common. The simplest hypothesis consistent with this, is that a random division of **B** between the species has taken place.

Consider a line of finite length. This may be broken at random into n parts by throwing $(n - 1)$ random points upon it. It would also be possible to divide the line successively by throwing n random pairs of points upon it. In the first case the division is into non-overlapping sections, in the second the sections overlap. MacArthur, whose paper may be consulted for references to the mathematical procedures involved, has given the expected distributions for the division of a line by these alternative methods (Fig. 1). He has moreover shown that with certain restrictions the distribution (I) which corresponds to non-intersecting specific biotops and so to non-intersecting realised niches, fits certain multispecific biological associations extremely well. The form of this distribution is independent of the number of dimensions in **B**. The alternative distribution with overlapping specific biotops predicts fewer species of intermediate rarity and more of great rarity than is actually found; proceding from the linear case (II), to division of an area or a volume, accentuates this discrepancy. Two very striking cases in which distribution I fits biological multispecific populations are given in Figures 2 and 3 from MacArthur and in Figure 4 from the recent studies of Dr. Alan Kohn (in press).

The limitation which is imposed by the theory is that in all large subdivisions of **B** the ratio of total number of individuals ($m = \sum_{i=1}^{n} N_i$) to

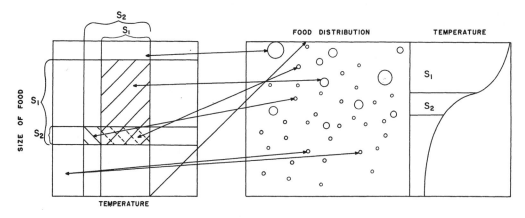

FIGURE 1. Two fundamental niches defined by a pair of variables in a two-dimensional niche space. Only one species is supposed to be able to persist in the intersection subset region. The lines joining equivalent points in the niche space and biotop space indicate the relationship of the two spaces. The distribution of the two species involved is shown on the right hand panel with a temperature depth curve of the kind usual in a lake in summer.

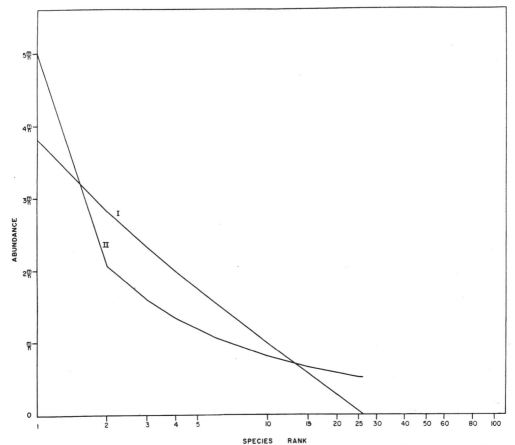

FIGURE 2. Rank order of species arranged per number of individuals according to the distributions I and II considered by MacArthur.

total number of species (n) must remain constant. This is likely to be the case in any biotop which is what may be termed *homogeneously diverse*, that is, in which the elements of the environmental mosaic (trees, stones, bushes, dead logs, etc.) are small compared with the mean free paths of the organisms under consideration. When a heterogeneously diverse area, comprising for instance stands of woodland separated by areas of pasture, is considered it is very unlikely that the ratio of total numbers of individuals to number of species will be identical in both woodland and pasture (if it occasionally were, the fact that both censuses could be added would not be of any biological interest). MacArthur finds that at least some bodies of published data which do not fit distribution I as a whole, can be broken down according to the type of environment into subcensuses which do fit the distribution. Data from moth traps and from populations of diatoms on slides submerged in rivers would not be expected to fit the distribu-

tion and in fact do not do so.[6] Such collection methods certainly sample very heterogeneously diverse areas.

The great merit of MacArthur's study is that it attempts to deduce operationally distinct differences between the results of two rival hypotheses, one of which corresponds essentially to the extreme density dependent view of interspecific interaction, the other to the opposite view. Although certain simplifying assumptions must be made in the theoretical treatment, the initial results suggest that in stable homogeneously diverse biotops the abundances of different species are arranged as if the realised niches were non-overlapping; this does not mean that populations may not exist under other conditions which would depart very widely from MacArthur's findings.

The problem of the saturation of the biotop. An important but quite inadequately studied aspect

[6] I am indebted to Dr. Ruth Patrick for the opportunity to test some of her diatometer censuses.

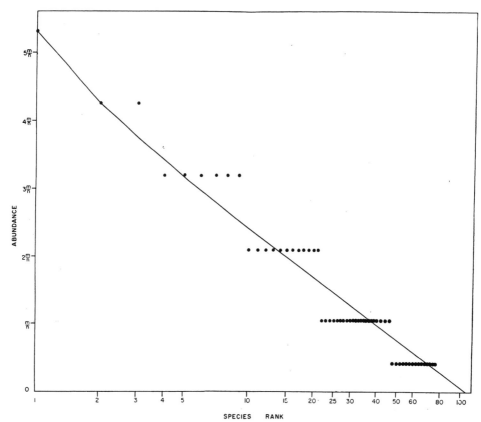

FIGURE 3. Rank order of species of birds in a tropical forest, closely following MacArthur's distribution I.

of niche specificity is that of the number of species that a given biotop can support. The nature of this problem can be best made clear by means of an example.

The aquatic bugs of the family *Corixidae* are of practically world wide distribution. Omitting a purely Australasian subfamily, they may be divided into the *Micronectinae* which are nearly always small, under 5 mm long and the *Corixinae* of which the great majority of species are over 5 mm long. Both subfamilies probably feed largely on organic detritus, though a few of the more primitive members of the *Corixinae* are definite predators. Some at least suck out the contents of algal cells, but unlike the other Heteroptera they can take particulate matter of some size unto their alimentary tracts. There is abundant evidence that the organic content of the bottom deposits of the shallow water in which these insects live is a major ecological factor regulating their occurrence. No *Micronectinae* occur in temperate North America and in the Old World this subfamily is

far more abundant in the tropics while the *Corixinae* are far more abundant in the temperate regions (Lundblad, 1934; Jaczewski, 1937). Thus in Britain there are 30 species of *Corixinae* and three of *Micronectinae* (Macan, 1956), in peninsular Italy 20 or 21 species of *Corixinae* and five of *Micronectinae* (Stickel, 1955), in non-Palaeartic India about a dozen species of *Corixinae* and at least ten species of *Micronectinae* (Hutchinson, 1940) and in Indonesia (Lundblad, 1934) only three *Corixinae* and 14 *Micronectinae*. A reasonable explanation of this variation in the relative proportions of the two subfamilies is suggested by the findings of Macan (1938) and the more casual observations of other investigators that *Micronecta* prefers a low organic subtratum; in tropical localities the high rate of decomposition would reduce the organic content.

In certain isolated tropical areas at high altitudes, notably Ethiopia and the Nilghiri Hills of southern India the decline in the numbers of *Micronectinae* with increasing altitudes, and so

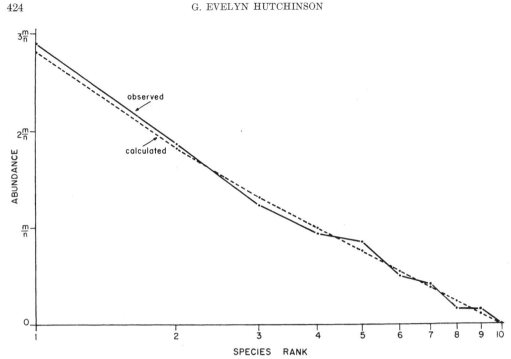

FIGURE 4. Rank order of species of *Conus* on a littoral bench in Hawaii (Kohn).

lower average water temperatures, is most noticable, but there is no increase in the number of *Corixinae*, presumably because the surrounding fauna is not rich enough to have permitted frequent invasion and speciation. Thus in the Nilghiri Hills between 2100 and 2300 m, intense collecting yielded three *Corixinae* of which two appear to be endemic, and one non-endemic species of *Micronecta*. Very casual collecting below 1000 m in south India has produced two species of *Corixinae* and five species of *Micronectinae*. The question raised by cases like this is whether the three Nilghiri *Corixinae* fill all the available niches which in Europe might support perhaps 15 or 20 species, or whether 'there are really empty niches. Intuitively one would suppose both alternatives might be partly true, but there is no information on which to form a real judgment. The rapid spread of introduced species often gives evidence of empty niches, but such rapid spread in many instances has taken place in disturbed areas. The problem clearly needs far more systematic study than it has been given. The addition and the replacement of species of fishes proceeding down a river, and the competitive situations involved, may provide some of the best material for this sort of study, but though much data exists, few attempts at systematic comparative interpretation have been made (*cf.* Hutchinson, 1939).

THE FUTURE OF COMPARATIVE DEMOGRAPHIC STUDIES

Perhaps the most interesting general aspect of the present Symposium is the strong emphasis placed on the changing nature of the populations with which almost all investigators deal. In certain cases, notably in the parthenogenetic crustacean *Daphnia* (Slobodkin, 1954), it is possible to work with clones that must be almost uniform genetically, but all the work on bisexual organisms is done under conditions in which evolution may take place. The emergence in Nicholson's experiments of strains of *Lucilia* in which adult females no longer need a protein meal before egg laying provides a dramatic example of evolution in the laboratory; the work reported by Dobzhansky, by Lewontin, and by Wallace, in discussion, shows how experimental evolution, for which subject the Carnegie Laboratory at Cold Spring Harbor was founded, has at last come into its own.

So far little attention has been paid to the problem of changes in the properties of populations of the greatest demographic interest in such experiments. A more systematic study of evolutionary change in fecundity, mean life span, age and duration of reproductive activity and length of post reproductive life is clearly needed. The most interesting models that might be devised would be those in which selection operated in favor of low fecundity, long pre-reproductive life and on any aspect of post-reproductive life.

There is in many groups, notably *Daphnia*, dependence of natality on food supply (Slobodkin, 1954) though the adjustment can never be instantaneous and so can lead to oscillations. In the case of birds the work of the Oxford school (Moreau, 1944; Lack, 1947a, 1954 and many papers quoted in the last named) indicates that in many birds natality is regulated by natural selection to correspond to the maximum number of young that can be reared in a clutch. In some circumstances the absolute survival of young is greater when the fecundity is low than when it is high. The peculiar nature of the subpopulations formed by groups of nestlings in nests makes this reasonable. Slobodkin (1953) has pointed out that in certain cases in which migration into numerous limited areas is possible, a high reproductive rate might have a lower selective. advantage than a low rate. Actually in a very broad sense the bird's nest is a device to formalise the numerous limited areas, the existence of which permits such a type of selection. It should be possible with some insects to set up population cages in which access to a large number of very small amounts of larval food is fairly difficult for a fertile female. If the individual masses of larval food were such that there was an appreciable chance that many larvae on a single mass would die of starvation while a few larvae would survive, it is possible that selection for low fecundity might occur. This experiment would certainly imitate many situations in nature.

The evolutionary aspects of the problem raised by those cases where there is a delay of reproductive activity after adult morphology has been achieved is much harder to understand. Some birds though they attain full body size within a year (or in the case of most passerines in the nest) are apparently not able to breed until their third or later year. It is difficult to see why this should be so. In any given species there may be good endocrinological reasons for the delay, but they can hardly be evolutionarily inevitable. The situation has an obvious *prima facie* disadvantage, since most birds have a strikingly diagonal survivorship curve after the first year of life and this in itself indicates little capacity for learning to live. One would have supposed that in the birds, mainly but not exclusively large sea birds, which show the delay, any genetic change favoring early reproduction would have a great selective advantage. Any experimental model imitating this situation would be of great interest.

The problem of possible social effects of long post-reproductive life, which can hardly be subject to direct selection, provides another case in which any hints from changes in demographic parameters in experiments would be most helpful. The experimental study of the evolutionary aspects of demography is certain to yield surprises. While we have Nicholson's work, in which

the amplitude of the oscillation in *Lucilia* populations. appear to be increased or at least not decreased as a result of the evolution he has observed, though the minima are less low and the variation less regular, we do not know if this sort of effect is likely to be general. Utida's elegant work on bean weevils appears to be consistent with some evolutionary damping of oscillations which would be theoretically a likely result.

The most curious case of a genetic change playing a regular part in a demographic process is certainly that in rodents described by Chitty. In view of the large number of simple ways which are now available to explain regular oscillations in a population, it is extremely important to heed Chitty's warning that the obvious explanation is not necessarily the true one. To the writer, this seems to be a particular danger in human demography, though the mysteries of variation of the human sex ratio, so clearly expounded by Colombo, should be a warning against over-simple hypotheses, for here no reasonable hypotheses have been suggested. Human demography relies too much on what psychologists call intervening variable theory. The reproducing organisms are taken for granted; when their properties change, either as the result of evolution or of changes in learned behaviour, the results are apt to be upsetting. The present "baby boom" is such an upset, and here a tendency to over-simplified thinking is also apparent. If, as appears clear at least for parts of North America, the present birth rate is positively correlated with economic position, it is easy to suppose that couples now have as many children as they can afford, just as most small birds appear to do. There is, however, a difference. If at any economic level a four child family was desired, but occasionally owing to the imperfections of birth control a five child family was actually achieved, we should not expect the fifth child to have a negligible expectation of life at birth, so that the total contribution to the population per family would be the same from a four and a five child family. Yet this is exactly what Lack and Arn (1947) found for the broods of the Alpine swift *Apus melba*. In man the criterion is never purely economic; it is not how large a brood can be reared, but how large a brood the parents think they can rear without undue economic sacrifice. Such a method of setting limits to natality is obviously extremely complicated. It involves an equilibrium between a series of desires, partly conscious, partly unconscious, and a series of estimates of present and future resources. There is absolutely no reason to suppose that the mean desired family size determined in such a way is a simple function of economics, uninfluenced by a vast number of other cultural factors. The assumption that a large family is *per se* a good thing is obviously involved; this may be accepted individually by most parents even though it is at

426   G. EVELYN HUTCHINSON

present a very dubious assumption on general grounds of social well being. Part of the acceptance of such an assumption is certain to be due to unconscious factors. Susannah Coolidge in a remarkable, as yet unpublished, essay,[7] "Population *versus* People," suggests that for many women a new pregnancy is an occasion for a temporary shifting of some of the responsibility for the older children away from the mother, and so is welcomed. She also suspects that it may be an unconscious expression of disappointment over, or repudiation of, the older children and so be essentially a repeated neurotic symptom. Moreover, the present highly conspicuous fashion for maternity, certainly a healthy reaction from the seclusion of upper-class pregnant women a couple of generations ago, is also quite likely fostered by those business interests which seem to believe that an indefinitely expanding economy is possible on a non-expanding planet.

An adequate science of human demography must take into account mechanism of these kinds, just as animal demography has taken into account all the available information on the physiological ecology and behaviour of blow flies, *Daphnia* and bean weevils. Unhappily, human beings are far harder to investigate than are these admirable laboratory animals; unhappily also, the need becomes more urgent daily.

REFERENCES

ANDREWARTHA, H. G., and BIRCH, L. C., 1954, The Distribution and Abundance of Animals. Chicago, University of Chicago Press. xv, 782 pp.
BETTS, M. M., 1955, The food of titmice in oak woodland. J. Anim. Ecol. *24:* 282–323.
BIRCH, L. C., 1957, The meanings of competition. Amer. Nat. *91:* 5–18.
BRIAN, M. V., 1953, Species frequencies from random samples in animal populations. J. Anim. Ecol. *22:* 57–64.
1956a, Segregation of species of the ant genus *Myrmica*. J. Anim. Ecol. *25:* 319–337.
1956b, Exploitation and interference in interspecies competition. J. Anim. Ecol. *25:* 339–347.
COOPER, D. M., and DOBZHANSKY, TH., 1956, Studies on the ecology of *Drosophila* in the Yosemite region of California. I. The occurrence of species of *Drosophila* in different life zones and at different seasons. Ecology *37:* 526–533.
CROMBIE, A. C., 1945, On competition between different species of graminivorous insects. Proc. Roy. Soc. Lond. *132*B: 362–395.
1946, Further experiments on insect competition. Proc. Roy. Soc. Lond. *133*B: 76–109.
1947, Interspecific competition. J. Anim. Ecol. *16:* 44–73.
DA CUNHA, A. B., EL-TABEY SHEKATA, A. M., and DE OLIVIERA, W., 1957, A study of the diet and nutritional preferences of tropical of *Drosophila*. Ecology *38:* 98–106.
DUMAS, P. C., 1956, The ecological relations of sympatry in *Plethodon dunni* and *Plethodon vehiculum*. Ecology *37:* 484–495.
FISHER, R. A., CORBET, A. S., and WILLIAMS, C. B., 1943, The relation between the number of species and the number of individuals in a ransom sample of an animal population. J. Anim. Ecol. *12:* 42–58.
GAUSE, G. F., 1934, The struggle for existence. Baltimore, Williams & Wilkins. 163 pp.
1935, Vérifications expérimentales de la théorie mathématique de la lutte pour la vie. Actualités scientifiques *277*. Paris. 63 pp.
HUTCHINSON, G. E., 1939, Ecological observations on the fishes of Kashmir and Indian Tibet. Ecol. Monogr. *9:* 145–182.
1940, A revision of the Corixidae of India and adjacent regions. Trans. Conn. Acad. Arts Sci. *33:* 339–476.
1941, Ecological aspects of succession in natural populations. Amer. Nat. *75:* 406–418.
1944, Limnological studies in Connecticut. VII. A critical examination of the supposed relationship between phytoplankton periodicity and chemical changes in lake waters. Ecology *25:* 3–26.
1951, Copepodology for the ornithologist. Ecology *32:* 571–577.
1953, The concept of pattern in ecology. Proc. Acad. Nat. Sci. Phila. *105:* 1–12.
HYNES, H. B. N., 1954, The ecology of *Gammarus deubeni* Lilljeborg and its occurrence in fresh water in western Britain. J. Anim. Ecol. *23:* 38–84.
1955, The reproductive cycle of some British freshwater Gammaridae. J. Anim. Ecol. *24:* 352–387.
JACZEWSKI, S., 1937, Allgemeine Zügeder geographischen Verbreitung der Wasserhemiptera. Arch. Hydrobiol. *31:* 565–591.
KOHN, A. J., The ecology of *Conus* in Hawaii. (Yale Dissertation 1057, in press.)
KURTÉN, B., 1957, Mammal migrations, cenozoic stratigraphy, and the age of Peking man and the australopithecines. J. Paleontol. *31:* 215–227.
LACK, D., 1947a, The significance of clutch-size. Ibis *89:* 302–352.
1947b, Darwin's finches. Cambridge, England. x, 208 pp.
1954, The natural regulation of animal numbers. Oxford, The Clarendon Press, viii, 343.
LACK, D., and ARN, H., 1947, Die Bedeutung der Gelegegrösse beim Alpensegler. Ornith. Beobact. *44:* 188–210.
LOTKA, A. J., The growth of mixed populations, two species competing for a common food supply. J. Wash. Acad. Sci. *22:* 461–469.
LUNDBLAD, O., 1933, Zur Kenntnis der aquatilen und semi-aquatilen Hemipteren von Sumatra, Java, und Bali. Arch. Hydrobiol. Suppl. *12:* 1–195, 263–489.
MACAN, T. T., 1938, Evolution of aquatic habitats with special reference to the distribution of Corixidae. J. Anim. Ecol. *7:* 1–19.
1956, A revised key to the British water bugs (Hemiptera, Heteroptera) Freshwater Biol. Assoc. Sci. Publ. *16:* 73 pp.
MACARTHUR, R. H., 1957, On the relative abundance of bird species. Proc. Nat. Acad. Sci. Wash. *43:* 293–295.
MOREAU, R. E., 1944, Clutch-size: a comparative study, with special reference to African birds. Ibis *86:* 286–347.
PEARSALL, W. H., 1932, The phytoplankton in the English lakes II. The composition of the phytoplankton in relation to dissolved substances. J. Ecol. *20:* 241–262.
PRESTON, F. W., 1948, The commonness, and rarity, of species. Ecology *29:* 254–283.
PROCTOR, V. W., 1957, Some factors controlling the distribution of *Haematococcus pluvialis*. Ecology, *in press*.
ROTHSCHILD, LORD, 1915, On the gigantic land tortoises of the Seychelles and Aldabra-Madagascar group with some notes on certain forms of the Mascarene group. Novitat. Zool. *22:* 418–442.
RUITER, L. DE, 1952, Some experiments on the camouflage of stick caterpillars. Behaviour *4:* 222–233.

---

[7] I am greatly indebted to the author of this work for permission to refer to some of her conclusions.

CONCLUDING REMARKS   427

SHINOZAKI, K., and URATA, N., 1953, Researches on population ecology II. Kyoto Univ. (not seen; ref. MacArthur, 1957).

SHORTEN, M., 1953, Notes on the distribution of the grey squirrel (*Sciurus carolinensis*) and the red squirrel (*Sciurus vulgaris leucourus*) in England and Wales from 1945 to 1952. J. Anim. Ecol. *22:* 134–140.

1954, Squirrels. (New Naturalist Monograph 12.) London 212 pp.

SKELLAM, J. G., 1951, Random dispersal in theoretical populations. Biometrika *38:* 196–218.

SLOBODKIN, L. B., 1953, An algebra of population growth. Ecology *34:* 513–517.

1954, Population dynamics in *Daphnia obtusa* Kurz. Ecol. Monogr. *24:* 69–88.

STEUSLOFF, V., 1943, Ein Beitrag zur Kenntniss der Verbreitung und der Lebensräume von *Gammarus*-Arten in Nordwest-Deutschland. Arch. Hydrobiol. *40:* 79–97.

STICKEL, W., 1955, Illustrierte Bestimmungstabellen der Wanzen. II. Europa. Hf. 2, 3. pp. 40–80 (Berlin, apparently published by author).

UENO, M., 1934, Yale North India Expedition. Report on the amphipod genus *Gammarus*. Mem. Conn. Acad. Arts Sci. *10:* 63–75.

VOLTERRA, V., 1926, Vartazioni e fluttuazioni del numero d'individui in specie animali conviventi. Mem. R. Accad. Lincei ser. 6, *2:* 1–36.

WINSOR, C. P., 1934, Mathematical analysis of the growth of mixed populations. Cold Spr. Harb. Svmp. Quant. Biol. *2:* 181–187.

THE
AMERICAN NATURALIST

Vol. XCIII   May–June, 1959   No. 870

HOMAGE TO SANTA ROSALIA
or
WHY ARE THERE SO MANY KINDS OF ANIMALS?*

G. E. HUTCHINSON

Department of Zoology, Yale University, New Haven, Connecticut

When you did me the honor of asking me to fill your presidential chair, I accepted perhaps without duly considering the duties of the president of a society, founded largely to further the study of evolution, at the close of the year that marks the centenary of Darwin and Wallace's initial presentation of the theory of natural selection. It seemed to me that most of the significant aspects of modern evolutionary theory have come either from geneticists, or from those heroic museum workers who suffering through years of neglect, were able to establish about 20 years ago what has come to be called the "new systematics." You had, however, chosen an ecologist as your president and one of that school at times supposed to study the environment without any relation to the organism.

A few months later I happened to be in Sicily. An early interest in zoogeography and in aquatic insects led me to attempt to collect near Palermo, certain species of water-bugs, of the genus Corixa, described a century ago by Fieber and supposed to occur in the region, but never fully reinvestigated. It is hard to find suitable localities in so highly cultivated a landscape as the Concha d'Oro. Fortunately, I was driven up Monte Pellegrino, the hill that rises to the west of the city, to admire the view. A little below the summit, a church with a simple baroque facade stands in front of a cave in the limestone of the hill. Here in the 16th century a stalactite encrusted skeleton associated with a cross and twelve beads was discovered. Of this skeleton nothing is certainly known save that it is that of Santa Rosalia, a saint of whom little is reliably reported save that she seems to have lived in the 12th century, that her skeleton was found in this cave, and that she has been the chief patroness of Palermo ever since. Other limestone caverns on Monte Pellegrino had yielded bones of extinct pleistocene Equus, and on the walls of one of the rock shelters at the bottom of the hill there are beautiful Gravettian engravings. Moreover, a small relic of the saint that I saw in the treasury of the Cathedral of Monreale has a venerable and

*Address of the President, American Society of Naturalists, delivered at the annual meeting, Washington, D. C., December 30, 1958.

petrified appearance, as might be expected. Nothing in her history being known to the contrary, perhaps for the moment we may take Santa Rosalia as the patroness of evolutionary studies, for just below the sanctuary, fed no doubt by the water that percolates through the limestone cracks of the mountain, and which formed the sacred cave, lies a small artificial pond, and when I could get to the pond a few weeks later, I got from it a hint of what I was looking for.

Vast numbers of Corixidae were living in the water. At first I was rather disappointed because every specimen of the larger of the two species present was a female, and so lacking in most critical diagnostic features, while both sexes of the second slightly smaller species were present in about equal number. Examination of the material at leisure, and of the relevant literature, has convinced me that the two species are the common European *C. punctata* and *C. affinis*, and that the peculiar Mediterranean species are illusionary. The larger *C. punctata* was clearly at the end of its breeding season, the smaller *C. affinis* was probably just beginning to breed. This is the sort of observation that any naturalist can and does make all the time. It was not until I asked myself why the larger species should breed first, and then the more general question as to why there should be two and not 20 or 200 species of the genus in the pond, that ideas suitable to present to you began to emerge. These ideas finally prompted the very general question as to why there are such an enormous number of animal species.

There are at the present time supposed to be (Muller and Campbell, 1954; Hyman, 1955) about one million described species of animals. Of these about three-quarters are insects, of which a quite disproportionately large number are members of a single order, the Coleoptera.[1] The marine fauna although it has at its disposal a much greater area than has the terrestrial, lacks this astonishing diversity (Thorson, 1958). If the insects are excluded, it would seem to be more diverse. The proper answer to my initial question would be to develop a theory at least predicting an order of magnitude for the number of species of 10^6 rather than 10^8 or 10^4. This I certainly cannot do. At most it is merely possible to point out some of the factors which would have to be considered if such a theory was ever to be constructed.

Before developing my ideas I should like to say that I subscribe to the view that the process of natural selection, coupled with isolation and later mutual invasion of ranges leads to the evolution of sympatric species, which at equilibrium occupy distinct niches, according to the Volterra-Gause principle. The empirical reasons for adopting this view and the correlative view that the boundaries of realized niches are set by competition are mainly indirect. So far as niches may be defined in terms of food, the subject has been carefully considered by Lack (1954). In general all the indirect evi-

---

[1] There is a story, possibly apocryphal, of the distinguished British biologist, J. B. S. Haldane, who found himself in the company of a group of theologians. On being asked what one could conclude as to the nature of the Creator from a study of his creation, Haldane is said to have answered, "An inordinate fondness for beetles."

dence is in accord with the view, which has the advantage of confirming theoretical expectation. Most of the opinions that have been held to the contrary appear to be due to misunderstandings and to loose formulation of the problem (Hutchinson, 1958).

In any study of evolutionary ecology, food relations appear as one of the most important aspects of the system of animate nature. There is quite obviously much more to living communities than the raw dictum "eat or be eaten," but in order to understand the higher intricacies of any ecological system, it is most easy to start from this crudely simple point of view.

FOOD CHAINS

Animal ecologists frequently think in terms of food chains, of the form *individuals of species* S_1 *are eaten by those of* S_2, *of* S_2 *by* S_3, *of* S_3 *by* S_4, etc. In such a food chain S_1 will ordinarily be some holophylic organism or material derived from such organisms. The simplest case is that in which we have a true *predator chain* in Odum's (1953) convenient terminology, in which the lowest link is a green plant, the next a herbivorous animal, the next a primary carnivore, the next a secondary carnivore, etc. A specially important type of predator chain may be designated Eltonian, because in recent years C. S. Elton (1927) has emphasized its widespread significance, in which the predator at each level is larger and rarer than its prey. This phenomenon was recognized much earlier, notably by A. R. Wallace in his contribution to the 1858 communication to the Linnean Society of London.

In such a system we can make a theoretical guess of the order of magnitude of the diversity that a single food chain can introduce into a community. If we assume that in general 20 per cent of the energy passing through one link can enter the next link in the chain, which is overgenerous (cf. Lindeman, 1942; Slobodkin in an unpublished study finds 13 per cent as a reasonable upper limit) and if we suppose that each predator has twice the mass, (or 1.26 the linear dimensions) of its prey, which is a very low estimate of the size difference between links, the fifth animal link will have a population of one ten thousandth (10^{-4}) of the first, and the fiftieth animal link, if there was one, a population of 10^{-49} the size of the first. Five animal links are certainly possible, a few fairly clear cut cases having been in fact recorded. If, however, we wanted 50 links, starting with a protozoan or rotifer feeding on algae with a density of 10^6 cells per ml, we should need a volume of 10^{26} cubic kilometers to accommodate on an average one specimen of the ultimate predator, and this is vastly greater than the volume of the world ocean. Clearly the Eltonian food-chain of itself cannot give any great diversity, and the same is almost certainly true of the other types of food chain, based on detritus feeding or on parasitism.

Natural selection

Before proceeding to a further consideration of diversity, it is, however, desirable to consider the kinds of selective force that may operate on a food chain, for this may limit the possible diversity.

It is reasonably certain that natural selection will tend to maintain the
efficiency of transfer from one level to another at a maximum. Any increase
in the predatory efficiency of the nth link of a simple food chain will however always increase the possibility of the extermination of the $(n-1)$th
link. If this occurs either the species constituting the nth link must adapt
itself to eating the $(n-2)$th link or itself become extinct. This process
will in fact tend to shortening of food chains. A lengthening can presumably occur most simply by the development of a new terminal carnivore link,
as its niche is by definition previously empty. In most cases this is not
likely to be easy. The evolution of the whale-bone whales, which at least
in the case of *Balaenoptera borealis*, can feed largely on copepods and so
rank on occasions as primary carnivores (Bigelow, 1926), presumably constitutes the most dramatic example of the shortening of a food chain. Mechanical considerations would have prevented the evolution of a larger rarer
predator, until man developed essentially non-Eltonian methods of hunting
whales.

Effect of size

A second important limitation of the length of a food chain is due to the
fact that ordinarily animals change their size during free life. If the terminal member of a chain were a fish that grew from say one cm to 150 cms in
the course of an ordinary life, this size change would set a limit by competition to the possible number of otherwise conceivable links in the 1-150
cm range. At least in fishes this type of process (metaphoetesis) may involve the smaller specimens belonging to links below the larger and the
chain length is thus lengthened, though under strong limitations, by can-
nibalism.

We may next enquire into what determines the number of food chains in a
community. In part the answer is clear, though if we cease to be zoologists
and become biologists, the answer begs the question. Within certain limits,
the number of kinds of primary producers is certainly involved, because many
herbivorous animals are somewhat eclectic in their tastes and many more
limited by their size or by such structural adaptations for feeding that they
have been able to develop.

Effects of terrestrial plants

The extraordinary diversity of the terrestrial fauna, which is much greater
than that of the marine fauna, is clearly due largely to the diversity provided
by terrestrial plants. This diversity is actually two-fold. Firstly, since terrestrial plants compete for light, they have tended to evolve into structures
growing into a gaseous medium of negligible buoyancy. This has led to the
formation of specialized supporting, photosynthetic, and reproductive structures which inevitably differ in chemical and physical properties. The ancient Danes and Irish are supposed to have eaten elm-bark, and sometimes
sawdust, in periods of stress, has been hydrolyzed to produce edible carbohydrate; but usually man, the most omnivorous of all animals, has avoided

almost all parts of trees except fruits as sources of food, though various in-dividual species of animals can deal with practically every tissue of many arboreal species. A major source of terrestrial diversity was thus introduced by the evolution of almost 200,000 species of flowering plants, and the three quarters of a million insects supposedly known today are in part a product of that diversity. But of itself merely providing five or ten kinds of food of different consistencies and compositions does not get us much further than the five or ten links of an Eltonian pyramid. On the whole the problem still remains, but in the new form: why are there so many kinds of plants? As a zoologist I do not want to attack that question directly, I want to stick with animals, but also to get the answer. Since, however, the plants are part of the general system of communities, any sufficiently abstract properties of such communities are likely to be relevant to plants as well as to herbi-vores and carnivores. It is, therefore, by being somewhat abstract, though with concrete zoological details as examples, that I intend to proceed.

INTERRELATIONS OF FOOD CHAINS

Biological communities do not consist of independent food chains, but of food webs, of such a kind that an individual at any level (corresponding to a link in a single chain) can use some but not all of the food provided by spe-cies in the levels below it.

It has long been realized that the presence of two species at any level, either of which can be eaten by a predator at a level above, but which may differ in palatability, ease of capture or seasonal and local abundance, may provide alternative foods for the predator. The predator, therefore, will neither become extinct itself nor exterminate its usual prey, when for any reason, not dependent on prey-predator relationships, the usual prey happens to be abnormally scarce. This aspect of complicated food webs has been stressed by many ecologists, of whom the Chicago school as represented by Allee, Emerson, Park, Park and Schmidt (1949), Odum (1953) and Elton (1958), may in particular be mentioned. Recently MacArthur (1955) using an ingenious but simple application of information theory has generalized the points of view of earlier workers by providing a formal proof of the increase in stability of a community as the number of links in its food web increases.

MacArthur concludes that in the evolution of a natural community two partly antagonistic processes are occurring. More efficient species will re-place less efficient species, but more stable communities will outlast less stable communities. In the process of community formation, the entry of a new species may involve one of three possibilities. It may completely dis-place an old species. This of itself does not necessarily change the sta-bility, though it may do so if the new species inherently has a more stable population (cf. Slobodkin, 1956) than the old. Secondly, it may occupy an unfilled niche, which may, by providing new partially independent links, in-crease stability. Thirdly, it may partition a niche with a pre-existing spe-cies. Elton (1958) in a fascinating work largely devoted to the fate of spe-cies accidentally or purposefully introduced by man, concludes that in very diverse communities such introductions are difficult. Early in the history of a community we may suppose many niches will be empty and invasion will proceed easily; as the community becomes more diversified, the process will be progressively more difficult. Sometimes an extremely successful invader may oust a species but add little or nothing to stability, at other times the invader by some specialization will be able to compete successfully for the marginal parts of a niche. In all cases it is probable that invasion is most likely when one or more species happen to be fluctuating and are under-represented at a given moment. As the communities build up, these opportunities will get progressively rarer. In this way a complex community containing some highly specialized species is constructed asymptotically.

Modern ecological theory therefore appears to answer our initial question at least partially by saying that there is a great diversity of organisms because communities of many diversified organisms are better able to persist than are communities of fewer less diversified organisms. Even though the entry of an invader which takes over part of a niche will lead to the reduction in the *average* population of the species originally present, it will also lead to an increase in stability reducing the risk of the original population being at times underrepresented to a dangerous degree. In this way loss of some niche space may be compensated by reduction in the amplitude of fluctuations in a way that can be advantageous to both species. The process however appears likely to be asymptotic and we have now to consider what sets the asymptote, or in simpler words why are there not more different kinds of animals?

LIMITATION OF DIVERSITY

It is first obvious that the processes of evolution of communities must be under various sorts of external control, and that in some cases such control limits the possible diversity. Several investigators, notably Odum (1953) and MacArthur (1955), have pointed out that the more or less cyclical oscillations observed in arctic and boreal fauna may be due in part to the communities not being sufficiently complex to damp out oscillations. It is certain that the fauna of any such region is qualitatively poorer than that of warm temperate and tropical areas of comparable effective precipitation. It is probably considered to be intuitively obvious that this should be so, but on analysis the obviousness tends to disappear. If we can have one or two species of a large family adapted to the rigors of Arctic existence, why can we not have more? It is reasonable to suppose that the total biomass may be involved. If the fundamental productivity of an area is limited by a short growing season to such a degree that the total biomass is less than under more favorable conditions, then the rarer species in a community may be so rare that they do not exist. It is also probable that certain absolute limitations on growth-forms of plants, such as those that make the development of forest impossible above a certain latitude, may in so acting, severely limit the number of niches. Dr. Robert MacArthur points out that the development of high tropical rain forest increases the bird fauna more than that of mammals, and Thorson (1957) likewise has shown that the so-called infauna show no increase of species toward the tropics while the marine epifauna becomes more diversified. The importance of this aspect of the plant or animal substratum, which depends largely on the length of the growing season and other aspects of productivity is related to that of the environmental mosaic discussed later.

We may also inquire, but at present cannot obtain any likely answer, whether the arctic fauna is not itself too young to have achieved its maximum diversity. Finally, the continual occurrence of catastrophes, as Wynne-Edwards (1952) has emphasized, may keep the arctic terrestrial community in a state of perennial though stunted youth.

Closely related to the problems of environmental rigor and stability, is the question of the absolute size of the habitat that can be colonized. Over much of western Europe there are three common species of small voles, namely *Microtus arvalis*, *M. agrestis* and *Clethrionomys glareolus*. These are sympatric but with somewhat different ecological preferences.

In the smaller islands off Britain and in the English channel, there is only one case of two species co-occurring on an island, namely *M. agrestis* and Clethrionomys on the island of Mull in the Inner Hebrides (Barrett-Hamilton and Hinton, 1911–1921). On the Orkneys the single species is *M. orcadensis*, which in morphology and cytology is a well-differentiated ally of *M. arvalis*; a comparable animal (*M. sarnius*) occurs on Guernsey. On most of the Scottish Islands only subspecies of *M. agrestis* occur, but on Mull and Raasay, on the Welsh island of Skomer, as well as on Jersey, races of Clethrionomys of somewhat uncertain status are found. No voles have reached Ireland, presumably for paleogeographic reasons, but they are also absent from a number of small islands, notably Alderney and Sark. The last named island must have been as well placed as Guernsey to receive *Microtus arvalis*. Still stranger is the fact that although it could not have got to the Orkneys without entering the mainland of Britain, no vole of the *arvalis* type now occurs in the latter country. Cases of this sort may be perhaps explained by the lack of favorable refuges in randomly distributed very unfavorable seasons or under special kinds of competition. This explanation is very reasonable as an explanation of the lack of Microtus on Sark, where it may have had difficulty in competing with *Rattus rattus* in a small area. It would be stretching one's credulity to suppose that the area of Great Britain is too small to permit the existence of two sympatric species of Microtus, but no other explanation seems to have been proposed.

It is a matter of considerable interest that Lack (1942) studying the populations of birds on some of these small British islands concluded that such populations are often unstable, and that the few species present often occupied larger niches than on the mainland in the presence of competitors. Such faunas provide examples of communities held at an early stage in development because there is not enough space for the evolution of a fuller and more stable community.

NICHE REQUIREMENTS

The various evolutionary tendencies, notably metaphoetesis, which operate on single food chains must operate equally on the food-web, but we also have a new, if comparable, problem as to how much difference between two species at the same level is needed to prevent them from occupying the same niche. Where metric characters are involved we can gain some insight into this extremely important problem by the study of what Brown and Wilson (1956) have called *character displacement* or the divergence shown when two partly allopatric species of comparable niche requirements become sympatric in part of their range.

I have collected together a number of cases of mammals and birds which appear to exhibit the phenomenon (table 1). These cases involve metric characters related to the trophic apparatus, the length of the culmen in birds and of the skull in mammals appearing to provide appropriate measures. Where the species co-occur, the ratio of the larger to the small form varies from 1.1 to 1.4, the mean ratio being 1.28 or roughly 1.3. This latter figure may tentatively be used as an indication of the kind of difference necessary to permit two species to co-occur in different niches but at the same level of a food-web. In the case of the aquatic insects with which I began my address, we have over most of Europe three very closely allied species of Corixa, the largest *punctata*, being about 116 per cent longer than the middle sized species *macrocephala*, and 146 per cent longer than the small species *affinis*. In northwestern Europe there is a fourth species, *C. dentipes*, as large as *C. punctata* and very similar in appearance. A single observation (Brown, 1948) suggests that this is what I have elsewhere (Hutchinson, 1951) termed a fugitive species, maintaining itself in the face of competition mainly on account of greater mobility. According to Macan (1954) while both *affinis* and *macrocephala* may occur with *punctata* they never are found with each other, so that all three species never occur together. In the eastern part of the range, *macrocephala* drops out, and *punctata* appears to have a discontinuous distribution, being recorded as far east as Simla, but not in southern Persia or Kashmir, where *affinis* occurs. In these eastern localities, where it occurs by itself, *affinis* is larger and darker than in the west, and superficially looks like *macrocephala* (Hutchinson, 1940).

This case is very interesting because it looks as though character displacement is occurring, but that the size differences between the three species are just not great enough to allow them all to co-occur. Other characters than size are in fact clearly involved in the separation, *macrocephala* preferring deeper water than *affinis* and the latter being more tolerant of brackish conditions. It is also interesting because it calls attention to a marked difference that must occur between hemimetabolous insects with annual life cycles involving relatively long growth periods, and birds or mammals in which the period of growth in length is short and of a very special nature compared with the total life span. In the latter, niche separation may be possible merely through genetic size differences, while in a pair of ani-

TABLE 1

Mean character displacement in measurable trophic structures in mammals (skull) and birds (culmen); data for Mustela from Miller (1912); Apodemus from Cranbrook (1957); Sitta from Brown and Wilson (1956) after Vaurie; Galapagos finches from Lack (1947)

	Locality and measurement when sympatric	Locality and measurement when allopatric	Ratio when sympatric
Mustela nivalis	Britain; skull ♂ 39.3 ♀ 33.6 mm.	(*boccamela*) S. France, Italy ♂ 42.9 ♀ 34.7 mm. (*iberica*) Spain, Portugal ♂ 40.4 ♀ 36.0 (*bibernica*) Ireland ♂ 46.0 ♀ 41.9	♂ 100:128 ♀ 100:134
M. erminea	Britain; " ♂ 50.4 ♀ 45.0		
Apodemus sylvaticus	Britain; " 24.8	unnamed races on Channel Islands 25.6–26.7	100:109
A. flavicollis	Britain; " 27.0		
Sitta tephronota	Iran; culmen 29.0	races east of overlap 25.5	100:124
S. neumayer	Iran; " 23.5	races west of overlap 26.0	
Geospiza fortis	Indefatigable Isl.; culmen 12.0	Daphne Isl. 10.5	100:143
G. fuliginosa	Indefatigable Isl.; " 8.4	Crossman Isl. 9.3	
Camarhyncbus parvulus	James Isl.; " 7.0 Indefatigable Isl.; " 7.5 S. Albemarle Isl.; " 7.3	N. Albemarle Isl. 7.0 Chatham Isl. 8.0	James 100:140:180 100:129
C. psittacula	James Isl.; " 9.8 Indefatigable Isl.; " 9.6 S. Albemarle Isl.; " 8.5	Abington Isl. 10.1 Bindloe Isl. 10.5	Indefatigable 100:128:162 100:127
C. pallidus	James Isl.; " 12.6 Indefatigable Isl.; " 12.1 S. Albemarle Isl.; " 11.2	N. Albemarle Isl. 11.7 Chatham Isl. 10.8	S. Albemarle 100:116:153 100:132
			Mean ratio 100:128

mals like *C. punctata* and *C. affinis* we need not only a size difference but a seasonal one in reproduction; this is likely to be a rather complicated matter. For the larger of two species always to be larger, it must never breed later than the smaller one. I do not doubt that this is what was happening in the pond on Monte Pellegrino, but have no idea how the difference is achieved.

I want to emphasize the complexity of the adaptation necessary on the part of two species inhabiting adjacent niches in a given biotope, as it probably underlies a phenomenon which to some has appeared rather puzzling. MacArthur (1957) has shown that in a sufficiently large bird fauna, in a uniform undisturbed habitat, areas occupied by the different species appear to correspond to the random non-overlapping fractionation of a plane or volume. Kohn (1959) has found the same thing for the cone-shells (Conus) on the Hawaiian reefs. This type of arrangement almost certainly implies such individual and unpredictable complexities in the determination of the niche boundaries, and so of the actual areas colonized, that in any overall view, the process would appear random. It is fairly obvious that in different types of community the divisibility of niches will differ and so the degree of diversity that can be achieved. The fine details of the process have not been adequately investigated, though many data must already exist that could be organized to throw light on the problem.

MOSAIC NATURE OF THE ENVIRONMENT

A final aspect of the limitation of possible diversity, and one that perhaps is of greatest importance, concerns what may be called the mosaic nature of the environment. Except perhaps in open water when only uniform quasi-horizontal surfaces are considered, every area colonized by organisms has some local diversity. The significance of such local diversity depends very largely on the size of the organisms under consideration. In another paper MacArthur and I (Hutchinson and MacArthur, 1959) have attempted a theoretical formulation of this property of living communities and have pointed out that even if we consider only the herbivorous level or only one of the carnivorous levels, there are likely, above a certain lower limit of size, to be more species of small or medium sized organisms than of large organisms. It is difficult to go much beyond crude qualitative impressions in testing this hypothesis, but we find that for mammal faunas, which contain such diverse organisms that they may well be regarded as models of whole faunas, there is a definite hint of the kind of theoretical distribution that we deduce. In qualitative terms the phenomenon can be exemplified by any of the larger species of ungulates which may require a number of different kinds of terrain within their home ranges, any one of which types of terrain might be the habitat of some small species. Most of the genera or even subfamilies of very large terrestrial animals contain only one or two sympatric species. In this connection I cannot refrain from pointing out the immense scientific importance of obtaining a really full insight into the ecology of the large mammals of Africa while they can still be studied under natural conditions. It is indeed quite possible that the results of studies on these wonderful animals would in long-range though purely practical terms pay for the establishment of greater reservations and National Parks than at present exist.

In the passerine birds the occurrence of five or six closely related sympatric species is a commonplace. In the mammal fauna of western Europe no genus appears to contain more than four strictly sympatric species. In Britain this number is not reached even by Mustela with three species, on the adjacent parts of the continent there may be three sympatric shrews of the genus Crocidura and in parts of Holland three of Microtus. In the same general region there are genera of insects containing hundreds of species, as in Athela in the Coleoptera and Dasyhelea in the Diptera Nematocera. The same phenomenon will be encountered whenever any well-studied fauna is considered. Irrespective of their position in a food chain, small size, by permitting animals to become specialized to the conditions offered by small diversified elements of the environmental mosaic, clearly makes possible a degree of diversity quite unknown among groups of larger organisms.

We may, therefore, conclude that the reason why there are so many species of animals is at least partly because a complex trophic organization of a community is more stable than a simple one, but that limits are set by the tendency of food chains to shorten or become blurred, by unfavorable physical factors, by space, by the fineness of possible subdivision of niches, and by those characters of the environmental mosaic which permit a greater diversity of small than of large allied species.

CONCLUDING DISCUSSION

In conclusion I should like to point out three very general aspects of the sort of process I have described. One speculative approach to evolutionary theory arises from some of these conclusions. Just as adaptative evolution by natural selection is less easy in a small population of a species than in a larger one, because the total pool of genetic variability is inevitably less, so it is probable that a group containing many diversified species will be able to seize new evolutionary opportunities more easily than an undiversified group. There will be some limits to this process. Where large size permits the development of a brain capable of much new learnt behavior, the greater plasticity acquired by the individual species will offset the disadvantage of the small number of allied species characteristic of groups of large animals. Early during evolution the main process from the standpoint of community structure was the filling of all the niche space potentially available for producer and decomposer organisms and for herbivorous animals. As the latter, and still more as carnivorous animals began to appear, the persistence of more stable communities would imply splitting of niches previously occupied by single species as the communities became more diverse. As this process continued one would expect the overall rate of evolution to have increased, as the increasing diversity increased the probability of the existence of species preadapted to new and unusual niches. It is reasonable to suppose that strong predation among macroscopic metazoa did not begin until the late Precambrian, and that the appearance of powerful predators led to the appearance of fossilizable skeletons. This seems the only reasonable hypothesis, of those so far advanced, to account for the relatively sudden appearance of several fossilizable groups in the Lower Cambrian. The process of diversification would, according to this argument, be somewhat autocatakinetic even without the increased stability that it would produce; with the increase in stability it would be still more a self inducing process, but one, as we have seen, with an upper limit. Part of this upper limit is set by the impossibility of having many sympatric allied species of large animals. These however are the animals that can pass from primarily innate to highly modifiable behavior. From an evolutionary point of view, once they have appeared, there is perhaps less need for diversity, though from other points of view, as Elton (1958) has stressed in dealing with human activities, the stability provided by diversity can be valuable even to the most adaptable of all large animals. We may perhaps therefore see in the process of evolution an increase in diversity at an increasing rate till the early Paleozoic, by which time the familiar types of community structure were established. There followed then a long period in which various large and finally large-brained species became dominant, and then a period in which man has been reducing diversity by a rapidly increasing tendency to cause extinction of supposedly unwanted species, often in an indiscriminate manner. Finally we may hope for a limited reversal of this process when man becomes aware of the value of diversity no less in an economic than in an esthetic and scientific sense.

A second and much more metaphysical general point is perhaps worth a moment's discussion. The evolution of biological communities, though each species appears to fend for itself alone, produces integrated aggregates which increase in stability. There is nothing mysterious about this; it follows from mathematical theory and appears to be confirmed to some extent empirically. It is however a phenomenon which also finds analogies in other fields in which a more complex type of behavior, that we intuitively regard as higher, emerges as the result of the interaction of less complex types of behavior, that we call lower. The emergence of love as an antidote to aggression, as Lorenz pictures the process, or the development of cooperation from various forms of more or less inevitable group behavior that Allee (1931) has stressed are examples of this from the more complex types of biological systems.

In the ordinary sense of explanation in science, such phenomena are explicable. The types of holistic philosophy which import *ad hoc* mysteries into science whenever such a situation is met are obviously unnecessary. Yet perhaps we may wonder whether the empirical fact that it is the nature of things for this type of explicable emergence to occur is not something that itself requires an explanation. Many objections can be raised to such a view; a friendly organization of biologists could not occur in a universe in which cooperative behavior was impossible and without your cooperation I could not raise the problem. The question may in fact appear to certain types of philosophers not to be a real one, though I suspect such philosophers in their desire to demonstrate how often people talk nonsense, may sometimes show less ingenuity than would be desirable in finding some sense in such questions. Even if the answer to such a question were positive, it might not get us very far; to an existentialist, life would have merely provided yet one more problem; students of Whitehead might be made happier, though on the whole the obscurities of that great writer do not seem to generate unhappiness; the religious philosophers would welcome a positive answer but note that it told them nothing that they did not know before; Marxists might merely say, "I told you so." In spite of this I suspect that the question is worth raising, and that it could be phrased so as to provide some sort of real dichotomy between alternatives; I therefore raise it knowing that I cannot, and suspecting that at present others cannot, provide an intellectually satisfying answer.

My third general point is less metaphysical, but not without interest. If I am right that it is easier to have a greater diversity of small than of large organisms, then the evolutionary process in small organisms will differ somewhat from that of large ones. Wherever we have a great array of allied sympatric species there must be an emphasis on very accurate interspecific mating barriers which is unnecessary where virtually no sympatric allies occur. We ourselves are large animals in this sense; it would seem very unlikely that the peculiar lability that seems to exist in man, in which even the direction of normal sexual behavior must be learnt, could have developed to quite the existing extent if species recognition, involving closely related sympatric congeners, had been necessary. Elsewhere (Hutchinson, 1959) I have attempted to show that the difficulties that *Homo sapiens* has to face in this regard may imply various unsuspected processes in human evolutionary selection. But perhaps Santa Rosalia would find at this point that we are speculating too freely, so for the moment, while under her patronage, I will say no more.

ACKNOWLEDGMENTS

Dr. A. Minganti of the University of Palermo enabled me to collect on Monte Pellegrino. Professor B. M. Knox of the Department of Classics of Yale University gave me a rare and elegant word from the Greek to express the blurring of a food chain. Dr. L. B. Slobodkin of the University of Michigan and Dr. R. H. MacArthur of the University of Pennsylvania provided me with their customary kinds of intellectual stimulation. To all these friends I am most grateful.

LITERATURE CITED

Allee, W. C., 1931, Animal aggregations: a study in general sociology. vii, 431 pp. University of Chicago Press, Chicago, Illinois.

Allee, W. C., A. E. Emerson, O. Park, T. Park and K. P. Schmidt, 1949, Principles of animal ecology. xii, 837 pp. W. B. Saunders Co., Philadelphia, Pennsylvania.

Barrett-Hamilton, G. E. H., and M. A. C. Hinton, 1911–1921, A history of British mammals. Vol. 2. 748 pp. Gurney and Jackson, London, England.

Bigelow, H. B., 1926, Plankton of the offshore waters of the Gulf of Maine. Bull. U. S. Bur. Fisheries 40: 1–509.

Brown, E. S., 1958, A contribution towards an ecological survey of the aquatic and semi-aquatic Hemiptera-Heteroptera (water-bugs) of the British Isles etc. Trans. Soc. British Entom. 9: 151–195.

Brown, W. L., and E. O. Wilson, 1956, Character displacement. Systematic Zoology 5: 49–64.

Cranbrook, Lord, 1957, Long-tailed field mice (Apodemus) from the Channel Islands. Proc. Zool. Soc. London 128: 597–600.

Elton, C. S., 1958, The ecology of invasions by animals and plants. 159 pp. Methuen Ltd., London, England.

Hutchinson, G. E., 1951, Copepodology for the ornithologist. Ecology 32: 571–577.

    1958, Concluding remarks. Cold Spring Harbor Symp. Quant. Biol. 22: 415–427.

    1959, A speculative consideration of certain possible forms of sexual selection in man. Amer. Nat. 93: 81–92.

Hutchinson, G. E., and R. MacArthur, 1959, A theoretical ecological model of size distributions among species of animals. Amer. Nat. 93: 117–126.

Hyman, L. H., 1955, How many species? Systematic Zoology 4: 142–143.

Kohn, A. J., 1959, The ecology of Conus in Hawaii. Ecol. Monogr. (in press).

Lack, D., 1942, Ecological features of the bird faunas of British small islands. J. Animal Ecol. London 11: 9–36.

    1947, Darwin's Finches. x, 208 pp. Cambridge University Press, Cambridge, England.

    1954, The natural regulation of animal numbers. viii, 347 pp. Clarendon Press, Oxford, England.

Lindeman, R. L., 1942, The trophic-dynamic aspect of ecology. Ecology 23: 399–408.

Macan, T. T., 1954, A contribution to the study of the ecology of Corixidae (Hemipt). J. Animal Ecol. 23: 115–141.

MacArthur, R. H., 1955, Fluctuations of animal populations and a measure of community stability. Ecology 3   533–536.

    1957, On the relative abundance of bird species. Proc. Nat. Acad. Sci. Wash. 43: 293–295.

Miller, G. S., Catalogue of the mammals of Western Europe. xv, 1019 pp. British Museum, London, England.

Muller, S. W., and A. Campbell, 1954, The relative number of living and fossil species of animals. Systematic Zoology 3: 168–170.

Odum, E. P., 1953, Fundamentals of ecology. xii, 387 pp. W. B. Saunders Co., Philadelphia, Pennsylvania, and London, England.

Slobodkin, L. B., 1955, Condition for population equilibrium. Ecology 35: 530–533.

Thorson, G., 1957, Bottom communities. Chap. 17 *in* Treatise on marine ecology and paleoecology. Vol. 1. Geol. Soc. Amer. Memoir 67: 461–534.

Wallace, A. R., 1858, On the tendency of varieties to depart indefinitely from the original type. *In* C. Darwin and A. R. Wallace, On the tendency of species to form varieties; and on the perpetuation of varieties and species by natural means of selection. J. Linn. Soc. (Zool.) 3: 45–62.

Wynne-Edwards, V. C., 1952, Zoology of the Baird Expedition (1950). I. The birds observed in central and southeast Baffin Island. Auk 69: 353–391.

Vol. XCV, No. 882   The American Naturalist   May–June, 1961

THE PARADOX OF THE PLANKTON*

G. E. HUTCHINSON

Osborn Zoological Laboratory, New Haven, Connecticut

The problem that I wish to discuss in the present contribution is raised by the very paradoxical situation of the plankton, particularly the phytoplankton, of relatively large bodies of water.

We know from laboratory experiments conducted by many workers over a long period of time (summary in Provasoli and Pintner, 1960) that most members of the phytoplankton are phototrophs, able to reproduce and build up populations in inorganic media containing a source of CO_2, inorganic nitrogen, sulphur, and phosphorus compounds and a considerable number of other elements (Na, K, Mg, Ca, Si, Fe, Mn, B, Cl, Cu, Zn, Mo, Co and V) most of which are required in small concentrations and not all of which are known to be required by all groups. In addition, a number of species are known which require one or more vitamins, namely thiamin, the cobalamines (B_{12} or related compounds), or biotin.

The problem that is presented by the phytoplankton is essentially how it is possible for a number of species to coexist in a relatively isotropic or unstructured environment all competing for the same sorts of materials. The problem is particularly acute because there is adequate evidence from enrichment experiments that natural waters, at least in the summer, present an environment of striking nutrient deficiency, so that competition is likely to be extremely severe.

According to the principle of *competitive exclusion* (Hardin, 1960) known by many names and developed over a long period of time by many investigators (see Rand, 1952; Udvardy, 1959; and Hardin, 1960, for historic reviews), we should expect that one species alone would outcompete all the others so that in a final equilibrium situation the assemblage would reduce to a population of a single species.

The principle of competitive exclusion has recently been under attack from a number of quarters. Since the principle can be deduced mathematically from a relatively simple series of postulates, which with the ordinary postulates of mathematics can be regarded as forming an axiom system, it follows that if the objections to the principle in any cases are valid, some or all the biological axioms introduced are in these cases incorrect. Most objections to the principle appear to imply the belief that equilibrium under a given set of environmental conditions is never in practice obtained. Since the deduction of the principle implies an equilibrium system, if such sys-

*Contribution to a symposium on Modern Aspects of Population Biology. Presented at the meeting of the American Society of Naturalists, cosponsored by the American Society of Zoologists, Ecological Society of America and the Society for the Study of Evolution. American Association for the Advancement of Science, New York, N. Y., December 27, 1960.

tems are rarely if ever approached, the principle though analytically true, is at first sight of little empirical interest.

The mathematical procedure for demonstrating the truth of the principle involves, in the elementary theory, abstraction from time. It does, however, provide in any given case a series of possible integral paths that the populations can follow, one relative to the other, and also paths that they cannot follow under a defined set of conditions. If the conditions change the integral paths change. Mere failure to obtain equilibrium owing to external variation in the environment does not mean that the kinds of competition described mathematically in the theory of competitive exclusion are not occuring continuously in nature.

Twenty years ago in a Naturalists' Symposium, I put (Hutchinson, 1941) forward the idea that the diversity of the phytoplankton was explicable primarily by a permanent failure to achieve equilibrium as the relevant external factors changed. I later pointed out that equilibrium would never be expected in nature whenever organisms had reproductive rates of such a kind that under constant conditions virtually complete competitive replacement of one species by another occurred in a time (t_c), of the same order, as the time (t_e) taken for a significant seasonal change in the environment. Note that in any theory involving continuity, the changes are asymptotic to complete replacement. Thus ideally we may have three classes of cases:

1. $t_c \ll t_e$, competitive exclusion at equilibrium complete before the environment changes significantly.
2. $t_c \simeq t_e$, no equilibrium achieved.
3. $t_c \gg t_e$, competitive exclusion occurring in a changing environment to the full range of which individual competitors would have to be adapted to live alone.

The first case applies to laboratory animals in controlled conditions, and conceivably to fast breeding bacteria under fairly constant conditions in nature. The second case applies to most organisms with a generation time approximately measured in days or weeks, and so may be expected to occur in the plankton and in the case of populations of multivoltine insects. The third case applies to animals with a life span of several years, such as birds and mammals.

Very slow and very fast breeders thus are likely to compete under conditions in which an approach to equilibrium is possible; organisms of intermediate rates of reproduction may not do so. This point of view was made clear in an earlier paper (Hutchinson, 1953), but the distribution of that paper was somewhat limited and it seems desirable to emphasize the matter again briefly.

It is probably no accident that the great proponents of the type of theory involved in competitive exclusion have been laboratory workers on the one hand (for example, Gause, 1934, 1935; Crombie, 1947; and by implication Nicholson, 1933, 1957) and vertebrate field zoologists (for example, Grinnell, 1904; Lack, 1954) on the other. The major critics of this type of approach, notably Andrewartha and Birch (1954), have largely worked with insects in the field, often under conditions considerably disturbed by human activity.

DISTRIBUTION OF SPECIES AND INDIVIDUALS

MacArthur (1957, 1960) has shown that by making certain reasonable assumptions as to the nature of niche diversification in homogeneously diversified[1] biotopes of large extent, the distribution of species at equilibrium follows a law such that the rth rarest species in a population of S_s species and N_s individuals may be expected to be

$$\frac{N_s}{S_s} \sum_{i=1}^{r} \frac{1}{S_s - i + 1} .$$

This distribution, which is conveniently designated as type I, holds remarkably well for birds in homogeneously diverse biotopes (MacArthur, 1957, 1960), for molluscs of the genus Conus (Kohn, 1959, 1960) and for at least one mammal population (J. Armstrong, personal communication). It does not hold for bird faunas in heterogeneously diverse biotopes, nor for diatoms settling on slides (Patrick in MacArthur, 1960) nor for the arthropods of soil (Hairston, 1959). Using Foged's (1954) data for the occurrence of planktonic diatoms in Braendegård Sø on the Danish island of Funen, it is also apparent (figure 1) that the type I distribution does not hold for such assemblages of diatom populations under quite natural conditions either.

MacArthur (1957, 1960) has deduced two other types of distribution (type II and type III) corresponding to different kinds of biological hypotheses. These distributions, unlike type I, do not imply competitive exclusion. So far in nature only type I distributions and a kind of empirical distribution which I shall designate type IV are known. The type IV distribution given by diatoms on slides, in the plankton and in the littoral of Braendegård Sø, as well as by soil arthropods, differs from the type I in having its commonest species commoner and all other species rarer. It could be explained as due to heterogeneous diversity, for if the biotope consisted of patches in each one of which the ratio of species to individuals differed, then the sum of the assemblages gives such a curve. This is essentially the same as Hairston's (1959) idea of a more structured community in the case of soil arthropods than in that of birds. It could probably arise if the environment changed in favoring temporarily a particular species at the expense of other species before equilibrium is achieved. This is, in fact, a sort of temporal analogue to

---

[1]A biotope is said to be *homogeneously diverse* relative to a group of organisms if the elements of the environmental mosaic relevant to the organism are small compared to the mean range of the organisms. A *heterogeneously diverse biotope* is divided into elements at least some of which are large compared to the ranges of the organisms. An area of woodland is homogeneously diverse relative to most birds, a large tract of stands of woodland in open country is heterogeneously diverse (Hutchinson, 1957, 1959).

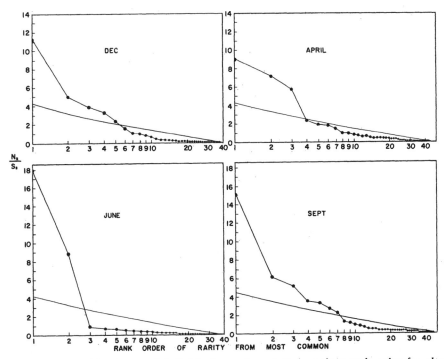

FIGURE 1. Abundance of individual species plotted against rank order for the planktonic diatoms of Braendegård Sø, for the four seasons, from Foged's data, showing type IV distributions. The unmarked line gives the type I distribution for a like number of species and individuals. The unit of population for each species is the ratio of total number of individuals (N_s) to total number of species (S_s).

heterogeneous diversity. Existence of the type IV distribution does not necessarily imply non-equilibrium, but if we assume niches are separated out of the niche-hyperspace with any boundary as probable as any other, we may conclude that either non-equilibrium in time or unexpected diversity in space are likely to underlie this type of distribution.

APPLICATION TO THE PLANKTON

Before proceeding to inquire how far plankton associations are either never in equilibrium in time or approach heterogeneous diversity in space in a rather subtle way, it is desirable to inquire how far ordinary homogeneous niche diversification may be involved. The presence of a light gradient in all epigean waters by day does imply a certain diversification, but in the epilimnia of lakes the chances of any organism remaining permanently in a particular narrow range of intensities is small in turbulent water. By day the stability of the epilimnion may well never be zero, but since what has to be explained is the presence of many species of competitors in a small volume of water, the role of small vertical variations is probably insignificant. A few organisms may be favored by peculiar chemical conditions at the surface film, but again this hardly seems an adequate explanation. The Langmuir spirals in the wind drift might also separate motile from non-motile forms or organisms of different densities to some extent but again the effect is likely to be small and transitory. It is hard to believe that in turbulent open water many physical opportunities for niche diversification exist.

SYMBIOSIS AND COMMENSALISM

The mathematical theory of competition permits the treatment of commensal and symbiotic relations by a simple change in sign of one or both of the competition functions. It can be shown (Gause and Witt, 1935) that under some conditions commensal or symbiotic species can occupy the same niche. There is a little evidence that occasionally water in which one alga has been growing may be stimulatory to another species (Lefèvre, Jacob and Nisbet, 1952; see also Hartman, 1960) though it is far more likely to be inhibitory. Since some phytoplankters require vitamins and others do not, a more generally efficient species, requiring vitamins produced in excess by an otherwise less efficient species not requiring such compounds, can produce a mixed equilibrium population. It is reasonably certain that this type of situation occurs in the phytoplankton. It is interesting to note that many vitamin-requiring algae are small and that the groups characteristically needing them (Euglenophyta, Crytophyceae, Chrysophyceae, and Dinophyceae) tend to be motile. The motility would give such organisms an advantage in meeting rare nutrient molecules, inorganic or organic. This type of advantage can be obtained by non-motile forms only by sinking in a turbulent medium (Munk and Riley, 1952) which is much more dangerous than even random swimming.

ROLE OF PREDATION

It can be shown theoretically, as Dr. MacArthur and I have developed in conversation, that if one of two competing species is limited by a predator, while the other is either not so limited or is fed on by a different predator, co-existence of the two prey species may in some cases be possible. This should permit some diversification of both prey and predator in a homogeneous habit.

RESULTS OF NON-EQUILIBRIUM CONDITIONS

The possibility of synergistic phenomena on the one hand and of specific predation on the other would probably permit the development of a somewhat diversified equilibrium plankton even in an environment that was essentially boundaryless and isotropic. It may, however, be doubted that such phenomena would ever permit assemblages of the order of magnitude of tens of species to co-occur. At least in homogeneous water in the open ocean there would seem to be no other alternative to a non-equilibrium, or as MacArthur (1960) would term it, an opportunistic community.

The great difficulty inherent in the opportunistic hypothesis is that since, if many species are present in a really variable environment which is controlling their competition, chance extinction is likely to be an important aspect of the process.[2] That this is not an important aspect of the problem, at least in some cases, is shown by the continual presence of certain dominant species of planktonic diatoms as microfossils in sediments laid down under fairly uniform conditions over periods of centuries or millenia. This is, for instance, clear from Patrick's (1943) study of the diatoms of Linsley Pond, in which locality *Stephanodiscus astrea*, *Melosira ambigua* and certain species of Cyclotella must have co-occurred commonly for long periods of time. It is always possible to suppose that the persistent species were continually reintroduced from outside whenever they became extinct locally, but this does not seem a reasonable explanation of the observed regularity.

IS THE PHYTOPLANKTON A VALID CONCEPT?

In view of the paradoxical nature of the phytoplankton, perhaps it is justifiable to inquire to what extent the concept itself has validity. In the ocean it is reasonably certain that the community is a self-perpetuating one, but in lakes it has long been regarded as largely an evolutionary derivative of the littoral benthos (for example, Wesenberg-Lund, 1908, pp. 323–325) and in recent years much evidence has accumulated to suggest that the derivation in some cases is not an evolutionary process in the ordinary sense of the word, but a process occurring annually, some individuals of a benthic flora moving at times into plankton. The remarkable work of Lund (1954, 1955) on Melosira indicates that the planktonic species of this genus become benthic, though probably in a non-reproductive condition, when turbulence is inadequate to keep them afloat. Brook (1959) believes that some of the supposed planktonic varieties of littoral-benthic desmids are non-genetic modifications exhibited by populations annually derived from the littoral. If most of the phytoplankton consisted of species with well-defined, if somewhat restricted, benthic littoral niches, from which at times large cultures in the open water were developed but perhaps left no descendants, much of our paradox would disappear. In the sea we should still apparently have to rely on synergism, predation and opportunism or failure to achieve equilibrium, but in fresh waters we might get still more diversity from transitory invasions of species which in the benthos probably occupy a heterogeneously diverse biotope like the soil fauna studied by Hairston (1959).

[2]The chance of extinction is always finite even in the absence of competition, but for the kind of population under consideration the arguments adduced, for instance, by Cole (1960) appear to the writer to be unrealistic. In a lake of area 1 km^2 or 10^6 m^2, in a layer of water only one meter deep, any organism present at a concentration of one individual per litre, which would be almost undetectibly rare to the planktologist using ordinary methods, would have a population N_0 of 10^9 individuals. If the individuals divided and the two fission products had equal chances of death or reproduction, so that in the expected case the population remained stable, the probability of random extinction (Skellam, 1955) is given by $p_e = [t/(1 + t)]^{N_0}$ where t is measured in generations. For large values of N_0 and t we may approximate by $t = -N_0/\ln p_e$. In the lake in question p_e would reach a value of 0.01 in 2.2×10^8 generations which for most phytoplankters would be a period of over a million years. Less than half a dozen lakes are as old as this, and all these are vastly larger than the hypothetical lake of area 1 km^2.

PARADOX OF THE PLANKTON   143

The available data appear to indicate that in a given lake district there is no correlation between the area of a lake and the number of species comprising its phytoplankton. This is apparent from Järnefelt's (1956) monumental study of the lakes of Finland, and also from Ruttner's (1952) fifteen Indonesian lakes. In the latter case, the correlation coefficient of the logarithm of the numbers of phytoplankton species on the logarithm of the area (the appropriate quantities to use in such a case), is -0.019, obviously not significantly different from zero.

It is obvious that something is happening in such cases that is quite different from the phenomena of species distribution of terrestrial animals on small islands, so illuminatingly discussed by Dr. E. O. Wilson in another contribution to this symposium. At first sight the apparent independence indicated in the limnological data also may appear not to be in accord with the position taken in the present contribution. If, however, we may suppose that the influence of the littoral on the species composition decreases as the area of the lake increases, while the diversity of the littoral flora that might appear in the plankton increases as the length of the littoral, and so its chances of diversification, increases, then we might expect much less effect of area than would initially appear reasonable. The lack of an observed relationship is, therefore, not at all inconsistent with the point of view here developed.

CONCLUSION

Apart from providing a few thoughts on what is to me a fascinating, if somewhat specialized subject, my main purpose has been to show how a certain theory, namely, that of competitive exclusion, can be used to examine a situation where its main conclusions seem to be empirically false. Just because the theory is analytically true and in a certain sense tautological, we can trust it in the work of trying to find out what has happened to cause its empirical falsification. It is, of course, possible that some people with greater insight might have seen further into the problem of the plankton without the theory that I have with it, but for the moment I am content that its use has demonstrated possible ways of looking at the problem and, I hope, of presenting that problem to you.

LITERATURE CITED

Andrewartha, H. G., and L. C. Birch, 1954, The distribution and abundance of animals. XV. 782 pp. Univ. of Chicago Press, Chicago, Ill.

Brook, A. H., 1959, The status of desmids in the plankton and the determination of phytoplankton quotients. J. Ecol. 47: 429–445.

Cole, L. C., 1960, Competitive exclusion. Science 132: 348.

Crombie, A. C., 1947, Interspecific competition. J. Animal Ecol. 16: 44–73.

Feller, W., 1939, Die Grundlagen der Volterraschen Theorie des Kampfes von Dasein in Wahrscheinlichkeitstheoretischer Behandlung. Acta Biotheoret. 5: 11–40.

Foged, N., 1954, On the diatom flora of some Funen Lakes. Folia Limnol. Scand. 5: 1–75.

144 THE AMERICAN NATURALIST

Gause, G. F., 1934, The struggle for existence. IX. 163 pp. Williams and Wilkins, Baltimore, Md.

———— 1935, Vérifications expérimentales de la théorie mathematique de la lutte pour la vie. Actual. scient. indust. 277: 1-62.

Gause, G. F., and A. A. Witt, 1935, Behavior of mixed populations and the problem of natural selections. Amer. Nat. 69: 596-609.

Grinnell, J., 1904, The origin and distribution of the chestnut-backed Chickadee. Auk 21: 364-382.

Hairston, N. G., 1959, Species abundance and community organization. Ecology 40: 404-416.

Hardin, G., 1960, The competitive exclusion principle. Science 131: 1292-1298.

Hartman, R. T., 1960, Algae and metabolites of natural waters. *In* The Pymatuning symposia in ecology: the ecology of algae. Pymatuning Laboratory of Field Biology, Univ. Pittsburgh Special Publ. No. 2: 38-55.

Hutchinson, G. E., 1941, Ecological aspects of succession in natural populations. Amer. Nat. 75: 406-418.

———— 1953, The concept of pattern in ecology. Proc. Acad. Nat. Sci. Phila. 105: 1-12.

———— 1957, Concluding remarks. Cold Spring Harbor Symp. Quant. Biol. 22: 415-427.

———— 1959, Il concetto moderno di nicchia ecologica. Mem. Ist. Ital. Idrobiol. 11: 9-22.

Järnefelt, H., 1956, Zur Limnologie einiger Gewässer Finnlands. XVI. Mit besonderer Berücksichtigung des Planktons. Ann. Zool. Soc. "Vancimo" 17(1): 1-201.

Kohn, A. J., 1959, The ecology of *Conus* in Hawaii. Ecol. Monogr. 29: 47-90.

———— 1960, Ecological notes on *Conus* (Mollusca: Gastropoda) in the Trincomalee region of Ceylon. Ann. Mag. Nat. Hist. 13(2): 309-320.

Lack, D., 1954, The natural regulation of animal numbers. VIII. 343 pp. Clarendon Press, Oxford, England.

Lefèvre, M., H. Jakob and M. Nisbet, 1952, Auto- et heteroantagonisme chez les algues d'eau dounce. Ann. Stat. Centr. Hydrobiol. Appl. 4: 5-197.

Lund, J. W. G., 1954, The seasonal cycle of the plankton diatom, *Melosira italica* (Ehr.) Kütz. subsp. *subarctica* O. Müll. J. Ecol. 42:151-179.

———— 1955, Further observations on the seasonal cycle of *Melosira italica* (Ehr.) Kütz. subsp. *subarctica* O. Müll. J. Ecol. 43: 90-102.

MacArthur, R. H., 1957, On the relative abundance of bird species. Proc. Natl. Acad. Sci. 45: 293-295.

———— 1960, On the relative abundance of species. Amer. Nat. 94: 25-36.

Munk, W. H., and G. A. Riley, 1952, Absorption of nutrients by aquatic plants. J. Mar. Research 11: 215-240.

Nicholson, A. J., 1933, The balance of animal populations. J. Animal Ecol. 2(suppl.): 132-178.

———— 1957, The self-adjustment of populations to change. Cold Spring Harbor Symp. Quant. Biol. 22: 153-173.

Patrick, R., 1943, The diatoms of Linsley Pond, Connecticut. Proc. Acad. Nat. Sci. Phila. 95: 53–110.

Provasoli, L., and I. J. Pintner, 1960, Artificial media for fresh-water algae: problems and suggestions. *In* The Pymatuning symposia in ecology: the ecology of algae. Pymatuning Laboratory of Field Biology, Univ. Pittsburgh Special Publ. No. 2: 84–96.

Rand, A. L., 1952, Secondary sexual characters and ecological competition. Fieldiana Zool. 34: 65–70.

Ruttner, F., 1952, Plankton studien der Deutschen Limnologischen Sunda-Expedition. Arch. Hydrobiol. Suppl. 21: 1–274.

Skellam, J. G., 1955, The mathematical approach to population dynamics. *In* The numbers of man and animals, ed. by J. G. Cragg and N. W. Pirie. Pp. 31–46. Oliver and Boyd (for the Institute of Biology), Edinburgh, Scotland.

Udvardy, M. F. D., 1959, Notes on the ecological concepts of habitat, biotope and niche. Ecology 40: 725–728.

Wesenberg-Lund, C., 1908, Plankton investigations of the Danish lakes. General part: the Baltic freshwater plankton, its origin and variation. 389 pp. Gyldendalske Boghandel, Copenhagen, Denmark.

THE INFLUENCE OF THE ENVIRONMENT

By G. Evelyn Hutchinson

DEPARTMENT OF BIOLOGY, YALE UNIVERSITY

The biosphere, or part of the earth within which organisms live, is a region in which temperatures range not far from those at which water is liquid. It receives a radiation flux of wavelength >3200 Å from the sun or has the products of photosynthesis made available by gravity as in the dark depths of the ocean. Numerous interfaces are present in most parts of the biosphere. It is geochemically characterized by atmophil elements in relative quantities such that both oxidized (Eh $\simeq 0.5$ volt) and quite reduced (Eh $\simeq 0.0$ volt) regions are both easily possible often within a few millimeters of each other as in lake sediments. Conditions for such a region on a planet are fairly critical, and would probably always involve loss of the initial gaseous phase with reformation of the atmosphere from frozen or chemically combined material.

Chemically, living organisms are mainly made of cosmically common, light, easily soluble atmophil and lithophil elements. The special properties of the several important elements, such as hydrogen bridge formation, the formation of long carbon chains, the possible existence of —COOH and of —NH$_2$ in the same molecule, the easily oxidized and reduced system —SH HS— $\rightleftharpoons$ —S—S—, the high energy phosphate bond and various other less striking properties exhibited by common biophil elements (of which phosphorus, with an odd atomic number between Si and S, is the rarest) are obviously fundamental. Without these properties life as we know it would be impossible.

The reduction of magnesium concentration in the earth's crust, as compared with the mantle, and the consequent approximate equalization of the amounts of Na, K, Mg, and Ca provide a geochemical rather than a purely chemical example of the "fitness of the environment" to use Henderson's phrase. However, it is difficult to be sure we know we are talking sense in this field without comparative instances, or whether we are involved in problems like the insoluble metaphysical question of childhood, Why am I not someone else? The exploration of the surface of Mars may give, long before the National Academy is celebrating another centenary, some welcome contrasting information. Meanwhile it is reasonable to suppose that extreme rarity and extreme insolubility, leading to a very low concentration of some elements within living tissues, do limit their functional importance.

If we consider an average mammalian liver cell of diameter about 23.4 μ, of volume, assuming a spherical form, of about 6700 μ^3, and of mass, if a density rather more than unity be assumed, of about 7×10^{-9} gm, we can obtain from published analyses[1] a rough idea of the mean number of atoms per cell as follows.

$>10^{14}$	**H, O**
10^{12}–10^{14}	**C, N**
10^{10}–10^{12}	**S,** **P, Na, K, Mg,** Cl, **Ca, Fe,** Si
10^8–10^{10}	**Zn,** Li, Rb, **Cu, Mn,** Al, Fe, Br
10^6–10^8	Sn, Ti, Mo, **Co, I,** Pb, Ag, B, Sr, Ni, V, Sc, Cd, **?Cr, Se**
10^4–10^6	U, Hg, Be ... $\rbrace$ 40 additional reactive natural elements probably in
10^2–10^4	.... $\rbrace$ these rows.
10^0–10^2	Ra

The elements known to have a function in mammals, other than in maintaining the integrity of skeletal structures (as do F, and possibly[2] Sr and Ba), are given in boldface type. It is evident that the probability of an element having a function decreases with decreasing concentration. When such a table was published twenty years ago, there were 9 elements in the 10^6–10^8 atoms per cell category with only cobalt functional; now there are analytic data for 14 and a presumption that chromium and selenium, which with iodine are now known to be functional, fall here. Evidently about a quarter of the elements with 10^6–10^8 atoms per cell may have a function. In the next two rows we might guess probabilities of 0.1–0.01 of function. This may imply one or two surprises. What is interesting is that although cobalt is enriched relative to nickel in liver, as in nearly all tissues of higher animals, over its concentration in the lithosphere, or for that matter in plants, it is still little more abundant than lead and less so than molybdenum. To use an element such as cobalt, the biochemistry of utilization must be reasonably specific. There are plenty of atoms of various kinds around in such concentrations that they could play a part as antimetabolites as well as significant functional roles in enzyme systems. It is possible that this sets the lower limits of concentration at which biochemically significant substances occur. There might be too many commoner accidental and potentially interfering materials around for any very important substance to work practically at 10^4 atoms or molecules per cell. The variety of elementary composition may thus set the standard of purity within which biochemical evolution has occurred.

We may roughly divide most of the biosphere into a purely liquid part, and solid-liquid and solid-gaseous parts, corresponding to (1) the open ocean, (2) the sea bottom, margin, and inland waters, and (3) the colonized land surfaces. In the first, it is possible that iron, which is almost insoluble under oxidizing conditions in inorganic aqueous systems, usually limits the amount of living matter, while in the last, water supply is the most important determinant. In the water-solid systems, including lakes and neritic marine environments, phosphorus, nitrogen, and other elements may be limiting.

The whole plant community can, since it is interconnected through the CO_2 and O_2 of the atmosphere, be regarded as an extremely inefficient (efficiency not much more than 0.1% in most cases in nature) photosynthetic machine. The details of the biochemistry of photosynthesis have been elucidated in recent years in most impressive studies by various investigators. We are, however, still rather ignorant of the quantitative details of the over-all biogeochemical process. It is evident that both the ocean[3] and the plant cover[4] of the continents play a major part in regulating the CO_2 content of the atmosphere, but this regulation is not sufficient to prevent a slow rise[5] due partly, but perhaps not entirely, to the production of the gas by combustion of fossil fuels.

The use of plant material by animals as food is ordinarily a much more efficient process than the photosynthetic capture of the radiation flux from the sun, and allows the existence of a considerable mass of and extraordinary diversity of animals. This diversity is however clearly in part due to the diversity and structural complexity of higher plants. A large part of contemporary ecological research is devoted to elucidation of the general principles that permit the coexistence of very large numbers of species together in a single locality.

The complexity of communities has fascinated naturalists from before Darwin, who described it classically. Only recently has it become apparent what a wealth of quantitative relationships can be seen in the complex structure. Remarkable and quite diverse types of theory, some of which have proved of considerable value in empirical studies, have been developed to deal with this sort of problem, though we still have an enormous amount to learn. While ordinarily the principle of competitive exclusion,[6] which can be phrased in abstract geometrical terms as the statement that two coexisting species do not occupy the same niche, is a good point of departure so long as it is not applied in too naive a manner. The claim may be made that the principle is inapplicable in practice because there would never be a possibility of demonstrating that the niche requirements of two species were, or were not, exactly the same. Actually what is involved is the question as to whether in competition two species could be so nearly equivalent that one would not replace the other in any reasonable time, such as the lifetime of an observer, the period of existence of relevant scientific records, or the period during which the average state of the habitat remained unchanged. The possibility that two competing species might be exactly matched contradicts what has been called the axiom of inequality,[6] that two natural bodies are never exactly the same. The possibility that in very large populations where the dynamics are essentially deterministic, they might be so nearly evenly matched that competition would proceed too slowly to detect, has been seriously suggested for phytoplankton associations, the multispecific nature of which seems otherwise paradoxical.[7] In this, as in other aspects of ecology, the role of time, or of the rates at which things can happen, has been inadequately studied.

In general the speed at which things happen in a very small organism will depend on physical processes of which ordinary diffusion is likely to be the most critical. In larger organisms the celestial mechanics of the solar system, giving days, tidal periods, lunar months, and years, can introduce apparently arbitrary rates into the life of an organism which will interact with rates set by small-scale physical processes. The extent of various kinds of biological clocks is one of the most important phenomena recently discovered in biology. The evidence from odd cases in which periodic processes occur pathologically[8] suggests that the full significance of the gearing of organisms, including ourselves, to the cycle of day and night is not even yet apparent. Moreover a case perhaps could be made for supposing that sometimes clocks can exist which, having evolved in relation to the environment, are no longer set to be synchronous with external periodicities. The human female reproductive cycle, shared by a number of primates, has long suggested a lunar month in spite of the lack of synchrony. Possibly it represents a lunar clock no longer set by the moon; an effect of moonlight on the reproductive cycle of some tropical mammals, including prosimians, has been noted.[9]

Even more extraordinary are the clocks regulating the three species of seventeen-year and the three of thirteen-year cicadas.[10] Here apparently synchrony is adaptive, so that three species bear the brunt of predation, with the different broods emerging so irregularly that there is little chance of a permanent increase in predator population.

Individual cases of a striking kind can, as these, be given at least hypothetical, though very plausible, explanations. We still lack, however, a really clear understanding of the relationship of rates of living and of evolution to the rates of physical change in the universe.

There are two extreme possible ways of evolution in relation to time. Since natural selection will go faster when generations succeed each other faster, one way is the evolution of progressively smaller and more rapidly reproducing organisms. However, the smaller an organism the less it can do. An alternative path gives large, slower reproducing organisms in which, when a nervous system capable of learning is developed, a premium is put on experience. Even in organisms such as plants, which do not learn in the ordinary sense of the word, a perennial can wait about at least metaphorically for a favorable season for reproduction. In the first case the time scale is set by the physical processes of diffusion; in the other extreme case, presumably by some function of the rate at which various things, such as learning, can occur.

In a varying environment, the time taken to learn about a seasonal or otherwise infrequent event will partly depend on the incidence of that event. Some sea birds seem to need several years' experience to learn how to get food for their chick or chicks.[11] The advantage of this learning must be great enough to offset the extra prereproductive mortality, which inevitably accompanies a delay in breeding.

It has recently been suggested that the great intraspecific competition on the limited feeding grounds near nesting sites puts a premium on expertness during the reproductive season and, until this has been acquired, attempts at breeding are wasteful and, to some extent, dangerous. Where mortality is lowest, possibly of the order of 3 per cent per annum in the albatrosses, the period prior to reproduction may be nine years, even though these immense birds have reached maximum size in their first year. In a case like this, learning can obviously only take place during the special period of reproduction, and expertness takes several years to acquire. But it is not clear why human learning of perennial activities should take so long when most of the individual events in our sensory and nervous systems take times measured in milliseconds, rather than months, while on the motor side, a good pianist can play ten notes a second if he has real cause to do so.

In the present and most legitimate excitement over the reading of genetic codes, it is important to remember that lexicography and grammar are not literature, even though the fixing of meaning to symbols and the rules of their ordering make literature possible. The literature of living organisms is very varied, and is perhaps most exciting in the epic or evolutionary forms in which organisms are continually changing in response to selection by a changing environment. Deduction from the possible molecular states of organisms is hardly likely to be an efficient way of exploration; an empirical approach to events is equally needed. A real ecology of time, relating the rates at which things happen in organisms, whether rapid physiological changes or the very slow changes of phylogenesis to the rates of the outside world, is so far only approached at the short-time physiological end. In the immediate future, as argon-potassium dating develops, it will be possible to study evolutionary rates of certain well-known phyletic lines, notably in the Tertiary mammals, with greatly increased precision. Details of variation in evolutionary rates will become accessible and should add enormously to our knowledge of organic change under long time spans. This aspect of biology is likely to be one of immense importance as the Academy moves into its second century of high scientific endeavor.

[1] Hutchinson, G. E., *Quart. Rev. Biol.*, **18**, 331 (1943); with additional data for Sc: Beck, G., *Mikrochemie ver. Mikrochim. Acta*, **34**, 62 (1948); for Be, B, Co, Hg: Forbes, R. M., A. R. Cooper, and H. H. Mitchell, *J. Biol. Chem.*, **209**, 857 (1954); for Si: Gettler, A. O., and C. J. Umberger, *Amer. J. Clin. Pathol.*, Tech. Sect., **9**, 1 (1945); for F: Gushchin, S. K., *Vopr. Pitaniya*, **19**, 71 (1960); for I: Gustun, M. J., *Vopr. Pitaniya*, **18**, 80 (1959); for As: Herman, M. A., T. J. Wiktor, and A. A. van Hee, *Bull. Agr. Congo Belge*, **51**, 403 (1960); for Br: Moruzzi, Giovanni, *Boll. Soc. Ital. Biol. Sper.*, **11**, 725 (1936); and for Cd: Voĭnar, A. O., *Akad. Nauk SSSR, Tr. Konf. Mikroelement*, **1950**, 580 (1952).

[2] Rygh, O., *Bull. Soc. Chim. Biol.* (Paris), **31**, 1052 (1949).

[3] Revelle, R., and H. E. Suess, *Tellus*, **9**, 18 (1957); Bolin, B., and E. Eriksson, in *The Atmosphere and the Sea in Motion* (Rockefeller Institute Press, 1959), p. 130; Eriksson, E., and P. Welander, *Tellus*, **8**, 155 (1956).

[4] Lieth, H., *J. Geophys. Res.*, **68**, 3887 (1963); Bolin, B., and C. D. Keeling, *J. Geophys. Res.*, **68**, 3899 (1963).

[5] Callendar, G. S., *Tellus*, **10**, 243 (1958); Bolin, B., and C. D. Keeling, *loc. cit.*

[6] Hardin, G., *Science*, **131**, 1292 (1960).

[7] Riley, G. A., in *Marine Biology 1*, Am. Inst. Biol. Sci. (1963), p. 70; Hutchinson, G. E., *Am. Naturalist*, **95**, 137 (1961).

[8] Richter, Curt P., these Proceedings, **46**, 1506 (1960).

[9] Cowgill, U. M., A. Bishop, R. J. Andrew, and G. E. Hutchinson, these Proceedings, **48**, 238 (1962); Harrison, J. L., *Bull. Raffles Mus.*, **24**, 109 (1952).

[10] Alexander, R. D., and T. E. Moore, *Misc. Publ. Mus. Zool., Univ. Mich.*, 1 21, 1 (1962); Dybas, H. S., and M. Lloyd, *Ecology*, **43**, 444 (1962).

[11] Ashmole, N. P., *Ibis*, **103b**, 458 (1963).

THE EVOLUTION OF LIVING SYSTEMS

By Ernst Mayr

MUSEUM OF COMPARATIVE ZOOLOGY, HARVARD UNIVERSITY

The number, kind, and diversity of living systems is overwhelmingly great, and each system, in its particular way, is unique. In the short time available to me, it would be quite futile to try to describe the evolution of viruses and fungi, whales and sequoias, or elephants and hummingbirds. Perhaps we can arrive at valid generalizations by approaching the process in a rather unorthodox way. Living systems evolve in order to meet the challenge of the environment. We can ask, therefore, what *are* the particular demands that organisms have to meet? The speakers preceding me have already focused attention on some of these demands.

The first challenge is to cope with a continuously changing and immensely diversified environment, the resources of which, however, are not inexhaustible. Mutation, the production of genetic variation, is the recognized means of coping with the diversity of the environment in space and time. Let us go back to the beginning of life. A primeval organism in need of a particular complex molecule in the primordial "soup" in which he lived, gained a special advantage by mutating in such a way that, after having exhausted this resource in his environment, he was able to synthesize the needed molecule from simpler molecules that were abundantly available. Simple organisms such as bacteria or viruses, with a new generation every 10 or 20 minutes and with enormous populations consisting of millions and billions of

Thoughts on Aquatic Insects

G. Evelyn Hutchinson

Aquatic adults seldom have terrestrial juveniles. Respiratory problems may lead to tropical species being smaller than related temperate ones. Suctorial feeding on higher plants seems unknown. Cryptic coloration is common, aposematic, and epigamic rare. Flightlessness is common, but potentially flying forms must be generally available; this restricts tracheal gills to juveniles. *(Accepted for publication 12 January 1981)*

KINDS OF LIFE HISTORY

Among the insects of inland waters there are certain peculiar situations which, though often obvious, at first seem hard to explain.

We may group the animals under consideration according to whether they are terrestrial or aquatic in active immature and predominantly trophic stages or in the mature and primarily reproductive climax of their lives. The following matrix shows the orders of magnitude of the four possible classes:

	Adult terrestrial	Adult aquatic
Immature terrestrial	n. 10^5	n. 10^2
Immature aquatic	n. 10^4	n. 10^3

This reveals the fact that it is easier to be immature in the water than to be mature. The category of the aquatic mature stage that has grown up terrestrially is curious-

Hutchinson is Sterling Professor of Zoology (emeritus) and senior research biologist, Department of Biology, Yale University, New Haven, CT 06511. This paper is derived from a lecture given on 12 September 1980 to Ruth Patrick's course on the biology of streams, School of Forestry and Environmental Studies, Yale University. He is greatly indebted to her and to the school for giving him the opportunity to present this material, which is based on a far more extended treatment (with a much fuller set of references) to appear in the fourth volume, on zoobenthos, of his *Treatise on Limnology*. © 1981 American Institute of Biological Sciences. All rights reserved.

ly rare, being exemplified only by certain beetles of the families Hydraenidae and Dryopidae (Hinton 1955). Moreover, though the aquatic bugs and beetles together include almost all the members of the adult aquatic and immature aquatic categories, the two orders differ in their relation to water: The endopterygote Coleoptera have a pupa that is nearly always formally terrestrial. It lies in a dry cocoon in a cavity excavated above water-level in the bank of the water; the larva lived in it, and all being well, the adult may return to it.

RESPIRATION IN AQUATIC HABITATS

For a member of a terrestrial group entering freshwater, respiration probably poses a greater problem than does any other vital function. Small aquatic insects (and other animals less than about 1 mm in diameter) are apt to breathe by diffusion over the entire thin, but largely unmodified, body wall. There may be specialized hemolymph sinuses below this wall. When the insect is a relatively long and thin cylinder, one may regard the whole body as a gill.

Many variations on this general theme are found in the Diptera, which commonly have blood gills or special diverticula of the body wall containing sinuses. In many cases, however, these have now been shown to be organs taking up chloride ions rather than oxygen. In some chironomids, the hemolymph contains hemoglobin. Comparable respiration involving the whole surface is common in

the first one or two nymphal instars of many aquatic insects such as the mayflies, the Hemiptera, and the aquatic moth *Acentropus*, though later these develop other respiratory mechanisms. The unique pair of genera, *Idiocoris* and *Paskia*, the smallest of all known aquatic Hemiptera, are apparently the only free-living adult apneustic insects; they are limited to the stony littoral of Lake Tanganyika (Esaki and China 1927).

The large aquatic insects breathe either by tracheal gills, outgrowths of the body wall that are richly supplied with trachea, or spiracles that may be fed from bubbles and replaced periodically by visits to the surface or to photosynthesizing plants. Such bubbles or air-stores are held in various ways. Immature insects retain air-stores usually in bunches of hair; in mature insects, the elytra of beetles and the equivalent hemelytra of bugs cover a large dorsal air-store. There may also be very fine hairs carrying air on the surfaces of various parts of the body, which give the insect its "silvery-light" appearance, as Mary Ball (Ball 1846) called it 140 years ago. From the air-stores, air is usually passed into the tracheal system much as in terrestrial insects.

Since the air in the air-store is in contact with water, it can act as a "physical gill"—a universal, though rather odd, appellation (Ege 1915, Popham 1964, Thorpe 1950). As respiration takes place, oxygen is removed; the equivalent carbon dioxide produced is much more soluble than the oxygen and quickly passes into the water, so that the bubble constituting the air-store contracts. The partial pressure of nitrogen in the bubble therefore increases, so nitrogen diffuses out. Since oxygen is lost by respiration, the partial pressure of that gas decreases, and oxygen diffuses into the bubble, at least from saturated water.

Ideally, the process can continue until all the nitrogen has been lost; but in actual cases, as the bubble becomes small, the oxygen content becomes dangerously low, and the air-store must be replaced at the surface, well before it disappears. The effectiveness of the process depends on the oxygen concentration in the water, which is related inversely to the temperature. It also depends on the exposed surface of the bubble, which, in a large animal, is smaller relative to the mass of metabolizing tissue than in a small one. Thus, the physical gill works best for small animals at low temperatures, but the effectiveness varies greatly with the morphology and physiology of the organism under study. Under favorable circumstances, the mechanism of the physical gill can certainly increase severalfold the period between surfacing, thereby reducing the risk of predation.

High temperatures not only decrease the physical gill's effectiveness, but also increase the metabolic demand for oxygen. Very little, unfortunately, is known about respiration in tropical aquatic insects, though Miller (1964) has clear evidence that the physical gill can work for *Enithares sobria* Stål, a backswimmer widespread in Central Africa. In the Corixidae, the larger Corixinae are predominantly temperate with a maximum number of species in the United States, whereas the smaller Micronectinae are largely tropical with the richest local fauna in Sri Lanka. In tropical uplands, such as the Nilgiri Hills in India and the mountains of Ethiopia, Micronectinae, as in temperate regions, tend to be rarer than Corixinae. These features of the distributions of the Corixidae are, thus, in line with small species being better adapted to warm climates, perhaps because they have greater ease in breathing.

Two remarkable respiratory specializations are known in particular groups of submerged insects. A few species of naucorid bugs and elminthid beetles have evolved an extreme development to the physical gill. A large area of the body is covered with a pile of very fine short hydrophobe hairs. In the Palaearctic naucorid *Aphelocheirus*, there may be 2.5 million such hairs (about 6μ long) per mm^2 (Thorpe 1950). At ordinary subaqueous pressures, surface forces prevent water from penetrating the pile, which remains the support for a thin dry gaseous layer, acting as a physical gill, from which specialized spiracles can receive air.

The insects that have perfected this respiratory plastron live in habitats where continual disturbance, or the photosynthetic activity of calcicole algae, keeps the water saturated with oxygen. Though in a perfectly developed case, such as that of *Aphelocheirus*, the plastron obviates the necessity for surfacing, it is evidently not a mechanism that can be effective in a wide variety of habitats.

A second specialization of the airstore is found in the lesser backswimmers of the subfamily Anisopinae in the family Notonectidae: the genera *Anisops* living in the warm temperate and tropical Old World and *Buenoa* in the New. These insects have an abdominal hemoglobin organ, richly supplied with tracheae. The hemoglobin in the cytoplasm of the cells of the organ is charged with oxygen at surfacing and slowly liberates the gas as it is used in respiration. The insect thus can adjust its density, remaining poised at the same depth for some minutes (Miller 1964).

Most adult insects living on the bottom or among weeds, if they use an air-store for respiration, will have to make frequent visits to the surface, except perhaps when dormant in winter. Such expeditions are energetically wasteful and limit the depths that can be colonized. Moreover, surfacing is bound to be a conspicuous act, increasing the dangers of predation. In spite of these disadvantages, no adult aquatic insect respires by tracheal gills, though such structures are obviously highly successful in the physiology of mayfly and stonefly nymphs and in more elaborate ways in those of dragonflies and damsel flies, in the larvae of alderflies and caddisflies, and in those of some beetles and two-winged flies.

A TREND TO SELECTIVE OMNIVORY

The feeding habits of aquatic insects are much more varied than their modes of respiration. Usable energy and the elements required in nutrition, other than hydrogen and oxygen, are obtained from many sources in a variety of ways.

The earliest putative aquatic insects, the Protodonata, were presumably carnivores, as are their immediate successors, the Odonata. The largest Protodonata, giants with a wingspan of up to 75 cm, must have had proportionately large nymphs of great repacity, which probably took a number of years to reach maturity. In their late instars, they may well have eaten fishes.

The oldest mayfly nymph of which the mouth parts are known in detail is *Kuka-*

lova americana Desmoulins from the Lower Permian of Oklahoma. Because the mandibles were very large and bore teeth, Kukalová (1968) initially regarded the insect as a carnivore. Later, Hubbard and Kukalová-Peck (1979) compared the mouth parts with some modern generalist feeders. Though the evidence is still inadequate, it seems that the mayflies may exemplify a trend detectible in other freshwater invertebrates that have evolved a kind of selective omnivory. For such animals, the haptobenthic algae and bacteria with the slime secreted by them, together with adherent detritus and associated small animals, constitute a universal pabulum covering all surfaces, at least in the shallower parts of temperate freshwaters (except perhaps the most oligotrophic). A change from carnivory to partial reliance on such a source of food, together with any other plant or animal matter that may be available, seems to characterize many of the small Corixidae and some species of Haliplidae.

Development of a like kind of omnivory is known in the Lymnaeidae among the watersnails. In all cases, some degree of selectivity is retained, though the range over which it is exercised is greatly extended. Some animals eat most algae other than *Cladophora*, or most angiosperms other than *Elodea*. These two plants are apparently largely inedible (and so most successful). The former is apparently protected by lauric, myristic, and palmitoleic acids (Lalonde et al. 1979, Larson 1981), which render the plant poisonous, or at least inhibitory in various ways to mosquito larvae, gastropod mollusks, and crayfish.[1] Unfortunately, except for *Nuphar* and *Nelumbo* (which are full of strange alkaloids), little is known of secondary compounds of a defensive kind in freshwater vascular plants. The existing data (or lack of it) suggest that such compounds are of less significance than on land, but this well may be an illusion based on ignorance. Even the most inedible plant will ultimately add to the food reserves, producing, when it dies, detritus and nutrients for algae and bacteria.

Generally, almost any moving food may be pursued by predatory aquatic insects, though the size and ease of handling make a great difference to what is actually taken. There are very few cases of food being rejected: *Notonecta undu-*

[1]Ruth Patrick, School of Forestry and Environmental Studies, Yale University, personal communication, November 1980.

lata Say refused a pink phyllopod of uncertain identity; *N. uhleri* Kirkaldy will not eat *Buenoa;* and a few of the Belostomatidae are specialized as predators on gastropod snails.

Most waterbugs are still suctorial carnivores; this is probably true of many corixids, even though others feed on haptobenthos. Indeed, fully aquatic insects have not developed the habit of sucking the fluid contents from vascular waterplants, though corixids and some larval haliplids may suck out all the contents of the cells of filamentous algae. There are no freshwater analogues to the vast array of Myridae in the Heteroptera, nor to the whole of the Homoptera. In the latter suborder, a few aphids are found on the aerial parts of emergent waterplants, but not under water. The absence of aquatic insects that suck higher plants may be of great antiquity. The Paleoptera, the more primitive group of flying insects, seems to have consisted in the late Palaeozoic of mandibulate orders with aquatic nymphs, which survive as mayflies and dragonflies, and extinct terrestrial haustorial orders (Palaeodictyoptera, Megasecoptera, and Diaphanopterodea), the sucking mouthparts of which are apparently derived from mayfly-like trophi (Hubbard and Kukalová-Peck 1979, Kukalová-Peck 1978).

CRYPTIC COLOR AND CHEMICAL DEFENSES

Aquatic insects are, of course, themselves subject to predation. Macan (1966) particularly has shown how important the littoral vegetation can be in determining if such insects can coexist with fish. The matter is clearly very complicated, depending largely on the species of predator and prey as well as the nature of the protective environment.

Striking reactions to predation have evolved. Popham (1941) found that most corixids tend to match the albedo of their background. The process is ontogenetic, depending on the background to which the insect is exposed three days before a moult. The corixids are so imprinted on the background that they become restless when placed on one that they do not match, even though they probably have no way—because of the geometry of their eyes and dorsal surfaces—of directly comparing their own coloration with that of the background on which they are perching. If, after a migratory flight, they enter a body of water with an inappropri-

ately colored bottom, they leave. Both in the laboratory and in nature, the matching can be shown to be effective against predation (Popham 1942, 1944).

A similar tendency to match a light or dark background occurs in some waterbeetles, notably the Haliplidae.[2] However, in holometabolous insects, which may have different larval and adult habitats, the variation in the adults is probably determined genetically rather than environmentally.

Many beetles secrete toxic secondary compounds in their pygidial or thoracic glands (Miller and Mumma 1976, Schildknecht 1970). In the Dytiscidae, at least in the subfamilies containing the larger species (Dytiscinae and Colymbetinae), the thoracic glands produce an astonishing array of steroids related to or identical with mammalian sex hormones. These are repellent, anaesthetic, or, in sufficient dosage, lethal to fish. In the Gyrinidae, norsesquiterpenes, produced in the pygidial glands, act similarly. The other families of waterbeetles have not been adequately studied, and the waterbugs, as well as the beetles, produce odors that may be repellent, but relatively little work has been done on their chemistry.

BREEDING, ASYMMETRY, AND SONG

There are certain very peculiar specializations related to reproduction. For the most part, they do not yet fit in with the other aspects of the life of insects underwater to make a unified story. Hungerford (1919) indicates that the male of *Notonecta irrorata* Uhler habitually takes up a position to the left of the female in mating, even though his exoskeleton and genitalia are not asymmetrical. In some naucorids, notably the common European *Ilyocoris cimicoides* (Linnaeus), the muscles of the male abdomen are asymmetrical, even though the exoskeleton is not. In some tropical members of the family (*Macrocoris, Neomacrocoris*), the seventh tergite of the male has asymmetrical lobes on its posterior margin, but their function is unknown. The Anisopinae have asymmetrical parameres or claspers, though the rest of the abdominal bodywall appears symmetrical. In the Corixidae, the whole of the abdominal exoskeleton from the fifth tergite backwards is extremely asymmetrical.

In most Corixidae, the male bears on the right side of the sixth abdominal tergite a condensed group of spines, usually forming several regular rows, called the strigil. In the very primitive Australian *Diaprepocoris,* the strigil is double, as if the left member of a pair of lobes had moved across the body to engage with the right, perhaps making a sound-producing organ. In the small and largely tropical Micronectinae, the strigil is developed as a row of short, wide spines on the right of the sixth abdominal tergite of the male. It is certainly a sound-producing organ, plucked by a plectrum formed from the right posterior margin of the fifth tergite (King 1976).

In the Corixinae, a well-developed strigil is present in most species. Mary Ball (Ball 1846), who first heard Corixidae stridulate in 1840, recorded abdominal movements, as in *Micronecta*, accompanying one type of sound made by *Sigara dorsalis* (Leach). No one else has seen any part other than the anterior legs being moved in relation to sound production in the Corixinae. In this subfamily, the strigil is probably used to increase the grip of the male on the female during coitus, though in many cases the structure looks like a sound-producing organ. In the Corixinae, the known stridulatory structures are a patch of short, pointed spines forming a *pars stridens* on the anterior femur of the males of many and the females of a few species, and a *plectrum*, a ridge on the edge of the head on which the *pars stridens* is rubbed (Jansson 1972). The sounds are mainly used in courtship, but they also may be employed to achieve a certain degree of spacing between males in a sort of territorial arrangement. Very remarkable differences in behavior on the part of closely allied species have been recorded by Jansson (1979); the comportment of the male of *Arctocorisa germari* (Fieber), which always waits for a reply from the female he is courting, contrasts with the activity of *A. carinata* (Sahlberg), which often engages in rape (Jansson 1979).

Though in most genera the strigil lies on the right side of the sixth abdominal tergite, throughout the family occasional sinistral specimens are recorded in a number of species. A few are regularly dimorphic, and others (*Corixa, Trichocorixa, Heliocorisa*) practically always have the strigil on the left. When this inversion occurs, it affects only the muscles and exoskeleton of the abdomen; other asymmetries remain unchanged. In one of the dimorphic species, *Ahuautlea mexicana* de la Llave, the homozygous

---

[2]E. J. Pearce, personal communication.

sinistral is lethal (Peters 1960). (There must be intense selection of heterozygotes bearing the sinistral gene.)

Very elaborate sound-producing mechanisms used in courtship are also found in the small backswimmers of the subfamily Anisopinae: *Anisops* in the Old World, *Buenoa* in the New. The few of the 175 species that have been studied differ markedly in the details of their behavior. The Anisopinae use a tibial comb on the anterior leg against a pair of prong projections for the beak as their main vocal organs; as well as sound, a characteristic pattern of movement varying from species to species is involved in courtship (Goertz 1963, Wilcox 1975).

Sound production is also recorded but hardly studied in a few other waterbugs and waterbeetles. Where it is used in an elaborate way in courtship, as in the Corixidae and Anisopinae, it seems characteristic of rather gregarious species ordinarily exhibiting little intraspecific aggression, unlike the potentially cannibalistic and very well-spaced Notonectini. This needs further study, as does the possibility that underwater auditory stimuli are better epigamic signals than are visual ones. Of course, they may be dangerous if there are predators around with receptor organs for sound. As in fishes (Popper and Coombes 1980), pitch plays little or no part in communication, which is primarily based on the length and spacing of trains of pulses.

FLIGHT AND FLIGHTLESSNESS

Finally, there is a systematic difference between the flightless forms of terrestrial bugs, usually apterous with vestigial elytra, and the flightless waterbugs and waterbeetles, in which the hemelytra or elytra are almost always large enough to cover a significant dorsal air-store. Only when there is a respiratory plastron, as in *Aphelocheirus*, does a condition develop comparable to what is found on land.

Very often in those Corixinae in which neither hemelytra nor wings are reduced, the insect is flightless because the indirect flight muscles that normally fill the thorax fail to develop at the last moult (Young 1965). This failure is usually temperature-dependent; only over about 15° C does the adult insect emerge as a flying form (Scudder 1976, Scudder and Meredith 1972). In the Micronectinae of temperate countries, nearly all specimens are flightless, about 2% having fully developed wings. The nature of the determination is unknown, though the

existence of the rare flying forms is important in dispersal (Wróblewski 1958). In *Micronecta* in tropical lands, many species are always—and most are frequently—winged, suggesting that in areas in which shallow water frequently dries up, flight is an essential function.

Flightless Corixidae with undeveloped muscles live longer and are more fertile under suboptimal, but not under optimal conditions, than fully developed specimens. When conditions are poor, the economy of not having to support the basal metabolism of a thorax full of flight musculature evidently better adapts the flightless forms, except that they are unable to fly away (Young 1965).

Less is known about the pterygodimorphism of beetles (Jackson 1956) than of bugs, but there is some evidence that the condition is genetic (Angus 1970).

THE DANGERS OF BEING TOO AQUATIC

In transient environments, which nearly all bodies of freshwater ultimately are, some opportunity for dispersal is necessary for the continued existence of a species. The nearer that this opportunity is to the reproductive phase in the life history, the more effective it will be. For any generation, the population will be greatest when present as newly fertilized eggs before any mortality has occurred and with the maximal chance of at least a few individuals being delivered into some unusually favorable environment. Since, to a flying insect, the period of flight is also the period of maximum potential dispersal, aeroterrestrial life would be expected at the time of reproduction. This would explain the rarity of species with terrestrial larvae and aquatic adults.

The absence of any adult insects with tracheal gills at first seems peculiar because any kind of respiration involving surfacing is always wasteful and necessitates the use of energy in moving to or from feeding grounds to the surface. The range of depths that can be occupied is clearly limited; nearly all waterbugs and waterbeetles never descend below a few meters. Moreover, surfacing is a dangerous activity not only to insects in torrential waters for mechanical reasons, as Parsons (1974) has pointed out, but also because, by making insects move conspicuously, it makes them more available to predators.

An enormous number of immature insects of all sorts of shapes, sizes, and

textures have tracheal gills. Why, when such organs would seem advantageous, do no adults? Perhaps, if an insect is to have gills, it must either give up flight or develop some mechanism for keeping the gills moist on land and in the air. Land crabs can do this, walking even on dry coral islands, as the branchial cavity provides a way to keep the gills moist. But there is no obvious preadaptation that would permit the development of such a shield in either order containing adult aquatic insects. Though many species of aquatic insects do not fly, very little genetic or environmental change would probably be needed to convert almost any actual nonflying species into one that can fly. Whatever may cause the appearance of fully developed macropterous forms in up to about 2% of the population of the Palaearctic species of *Micronecta*, the chance of any individual being able to fly is small. But the chance of there being some individuals, in a drying habitat, that could fly away to a more permanent body of water is great. Wróblewski (1958) has given actual examples that indicate that such behavior would be quite possible in nature. In an extreme case of genetically determined wing development, back-mutation might be successfully balanced against the selection in favor of a flightless form. This would permit survival of the species so long as aerial respiration remained possible. If tracheal gills had been evolved, the species might be completely unable to look for new habitats as the water dried up. Thus, the only adult aquatic insects without open spiracles are the minute *Idiocoris* and *Paskia*, which live in one of the two lakes that are least likely to have dried out completely for many millions of years.

Throughout the Mesozoic, the rise of the angiosperms must have permitted the development of a modern marginal flora, which provides protection against excessive predation by fish and a place where insect predators may lurk unseen while waiting for the appearance of their own prey. The early immature aquatic insects seem to have had gill-plates, producing a simple current used in dermal respiration. Kukalová-Peck (1978), reviewing the many theories put forward as to the origin of the insect wing, makes a clear case for regarding wings and the gill-plates of Ephemeroptera as serially homologous, e.g., in an organism such as the Permian mayfly nymph, *Kukalova americana* (Kukalová 1968). She is, however, wary about attributing an aquatic habitat to the ancestor of all the

Pterygota. The earliest such ancestor presumably continued to moult throughout life without any sharp break in form at a particular metamorphic moult. There are several hypotheses as to the value of a series of lobes arising from the first thoracic segment nearly to the end of the abdomen, some not involving an aquatic stage. However, the development of a clear-cut metamorphic moult at which the lobes on the second and third thoracic segments become functional wings is consistent with a change of environment from water to air, as in the penultimate moult of a mayfly. So there may be an evolutionary connection here.

From the serial homology so obvious in the *Kukalova* nymph, we may reasonably expect at least the first aquatic insects to have had gill-plates, if not tracheal gills. As the phanerogamic littoral vegetation developed, underwater bugs or Nepomorpha (Popov 1971)—and doubtless some of the families of waterbeetles—evolved. A second kind of aquatic respiration with an airstore permitted adults as well as nymphs or larvae to live underwater, and such insects were clearly prone to occupy some of the new niches offered by a rich marginal zone of angiosperms. The extensive haptobenthos associated with these plants would be a convenient source of food, perhaps better than the plants themselves. The late appearance of the diatoms suggests that the haptobenthos, which must have existed as a layer of colorless and blue-green bacteria far into the Precambrian, underwent considerable further evolution throughout the Palaeozoic. The development of a community capable of supporting a fairly rich microfauna would have added to the nutrient value of the so-called universal freshwater pabulum.

THE DINGINESS OF FRESHWATER INVERTEBRATES

One further aspect of this story seems ripe for investigation. Freshwater animals are notoriously dingy in color, as E. M. da Costa (1776), noticed more than 200 years ago. One of the brightest colors found in limnetic animals is the brilliant red of some planktonic copepods, which Hairston (1976) has shown has a purely physiological function, providing protection against excessive short-wave solar radiation. Among larger invertebrates, a crimson polyclad in Borneo and a green opisthobranch on

some islands in the western Pacific are the most brilliant, but they are of relatively recent marine origin. Some leeches and some mayfly nymphs have charming color patterns, but they are relatively subdued. Among mollusks the most decorative are certain Neritidae, notably in the genus *Theodoxus,* which, like their marine allies, have porphyrin pigments in the shells (Comfort 1950).

The only reasonably clear cases of aposematic coloring are provided by the scarlet water mites, which can be unpalatable (Elton 1922, Popham 1948).[3] Ellis and Borden (1970) found that blood worms were the least preferred kind of food in a considerable selection offered to *Notonecta undulata,* suggesting that chironomid larvae, having acquired hemoglobin for respiration, have evolved some impalatability as an appropriate counterpoise to their increased conspicuousness when discovered by a predator nosing about on a shallow bottom.

The very striking series of poisonous secondary compounds in the Dytiscidae and Gyrinidae go unadvertised, doubtless because the advantages of warning coloration are more than offset by the disadvantage a predator has in being conspicuous. However, at the surface a scintillating school of whirlgig beetles may be conspicuous enough to discourage experienced predators such as birds.

On land a large number of distasteful species owe their unpalatability to secondary compounds derived from their food plants. Such herbivorous species can safely develop warning coloration. In freshwater the apparent rarity of plants containing such compounds would probably rule out this particular strategy. It would be interesting to know whether the species of *Donacia* feeding on Nympheaceae sequester alkaloids from their food plants, as the beaver feeding on the same plants appears to do.

There may be further reasons for the drabness of most freshwater animals. In quiet water, eddy diffusivity is likely to be lower than in air, so that a repulsive odor may be as effective a localized and permanent warning as a bright color, except perhaps for very small mobile animals, such as watermites. The work of Popham (1948) suggests that poor color vision and poor learning capacity may also play a part. Most aposematic coloration on land is directed against birds with superior color vision and adequate learning capacity.

---

[3]C. Kerfoot, personal communication.

Except among fishes, there is no certain development of epigamic coloration in freshwater animals. The backswimmers of the western American subgenus *Erythronecta* show some sexual dimorphism in the color of their hemelytra, the black and red females being the most brilliant of all aquatic insects. Though there is evidence that in *Notonecta undulata* males find their mates by sight (Clark 1928) and that part of the compound eye of *N. glauca* Linnaeus is sensitive to differences in color (Rokohl 1942), it is by no means certain that red and black hemelytra, seen from below against the sky and obscured by a silvery air-store, would provide a very powerful sexual stimulus. Except in the largest and clearest lakes, in which most of the more spectacular freshwater fishes having epigamic colors are found, the water may have always had just sufficient brown organic stain to dull the kind of brilliance associated with mating behavior on land.

None of these explanations is probably adequate by itself to account for the chromatic characteristics of the freshwater fauna. Taken together, they doubtless provide, in some cases, a partial explanation. Certainly, our knowledge of aquatic insects is still not great enough to enable us to think about them satisfactorily.

ACKNOWLEDGMENTS

I wish to acknowledge the help of three friends: Ruth Patrick has given me numerous and valuable pieces of information; Charles L. Remington has put his enormous store of entomological learning at my disposal; and Nancy Knowlton has been a lively ecological critic, saving me from a serious error in my discussion of underwater coloration.

REFERENCES CITED

Angus, R. B. 1970. Genetic experiments on *Helophorus* F. (Coleoptera, Hydrophilidae). *Trans. R. Entomol. Soc. Lond.* 122: 257–276.

Ball, R. 1846. [All but introductory paragraph is by Mary Ball, though she is not mentioned by name] *Corixa striata,* Curtis. *Ann. Mag. Nat. Hist.* 17: 135–136.

Clark, L. B. 1928. Seasonal distribution and life history of *Notonecta undulata* in the Winnipeg region, Canada. *Ecology* 9: 383–403.

Comfort, A. 1950. Biochemistry of molluscan shell pigments. *Proc. Malacol. Soc. Lond.* 27: 79–85.

da Costa, E. M. 1776. *Elements of Conchology.* B. White, London.

Ege, R. 1915. On the respiratory function of the air store carried by some aquatic insects. *Zeit. Allg. Physiol.* 17: 81–124.

Ellis, R. A., and J. H. Borden. 1970. Predation by *Notonecta undulata* (Heteroptera: Notonectidae) on larvae of the yellow fever mosquito. *Ann. Entomol. Soc. Am.* 63: 963–973.

Elton, C. 1922. On the colours of water-mites. *Proc. Zool. Soc. Lond.* 1922(2): 1231–1239.

Esaki, T., and W. E. China. 1927. A new family of aquatic Hemiptera. *Trans. Entomol. Soc. Lond.* 1927(II): 279–295.

Goertz, M. E. 1963. Sound production in two sympatric species of aquatic Hemiptera *Buenoa macrotibialis* Hungerford and *Buenoa confusa.* Unpublished M.A. thesis, Dartmouth College, Hanover, NH.

Hairston, N. G. 1976. Photoprotection by carotenoid pigments in the copepod *Diaptomus nevadensis. Proc. Natl. Acad. Sci. USA* 73: 971–974.

Hinton, H. E. 1955. On the respiratory adaptations, biology, and taxonomy of the Psephenidae with notes on some related families (Coleoptera). *Proc. Zool. Soc. Lon.* 125: 543–568.

Hubbard, M. D., and J. Kukalová-Peck. 1979. Permian mayfly nymphs: new taxa and systematic characters. Pages 19–31 in J. F. Flannagen and K. E. Marshall, eds. *Advances in Ephemeroptera Biology.* Plenum Press, New York.

Hungerford, H. B. 1919. Biology and ecology of aquatic and semiaquatic Hemiptera. *Kansas Univ. Sci. Bull.* 11 (21): 1–256.

Jackson, D. J. 1956. The capacity for flight of certain waterbeetles and its bearing on their origin in the western Scottish Isles. *Proc. Linn. Soc. Lond.* 167: 76–96.

Jansson, A. 1972. Mechanisms of sound production and morphology of the stridulatory apparatus in the genus *Cenocorixa* (Hemiptera, Corixidae). *Ann. Zool. Fennici* 9: 120–129.

———. 1979. Reproductive isolation and experimental hybridization between *Arctocorisa carinata* and *A. germani* (Heteroptera, Corixidae). *Ann. Zool. Fennici* 16: 89–104.

King, J. M. 1976. Underwater sound production in *Micronecta batilla* Hale (Hemiptera: Corixidae). *J. Austral. Entomol. Soc.* 15: 35–43.

Kukalová, J. 1968. Permian mayfly nymphs. *Psyche* 75: 310–327.

Kukalová-Peck, J. 1978. Origin and evolution of insect wings and their relation to metamorphosis, as documented by the fossil record. *J. Morphol.* 156: 53–126.

Lalonde, R. T., C. D. Morris, C. F. Wong, L. C. Gardiner, D. J. Eckert, D. B. King, and R. H. Zimmerman. 1979. Response of *Aedes seriatus* larvae to fatty acids of *Cladophora. J. Chem. Ecol.* 5: 371–381.

Larson, R. 1981. Potential for biological controls of *Cladophora glomerata.* In a report of the Environmental Protection Agency. EPA, Washington, DC, in press.

Macan, T. T. 1966. The influence of predation on the fauna of a moorland fish pond. *Arch. Hydrobiol.* 61: 432–452.

Miller, J. R., and R. O. Mumma. 1976. Physiological activity of water beetle defensive agents. I. Toxicity and anaesthetic activity of steroids and norsesquitespenes administered in solution to the minnow *Pimephales promelas* Raf. *J. Chem. Ecol.* 21: 115–130.

Miller, P. L. 1964. The possible role of haemoglobin in *Anisops* and *Buenoa* (Hemiptera: Notonectidae). *Proc. R. Entomol. Soc. Lond.* 39 A: 166–175.

Parsons, M. C. 1974. Anterior displacement of metathoracic spiracle and lateral intersegmental boundary in the pterothorax of Hydrocorisa (Aquatic Hemiptera). *Zeit. Morph. Tiere* 79: 165–198.

Peters, W. 1960. Inheritance of asymmetry in a water-boatman (*Krizousacorixa femorata*). *Nature (Lond.)* 186: 737.

Popham, E. J. 1941. The variation in the colour of certain species of *Arctocorisa* (Hemiptera, Corixidae) and its significance. *Proc. Zool. Soc. Lond. Ser. A* 111: 135–172.

———. 1942. Further experimental studies of the selective action of predators. *Proc. Zool. Soc. Lond. Ser. A* 111: 105–117.

———. 1944. A study of the changes in an aquatic insect population, using minnows as predators. *Proc. Zool. Soc. Lond.* 114: 74–81.

———. 1948. Experimental studies of the biological significance of noncryptic pigmentation with special reference to insects. *Proc. Zool. Soc. Lond.* 117: 768–783.

———. 1964. The migration of aquatic bugs with special reference to the Corixidae (Hemiptera: Heteroptera). *Arch. Hydrobiol.* 60: 450–496.

Popov, I. A. 1971. Istoricheskoe razvitie poluzhestkokrylykh infraotriada Nepomorpha. Akad. Nauk. USSR., *Trudy Palaeont. Inst.* 120, (in Russian).

Popper, A. N., and S. Coombs, 1980. Auditory mechanisms in teleost fishes. *Am. Sci.* 68: 429–440.

Rokohl, R. 1942. Über die regionale Verschiedenhiet der Farbentüchtigkeit in zusammengesetzen Auge von *Notonecta glauca. Zeit. Vergl. Physiol.* 29: 638–676.

Schildknecht, H. 1970. The defensive chemistry of land and water beetles. *Angew. Chem.* (Int. Ed.) 9: 1–19.

Scudder, G. G. E. 1976. Field studies of the flight muscle polymorphism in *Cenocorixa* (Hemiptera, Corixidae). *Verh. Int. Verem. Limnol.* 19: 3064–3072.

Scudder, G. G. E., and J. Meredith. 1972. Temperature induced development in the indirect flight muscles of adult *Cenocorixa. Dev. Biol.* 29: 330–336.

Thorpe, W. H. 1950. Plastron respiration in aquatic insects. *Biol. Rev.* 25: 334–390.

Wróblewski, A. 1958. The Polish species of the genus *Micronecta* Kirk. (Heteroptera, Corixidae). *Ann. Zool.* (Warsaw) 17: 247–381.

Young, E. C. 1965. Flight muscle polymorphism in British Corixidae: ecological observations. *J. Anim. Ecol.* 34: 353–390.

Part

5 *Museums*

Experiencing Green Pigeons

Michael J. Donoghue and Jane Pickering

Museums, in G. Evelyn Hutchinson's words, "provide the marvelous and the beautiful."[1] And, as you will discover from the essays in this chapter, Hutchinson considered them crucial to the pursuit of scientific knowledge, frequently highlighting the importance of natural history collections and the rare emotional experiences that museums can provide. These writings are not just the offbeat musings of a polymath on art and antiquities. Instead, as we will argue, they are central to understanding how Hutchinson attained the scientific perspective that we so admire, especially his ability to seamlessly integrate ecological and evolutionary thinking. As you will also find, Hutchinson's deep belief in the importance of museums and collections stemmed from his appreciation of their aesthetic qualities, for in the end, as Sharon Kingsland documents in her essay in this volume, he viewed science as being fundamentally about the beauty of nature.

"A CONSTANT SOURCE OF DELIGHT"

Hutchinson's close relationship with museums began, as did his passion for collecting, as a child growing up in the rich intellectual environment of a Cambridge college, where eminent naturalists were frequent visitors and he had ready access to the university's great museums. He spent countless hours in the field, kindling his lifelong love of water bugs, butterflies, and many other animals. These collecting jaunts, among other things, provided the basis for his early scientific publications—he had published six papers on insects before his twentieth birthday.[2]

Especially in the winter months, "when little was stirring in the field," he frequently visited the university's museums. Later he recollected that he probably knew the specimens in the zoology museum by heart by the age of fourteen.[3] The displays in that museum taught him bio-

Hutchinson collecting *Philaenus spumarius* (the common froghopper or meadow spittlebug) at Cherryhinton Chalk Pit, Cambridge, as a student in 1920. Students were allowed to roam only on Sunday afternoons and, as Hutchinson wryly noted, his "Sunday-best" clothes were "tiresome and uneconomical" in the field (Hutchinson, 1979, p. 70). Yale University Archives.

logical principles that were, as he later recognized, "essential ingredients" of his life.[4] For example, it was there that he first encountered cryptic coloration and mimicry, and animals as bizarre and delightful as the dodo and the Surinam toad.

These childhood experiences firmly established Hutchinson's views on the importance of specimens for research, and in particular the role of museums in comparative biology and evolutionary ecology. This is perhaps most directly and forcefully expressed in "On Being a Meter and a Half Long," which he wrote in connection with the celebration of the two hundredth anniversary of the birth of the British scientist James Smithson, whose gift to the United States enabled the founding of the Smithsonian Institution. He argued that although publicists and professional administrators focus only on using the latest experimental techniques (for them, "it is considered far more important to be up-to-date than to be interesting and useful"), the synthesis (and enjoyment) of existing knowledge, particularly that based on "material objects," was of equal importance to science.[5] Earlier, in "A Note on the Functions of a University," he argued that such synthesis— or what he later termed the "extensive" study of nature—required "access to objects, manuscripts, books, pictures, records of all sorts, apparatus, chemicals, specimens, etc., which may be termed the material basis of scholarship."[6] At the time, of course, this view was highly unfashionable. Museums had played a key role in the development of the sciences in the eighteenth and nineteenth centuries, but with the rise of experimentalism in the late nineteenth and early twentieth centuries research tended to move away from museums, which in turn began to justify their existence mainly in terms of public education.[7]

Hutchinson understood this history. But, having grappled with difficult taxonomic issues surrounding water bugs and other organisms, and understanding firsthand that solutions to "deep and difficult genetical and ecological questions . . . depended on taxonomic distinctions," he consistently and strongly promoted the basic taxonomic research that depends fundamentally on museum collections.[8] Perhaps the most elemental expression of this support appeared in a letter regarding a search for a new director of the Yale Peabody Museum: "I feel very strongly that whatever the Director's main research interest, he or she should have real experience with taxonomic biology . . . so that a tradition is maintained that supports everyone whose work depends on A being A and not B by mistake."[9] He also recognized the renewal of systematics and evolutionary morphology in connection with the "modern synthesis" in evolutionary biology, which he saw as leading to a renaissance in the use of museums, particularly, again, at the Peabody Museum.[10]

Hutchinson's close connection with the Peabody Museum began from the moment he arrived at Yale, and that connection is perhaps the best illustration of his lifelong and multifaceted relationship with museums. He worked closely and productively with a number of the Peabody's curators, donated important entomological collections to the museum, frequently used Peabody specimens in his teaching, and supported the Peabody's educational endeavors in the New Haven community.[11] In recognition of this diverse activity Hutchinson was elected in 1975 as one of the first honorary curators of the Peabody, a special status created to recognize "a leading authority in the natural sciences . . . appointed to the Museum at large, not to a particular division." In 1981, he was awarded the Verrill Medal, the Peabody Museum's highest honor, and his name is right at home among the luminary recipients of that prize, including George Gaylord Simpson, G. Ledyard Stebbins, Ernst Mayr, and Theodosius Dobzhansky. In presenting the award, his former student and lifelong friend S. Dillon Ripley, who was then Secretary of the Smithsonian Institution, provided this wonderfully apt citation: "Refreshingly you have always believed in the value of specimens and museum collections as essential to learning. Yale's Peabody Museum is blessed by your multidimensional presence."[12]

Hutchinson (center) holds the Verrill Medal that S. Dillon Ripley (right) presented to him on behalf of the Peabody Museum. Karl M. Waage, then director, is on the left. The book in Secretary Ripley's hand is Hutchinson's account of his travels in Goa and Indian Tibet. Yale Peabody Museum Archives; copyright Yale University.

This "multidimensional presence" also included Hutchinson's well-documented admiration of the great works of art for which the museum is famous—most notably Rudolph Zallinger's mural *The Age of Reptiles*.[13] While Hutchinson clearly recognized the great scientific value of natural history collections, his essays highlighted the great significance he also attached to their aesthetic value, which he believed was as important as any perceived "relevance." It is to this aspect of his understanding of museums and their collections that we turn now.

"FEELINGS OF THIS SORT MOLD OUR LIVES"

Hutchinson believed that appreciating beauty in the world—recognizing "that the universe and its inhabitants can be extremely decorative"[14]—was as fundamental to science as it was to the enjoyment of life, and he himself took great pleasure in a very broad range of "material objects." For example, he declared in his autobiography that the "greatest" museum in Cambridge was the Fitzwilliam Museum of art and antiquities, and he made frequent visits there to enjoy its paintings.[15] In general, he made little distinction between art and natural history objects, focusing instead on the common emotional experiences that they can elicit.

He approached these issues most directly in "The Naturalist as an Art Critic," delivered as a lecture in 1963 to commemorate the 150th anniversary of the Academy of Natural Sciences of Philadelphia. Why, he asked, are some objects placed in an art museum and others in a natural history museum? What, after all, is the distinction between a work of art and an object of natural beauty? In the very first museum collections, natural and man-made objects were intermingled, or even variously conjoined (which he illustrated with an ostrich egg goblet), whereas moving toward the present these "have been sorted out, purified, or perhaps merely divided into categories convenient in administration."[16] This sorting is not, he asserted, about beauty, or symmetry, or other such aesthetic qualities. Instead it perhaps rests upon whether there is "some evidence of a message from, or expression of, the personality of another human being, the artist who made the work."[17] However, with the clever aid of fake paintings and artistic apes, Hutchinson quickly blurred any such distinction. Extending this thinking still further, he argued that "if the whole aspect of the work of a natural history museum is considered in this light, a taxonomically arranged set of diatom slides or a drawer of insects, no less than a habitat group or the magnificent fulgerite . . . are seen to have some of the properties of works of art."[18]

Beyond beauty there are a host of other aesthetic experiences, which were explored first by Immanuel Kant and more recently in connection with biodiversity by Kiester.[19] Hutchinson's writings feature some of the deep emotions inspired by natural history museums. For example, in reflecting that "many people entering a natural history museum for the first time must wonder, if only for a moment, whether a pterodactyl or dinosaur really could have lived," he puts his finger directly on the fine and shifting line that separates reality from make-believe, and the sublime disorientation of truly not knowing which side of that line you're on.[20]

He also provided vivid examples of the feelings associated with actually handling museum objects. He recalled, for example, being shown some of the "special treasures" of the Cambridge Museum of Ethnology and Archaeology on his tenth or eleventh birthday, and actually trying on a magnificent Hawaiian feather cloak. In South Africa he had a chance to handle some of anthropologist Raymond Dart's specimens: "The memory of having the original Australopithecus skull in my hands still thrills me."[21] But, for us, the most engaging account of such an experience comes from a speech that Hutch-

inson delivered in 1960 as part of a symposium titled "The Role of the Museum in Teaching and Research at Yale," held to mark the opening of the Oceanographic and Ornithological Wing (better known as the Bingham Laboratory) of the Peabody Museum. This essay was published later under the title "The Uses of Beetles" in *The Enchanted Voyage*.[22] In concluding he recounted this experience: "A few weeks ago, I happened to be in the Coe Memorial Room at the top of the new building where our enthusiastic ornithologists were arranging the collection of bird skins. Suddenly it dawned on me that I had never realized what an extraordinary number of pigeons are bright green. Most of you will also probably not have experienced any large number of green pigeons, though to the ornithologist they are a commonplace. Many of them have in addition minor decoration in a great variety of other colors, often of a rather startling kind. To me, this realization, though it had no apparent value in relation to anything else that I knew, gave me intense pleasure that I can still recall and re-experience. Feelings of this sort mold our lives, I think always enriching them."[23] Embodied in this story are themes that resonate throughout Hutchinson's writings. He was keenly aware of and attributed great significance to the incomparable, if fleeting, emotional experiences stimulated by natural objects, and he viewed museums as providing circumstances highly conducive to such experiences. These are of crucial importance, he felt, despite their having "no apparent value in relation to anything else." We believe this to be an exceptionally deep insight, and it certainly is one that has oriented our own thinking about museum experiences and their place in our lives.

"THE SMALL ROOTS OF MODERN SCIENCE"

While some aesthetic experiences pass without noticeable consequence, others form the basis of a more extended commitment, which can, in the right hands and at the right moment, yield profound insights. So it was for Hutchinson. His lifelong interest in water bugs and moths, which was founded first of all on aesthetic at-

A variety of "green pigeons" from the Peabody's ornithology collection.

tractions and the simple joys of collecting, inspired many of his most important ideas. Perhaps most notably, "Homage to Santa Rosalia," in which he contemplated niche differences among related species, was stimulated by his observations on the *Corixa* water bugs that he collected from a small pond just below the sanctuary of Santa Rosalia on Monte Pellegrino, near Palermo, Italy.[24]

As for the Lepidoptera, he observed that they "carry all the major problems of evolutionary biology set out in colored two-dimensional diagrams on their wings."[25] Indeed, he used them often in his writings, as for example in one of his best-known short essays, mysteriously entitled "The Cream in the Gooseberry Fool."[26] This work illuminated a pathway from an aesthetics-based collecting impulse to a major scientific insight, and at the same time it provided an astonishing tribute to the importance

of museum specimens and amateur naturalists. The human star of the "Gooseberry" tale is one Reverend Raynor (1854–1929), a keen amateur lepidopterist. The animal protagonist is the magpie moth, *Abraxas grossulariata,* known also as the gooseberry moth because the larvae feed on currant and gooseberry plants. Hutchinson had himself studied these organisms carefully as a child in Cambridge, later reflecting that "the fantastic variation exhibited by a small minority of the specimens of the species, an aristocratic variability remote from the daily life of the average currant moth, is always beautiful but also in turn puzzling, frustrating and challenging."[27]

Raynor deposited a series of *Abraxas* specimens in the zoological museum in Cambridge, documenting the results of his extensive breeding experiments with aberrant forms of the moth. Hutchinson found these annotated by the geneticist Leonard Doncaster, who recog-

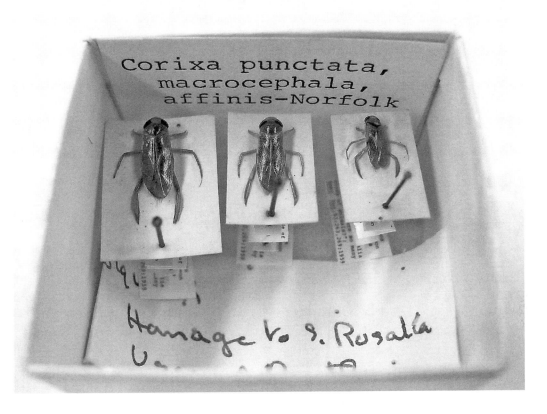

Corixa waterbugs, with Hutchinson's annotations, from the Peabody's entomology collection, showing the size differences discussed in "Homage to Santa Rosalia" (Hutchinson, 1959) and a part of the collection that he used in his research and in classroom demonstrations. These specimens, from Norfolk, England, were collected by Hutchinson when he was a student.

nized in them the first example of what we now call sex-linked inheritance. Raynor and his moths later came to the attention of yet another renowned geneticist, Huia Onslow, who then used *Abraxas* to demonstrate the Mendelian basis of what appeared to be blending inheritance. So, in the end, these studies underpinned the modern chromosomal theory of inheritance, the work of E. B. Ford on natural selection in the wild, and "all that has happened subsequently in genetics."[28]

The basic message is a simple one: collections, carefully made and maintained, can and actually do inspire great discoveries, but often quite "accidentally." Raynor did not, after all, set out to prove or disprove the existence of sex-linked inheritance, he simply delighted in what Hutchinson called "the protean beauty of the magpie moth."[29] The meaning of the essay's title is trickier to decipher. A play on the popular dessert, it refers to the amateur lepidopterists who gathered to purchase some of Raynor's *Abraxas* specimens when they were put up for auction in 1907. "Gooseberry fools" purchased them, at quite high prices, solely as curiosities, while only the "cream" among them appreciated their true scientific significance.

The deeper theme of the "Gooseberry" essay—that big ideas are so often stimulated by small, seemingly inconsequential observations—appears repeatedly in Hutchinson's essays. As Kingsland describes in this volume, Hutchinson firmly believed that all observations, however small, could be of great importance. For example, consider this lovely point from "Science Has Been Liberal Handed . . .": "Only to the

Gooseberry moths (*Abraxas grossulariata*) in the Peabody's entomology collection, with Hutchinson's annotations. These were purchased in London by the curator of entomology, Charles Remington, specifically for Hutchinson's work, including his writing of "The Cream in the Gooseberry Fool" (Hutchinson, 1965).

inhabitants of a university would it seem natural that an article with the title 'The origin and distribution of the chestnut-backed chickadee' should contain a conclusion which, we now realize after nearly three-quarters of a century, describes the basic role that competition plays in the distribution of all living things, plant or animal, friend or foe of man."[30] "Conjectures Arising in a Quiet Museum," in which he considered the consequences of incomplete penetrance for adaptive evolution, presented yet another example from his own work with specimens.[31] This essay begins, "These speculations arose largely from visits to Tring to gaze at the huge series of *Abraxas grossulariata*," and it continues, "Now that I think I see what some of it may mean, the meaning tends to be related to other

organisms, watersnails, stoats, and even man himself."[32]

As these last remarks begin to demonstrate, Hutchinson was, largely by virtue of his museum background, a superb comparative biologist, who also very clearly recognized threats to biodiversity and the great need to preserve it. The following passage, taken from "Fifty Years of Man in the Zoo," provides a flavor of his natural ability to draw together his exceptional knowledge of the diversity of life in framing and then extending an argument: "The themes that have been developed have been illuminated by studies on butterflies, deer mice, robins, a lioness, and by implication all the primates living as well as fossil. It would have been quite possible to develop other themes involving snails, waterbugs, birds of paradise, dolphins, giraffes, and rhinoceroses. We are only at the beginning of this kind of study. . . . Without this diversity it will be immeasurably more difficult to understand ourselves."[33] Taking an even broader perspective, we can see the deep imprint of Hutchinson's museum experiences in his blending of ecological and evolutionary thinking into what we now call "evolutionary ecology." That discipline today is perhaps not quite so interesting as it was in Hutchinson's hands. What is missing in many modern practitioners is Hutchinson's hands-on knowledge of museums and collections and, consequently, his deep-seated appreciation of the intricacies of biological diversity, and his ever-present awareness of the influences of deep evolutionary history. Consider, for example, this passage from *The Ecological Theater and the Evolutionary Play:* "It is evident that at any level in the structure of the biological community there is a set of complicated relations between species, which probably tend to become less important as the species become less closely allied. These relations are of the kind which insure niche separation. They are probably balanced by another set of relationships expressing the fact that organisms of common ancestry are more likely to inherit a common way of life."[34] Such formulations, which came so naturally to Hutchinson (as they also did to Charles Darwin), appear to have become increasingly difficult as ecology and evolutionary biology have tended to separate from one another. It is only over the past several years that we are undertaking a serious integration of the study of phylogeny with community ecology issues or with the analysis of global biogeographic patterns.[35] For example, Cavender-Bares and her colleagues provided an outstanding study on oak trees of the balance that Hutchinson so thoroughly appreciated.[36]

"A CERTAIN DEGREE OF LOVABLENESS"

Reading the body of Hutchinson's work represented in this chapter, we are struck not only by his deep love and knowledge of museums and the unique roles that they play in human experience, but also by the ways in which he embraced these things in his own science. He seems to have been entirely comfortable in discussing and using emotional and aesthetic experiences to enrich his work. One has the sense that scientists are anxious to distance themselves as much as possible from discussions of emotional experiences and the like, considering them "unscientific." But for Hutchinson these boundaries just didn't seem to matter, for, as he put it, "ultimately the values of pure science and of the fine arts are identical."[37] In fact, it is the very blurring of these boundaries that we associate with his peculiar genius—the style of thinking and writing that produced his most innovative work.

So, far from being a sideline in Hutchinson's life that just happened to have brought forward some of his finest writing, his museum experiences are critical to properly understanding the development of his key scientific contributions and, in particular, the beginnings of a deeply satisfying (if still quite underdeveloped) integration of ecology and evolutionary biology. Evelyn Hutchinson believed that the goal of inductive knowledge was to produce beautiful conceptual schemes and thereby to increase the "lovableness" of the universe. Owing in no small measure to his museum experiences, he surely achieved this end quite perfectly.

A NOTE ON THE FUNCTIONS
OF A UNIVERSITY

THESE NOTES are written on the basic assumption that the professed ideals of our culture, of Hellenic, Judao-Christian, and Anglo-Saxon derivation, are reasonably satisfactory, but that present in that culture are a number of inconsistent elements which prevent an approach to the realization of the ideals.

Until a great deal more knowledge is developed and disseminated, such inconsistencies will remain. Ethical problems therefore are not considered in this statement. The primary ethical orientation of the culture is taken for granted, and the first task of the human intellect is regarded as that of finding ways to remove the inconsistencies. Problems of pure aesthetics, as opposed to the intellectual handling of aesthetic elements, are purposely avoided, because the writer, in common with other men, knows of no satisfactory way to handle them. It is, however, probable that a point of view parallel to that developed with regard to intellectual values could be developed with regard to pure appreciation of works of art. Actually it is impossible to distinguish in many cases between what are referred to below as the pleasures of learning and the pleasures arising from aesthetic experiences. Pure rote learning, as in the study of languages, is not discussed, and in general it may be considered as one of a number of methods for gaining experience of the material basis of scholarship.

The university may be regarded as the institutional mechanism

144

THE FUNCTIONS OF A UNIVERSITY

for developing and transmitting the part of our culture that consists of complex generalizations. Statements of all particular experiences of life are to be included as instances of such generalizations.

This is explicitly recognized in the learning and teaching of the deductive sciences (mathematics, logic). It is less easily recognized in the areas of the natural sciences, where a multitude of facts may bewilder the student and may appear useless and pedantic to the outside world. These facts are, however, valued primarily as the raw material for inductive generalization. In history it is true that some authorities deny the reality of any general propositions, but if this is accepted the study of history must not claim to be of the least practical significance in the world of affairs. We can learn only by experience. The university is a mechanism by which past experience can continue to be at the service of living men and women. If the experience does not permit a judgment of the form, "if a and b and c . . . we may with some probability expect x but not y," then we have learnt nothing of practical value. It is a matter of particular importance that the students and teachers of the humanities realize this fact if they wish to make claims that their subjects prepare men for life.

The intellectual activity of the university differs from that of the lower educational levels in being more difficult. At the same time this intellectual activity is more endowed with the quality that gives rise to intellectual excitement. The discovery of this quality, the history of which does not seem to have been much considered, represents one of the great landmarks in human development. In spite of what has recently been written to the contrary, it is hardly to be doubted that most of human progress, insofar as it depends on purely intellectual factors, is ultimately based on the mental pleasure arising from seeing new facts as special cases of generalizations or new generalizations arising from old and new facts. The pleasure in analysis and intellectual

145

THE ITINERANT IVORY TOWER

resynthesis of experience, whether of nature or man, is one of the mainsprings of our culture, and therefore one of the primary functions of the university is to encourage the appreciation of this pleasure. Advance in our environmental relations insistently demands more knowledge and more experience. The labor of gaining the necessary intellectual experience would require more than heroic fortitude if it were not for the excitement and satisfaction that accompany the task. The existence of this satisfaction is, however, something that has to be discovered and learnt.

In all intellectual activity that is not purely deductive it is necessary to have access to objects, manuscripts, books, pictures, records of all sorts, apparatus, chemicals, specimens, etc., which may be termed the material basis of scholarship. In view of the impossibility of knowing what is likely to prove useful, it has generally been regarded as an essential function of a university to maintain as large a library as possible, containing as great a part of the world's serious literature, and also of such seemingly ephemeral records, as might conceivably be of value. Recently it has become a normal library function to make available through loans, photostats, and microfilm such rare material as is needed elsewhere. In view of the destruction of libraries and collections in both Europe and Asia, duplication and dissemination of the material basis of scholarship will become an essential and important function of the university in the near future.

Too little attention is paid today to the nonliterary aspects of the material basis of scholarship. There is an unfortunate tendency to regard university museums only as means of interesting the public in the arts and sciences, rather than as a potential storehouse of facts on which new generalizations can be built.

The university should be regarded primarily as a place of learning, and not as a place of teaching. Margaret Mead [1] has

1. Our educational emphases in primitive perspective, *Am. J. Sociology, 48,* 633–9, 1943.

THE FUNCTIONS OF A UNIVERSITY

made essentially this distinction in an article in which she points out that in most cultures the pupil goes to the learned or expert man or woman to acquire his or her skill, the emphasis being placed on learning, whereas in our culture the pupil is exposed to an education, which "takes" only if the teaching is effective. Dr. Mead points out that a synthesis of these antitheses is required. The synthesis in the university seems to the present writer to require the teacher primarily to show by example how learning is accompanied by an intense mental excitement. The antithesis between teaching and research has often been acrimoniously debated. There is, however, no antithesis between learning and research, because if the teacher is not learning himself he can never teach by example. He should be encouraged to arrange his formal teaching in such a way as to bring out the inherent excitement in the relations of the subject matter. It may be urged that many students can feel little or nothing of this, but that they must nevertheless be educated as a measure of safety in a democracy. It is not certain, however, that the appreciation of learning is given a high enough value in our culture to permit every reasonably intelligent student to be brought up in a tradition that fosters such appreciation. In a general sense it is the basic function of the university to emphasize, as vigorously as possible, that intellectual activity is one of the great pleasures of life, for in so doing the university performs the fundamental duty of encouraging us to know enough to set our house in order.

THE USES OF BEETLES[1]

*Beetles serve for divers uses, for they both
profit our mindes, and they cure some
infirmities of our bodies.*

A natural history museum, considered in terms of its purposes,
may be defined as an institute for the study of the diversity of nature.
This is true even if the student is a four-year-old child who can
learn that although monkeys and elephants are very different, in
each both front and back legs are organized on the same principle,
"one bone, two bones and a lot of little bones." It is equally true
of the paleontologist using the same principle of homology in
attempting to solve the still obscure problem of the origin of the
five-toed limb. It was such diversity in unity that Charles Darwin
perceived when he compared his skins of Galapagos mockingbirds
from the different islands and found that their beaks differed, writing
in his notebook in 1835, "When I see these Islands in sight of each
other and possessed of but a scanty stock of animals tenanted by these
birds but slightly differing in structure and filling the same place in
nature, I must suspect they are only varieties. . . . If there is the slightest
foundation for these remarks, the Zoology of Archipelagoes will be
well worth examination; for such facts would undermine the stability

1. Contribution to a symposium on "The Role of the Museum in Teaching
   and Research at Yale," held by the Yale Alumni Board on the occasion of
   the opening of the Oceanographic and Ornithological Wing of the Peabody
   Museum.

THE USES OF BEETLES

of species." This was the germ of his interest in evolution, the full significance of which we have doubtless not yet appreciated.

Several ancient peoples, notably the Egyptians and the Aztecs, kept a variety of wild animals for pleasure. Aristotle, the most significant pure biologist of the ancient world, is said to have received collections that Alexander the Great sent back from his campaign to his old teacher. Little, however, can be said about the history of natural history in the ancient world. The immediate precursors of the collections in the buildings, the opening of which we are attending, do not go back much beyond four hundred years. This four hundred years of history is, however, very illuminating.

The history of natural history in the Western World may be subsumed under three periods. The first or medieval, from the twelfth century to the early sixteenth century, saw the beginning of the study of Greek and Arabic texts, and the production of a few original works such as the biological writings of Albertus Magnus and the famous treatise of Frederick II on falconry, the relation of which to earlier Arabic works apparently requires further study. The main significance of the medieval period does not lie in these sporadic literary contributions, which were sparse in comparison with the ever-popular and almost entirely fabulous bestiaries based on the "Physiologus." What is really important in the medieval period is the development of an iconography of animals and plants by sculptors and illuminators. We cannot talk about the origin of species, though many people have of course tried, until we know what we mean by species. The sculptors who correctly distinguished, in the Chapterhouse at Southwell, between the two British oaks, were laying the foundation of this knowledge. Somewhat later, illuminators often introduced plants, insects, and birds into the decorative borders of manuscripts. The most astonishing of such decorations, full of insects, spiders, and shells, with a quite recognizable carpet beetle, have been studied by A.C.Crombie of Oxford.[2]

2. "Cybo d'Hyères: A Fourteenth-century Zoological Artist," *Endeavour*, *11* (1952), 183–87.

THE ENCHANTED VOYAGE

They date from the fourteenth century and are ascribed to Cybo
d'Hyères, a Genoese who seems to have worked in Provence, in a
region which was undoubtedly somewhat more influenced by
Gothic natural iconography than was much of Italy. There are
numerous later examples; a particularly lovely one, a late fifteenth-
century "Book of Hours," formerly at Aldenham House and recently
from the library of Yale's great benefactor Louis Rabinowitz, is in
the Rare Book Room of the Sterling Library. In this exquisite work,
believed to have been executed in Ghent or Bruges and certainly
dating from the late fifteenth century, the flowers and fruits, which
doubtless kept still, are all naturalistic; the birds, which did not,
are mainly the types that later turn up on painted china or lacquered
screens, though a hoopoe, a goldfinch, and a great-tit are clearly
recognizable. The butterflies, of intermediate activity, are fairly
realistic, although the artist responsible had some ideas of his own.
He depicted a clouded yellow (*Colias croceus*), the European ally of
our sulphur butterfly, on several pages (Plate 3, *above*), a wall
butterfly (*Pararge megera*) on folio 44v and folio 96r, and an Apollo
butterfly (*Parnassius apollo*) on folio 108r (Plate 3, *middle*); but there
are also some strangely intermediate forms and at least one blue and
yellow species (folios 54r and 54v) whose habitat can only have been
the artist's imagination. The presence of the Apollo butterfly is very
interesting, for it is an Alpine form that could not have occurred
near Ghent or any other Flemish center of the illuminator's art,
save as a very rare straggler. It suggests that a set of illustrations of
butterflies for the use of illuminators and other decorators may
have been circulated. This is probably more likely than the transfer
of actual and rather fragile specimens. At least one ornithological
manuscript[3] surviving from the fourteenth century probably

3. In the Pepysian Library, Magdalene College, Cambridge, and reproduced
   in part in *Illustrated London News*, *235* (Dec. 12, 1959), pt. 4, facing p. 853. In
   the caption to this colored reproduction, David Lack points out that one of
   the birds illustrated is a red-legged partridge, not native to Britain, though
   there is evidence from a few bird names that the manuscript was made or at
   least annotated in England.

PLATE 3. *Above: Colias croceus*, deep yellow with dark borders to wings; *left*, contemporary specimen ($\times 2/3$), *right*, miniature Aldenham-Rabinowitz "Book of Hours" folio 100r ($\times 2$).

Middle: Parnassius apollo, white with black markings and crimson ocellus on hind wings; *left*, contemporary specimen ($\times 2/3$), *right*, miniature from Aldenham-Rabinowitz "Book of Hours" folio 108r ($\times 2$).

Below, left to right: Corixa punctata, *C. macrocephala*, and *C. affinis* (all $\times 2$ and all British specimens).

THE ENCHANTED VOYAGE

represented such a collection of bird drawings. Along with illustrated herbals, which have a good Greek ancestry, such treasuries of illustrations of birds and perhaps insects may have been the forebears of the monographs all systematists now use.

Within a hundred years of this work, we have clear evidence that several people were collecting plants and animals, writing about them, drawing them, even attempting to publish about them. The second period in Western natural history had begun. The origin of such collections poses certain interesting problems on which too little research has been done.

The storage of dry drugs in comparably shaped but diversely labeled pots in an apothecary's shop gives some of the diversity in unity that satisfies a collector, as do the spices in neat boxes in a kitchen. The great collecting craze of the Middle Ages was for relics. At least in countries touched by the Reformation, the collection of both works of art and, later, of natural curiosities may have taken the place of the vast accumulations of minute fragments of a whole army of saints, neatly labeled, authenticated, and stored in glazed ornamental containers. A further stimulus to collecting doubtless came from natural rarities such as narwhal's tusks, doing duty for unicorns' horns, and bezoar stones, which were collectors' items, as well as more or less magical drugs.

Sometime during the sixteenth century, the esthetic interest in organisms, the practical interest in drugs, the learned interest in the writings of the ancients that dealt with natural history, and the collectors' enthusiasm previously directed toward books, relics, and *objets d'art* must have united to produce the first systematic natural history collections. Gesner was making collections of plants before 1560, Samuel Quickelberg published a catalogue of his collections in 1565, and there was a storehouse of natural rarities at the court of Saxony. William Penny, who died in 1588, had a "treasury of insects" and left a mass of manuscripts, which later were edited by Thomas Moffet as the "Theatrum Insectorum," from which the epigraph at the beginning of this chapter is taken.

94

THE USES OF BEETLES

The material brought back from the distant parts of the world that became known from the late fifteenth century onward added greatly to the interest of these and subsequent collectors. It is, moreover, not impossible that knowledge of the diversity of the fauna of Mexico, in which region species after species had a native name, stimulated a systematic attitude in the minds of anyone who had seen what was published of Hernandez' great work on Mexican natural history.[4]

It is important to notice that throughout the whole of the second or Renaissance period of natural history, which lasted from sometime in the sixteenth century to 1859, there is always a strong religious motive for the study. "He that beholds the forms, clothing, elegancy and rich habits of the Butterflies, how can he choose but admire the bountiful God, who is the Author and giver of so rich treasure." In a like vein, Linnaeus believed that his "Systema Naturae" represented valid insight into the thoughts of the Creator. This attitude persisted so long as people believed that they were created to praise God.

The third or modern period of natural history has lasted just a century. It has been dominated by the idea of evolution, a scientific idea that has had a far more powerful effect on men's minds than have any of the physical sciences, however much they may have altered our material environment. Evolution as we know it was born in the mind of Charles Darwin when collecting museum specimens. Much of the later development of the theory has been due to museum workers. A large part of the justification of any university museum lies in the contribution that can be made in it to this kind of knowledge. We need to know far more than we do.

4. Francisco Hernandez, *Nova plantarum, animalium et mineralium mexicanorum historia a Francisco Hernandez Medico in Indiis praestantissimo primum compilata, dein a Nardo Antonio Reccho in volumen digesta, a Io. Terentio, Io. Fabro, et Fabio columna Lynceis notis, & additionibus longe doctissimis illustrata* (Rome, Sumptibus Blasii Deversini, & Zunobii Masotti Bibliopolarum, Typis Vitalis Mascardi, 1651).

THE ENCHANTED VOYAGE

Discoveries of any sort in science can be applied to human affairs, but it is evident that in general one needs to know vastly more in order to make the applications wisely and constructively than is needed to do it foolishly and destructively. That is one cause of our present predicament. We are presumably still evolving; people makes guesses that we are undergoing evolutionary loss of teeth, hair, or little toes. For most such guesses, there is not a particle of evidence. There is, however, much evidence to show that, for example, the peculiar characteristic of having a big toe shorter than the second toe, a character that apparently is slightly maladaptive, was regarded as esthetically ideal by the Greeks, and is known to be inherited, has spread through a large part of the population. No one knows why. We are full, body and mind, of such problems. In most cases, we can learn best about ourselves when guided by analogies from birds, fishes, butterflies, sea shells, and sponges.

There is yet another cause of our present discontent. The theory of evolution has often been applied unwisely, sometimes wickedly, to men's minds. A hundred years ago it became evident that each species was not made in Eden individually on the basis of some archetypal design kept in Heaven. When such an oversimple idea was abandoned, the religious meaning for the study of natural history, with much of the wonder and glory, officially disappeared. I do not think that this fundamental attitude really disappeared from the mind of the investigator; a love of the created world, a sort of natural piety, exists in the minds of nearly all naturalists. What happened was that the attitude became something to be held in private, in an apologetic manner, something regarded in the public world as sentimental, impractical, or unrealistic. I believe very strongly that, whatever religious beliefs an investigator may or may not have, this point of view is an essential ingredient of good work, and in our human predicament can be a saving grace. It is one of the attitudes that can help us from destroying our natural heritage, which we are doing at an alarming and ever-increasing rate. I would remind you that during nearly all the history of our

THE USES OF BEETLES

species man has lived in association with large, often terrifying, but always exciting animals. Models of the survivors, toy elephants, giraffes and pandas, are an integral part of contemporary childhood. If all these animals became extinct, as is quite possible, are we sure that some irreparable harm to our psychological development would not be done ? In a broader context, are we certain that we can really persist as humane human beings in a world in which natural and much artificial beauty is being continually replaced by ugliness or at best by neutral functional forms ?

A few weeks ago, I happened to be in the Coe Memorial Room at the top of the new building where our enthusiastic ornithologists were arranging the collection of bird skins. Suddenly it dawned on me that I had never realized what an extraordinary number of pigeons are bright green. Most of you will also probably not have experienced any large number of green pigeons, though to the ornithologist they are a commonplace. Many of them have in addition minor decoration in a great variety of other colors, often of a rather startling kind. To me, this realization, though it had no apparent value in relation to anything else that I knew, gave me intense pleasure that I can still recall and re-experience. Feelings of this sort mold our lives. I think always enriching them. It is for the further opportunity for such experience that I would like to thank all who are responsible for the splendid addition that is being formally presented to the University today.

THE NATURALIST AS AN ART CRITIC [1]

BY G. EVELYN HUTCHINSON

Sterling Professor of Zoology
Department of Biology, Yale University

During the early period of the formation of those large collections which ultimately became the bases of the public museums of Europe, such virtuosi and cognoscenti as collected objects of natural origin also usually collected human artifacts, both for their intrinsic value, beauty, and on account of their historic associations.

In the earliest inventory [2] of a great princely collection in Western Europe, that of the Duc de Berry, brother of Charles V of France, who was born in 1340 and died in 1416, there were a few odd natural history specimens mentioned, ostrich eggs, probably an elephant molar, tusks of wild boars, a bird's bone remarkable for its lightness, a porcupine quill and various pebbles which seem odd in a collection made up of an unbelievable number of precious stones, pearls, jewels, vessels and images of gold and silver and relics of the saints, almost all of which have disappeared, and of manuscripts, some of which are still among the glories of mediaeval French art.

Later collections, in the 16th century, were richer in natural history, and in fact almost exclusively biological collections were first made at that time. However, a number of the most famous were very mixed even at a much later date. The most striking examples are those of Elias Ashmole, actually largely assembled by his friend John Tradescant whom we commemorate in *Tradescantia,* which enriched the University of Oxford, and of Sir Hans Sloan, in part based on the cabinets of other collectors, which formed the basis of the British Museum in both its branches. Perhaps even in the 17th century such collections may have raised philosophical or moral problems. Jan van Kessel's painting (Figure 1) in Florence now called " Lo studio di un naturalista," though certainly amusing, must also have allegorical roots that I am not expert enough to excavate.[3] The naturalist whose study is depicted by van Kessel seems to have been interested in birds, caterpillars, strange and mythological plants such as the mandragora or mandrake, sur-

---

[1] A distinguished scientist's lecture given at the Annual Meeting of the Academy of Natural Sciences of Philadelphia on 29 April 1963.

[2] Guiffrey, J. 1894–6. Inventaires de Jean Duc de Berry (1401–1416). Paris, E. Leroux. I. (1894) CXCIV, 347 pp. II. (1896) 321 pp.

[3] In a very curious painting said to be the only known work of Giuseppe Crespi the younger, reproduced (Plate L) and discussed (pp. 306–307) by H. W. Janson (*Apes and Ape Lore in the Middle Ages and the Renaissance,* London, Warburg Institute 1952, 384 pp.), a monkey is depicted holding what looks like a shell to his ear, surrounded by a fantastic assemblage of instruments, natural history specimens and antiquities. Janson connects this picture from the late 18th century with the *Tractatus secundus de Naturae Simia* of Robert Fludd, 1618. Jan van Kessel's painting perhaps belongs in the same obscure tradition, though he also painted less problematic *singeries.*

(99)

Proc. Acad. Nat. Sci., Phila., Vol. 115, No. 5, 1963.

veying instruments, telescopes and coins. I would call your attention to the amount of jewellery that he amassed; this seems to have been one of the classes of object most favored by early collectors, partly no doubt as an investment as well as for its beauty.

FIGURE 1. J. van Kessel (1626–1679). Lo Studio di un naturalista. Pitti Palace, Florence. Courtesy of the Gabinetto Fotographico alla Soprintendenza, Uffizi, Florence.

Since we are celebrating the close of the hundred and fiftieth anniversary year of the oldest natural history museum in the United States, in a city that is also famous for its art collection, it has seemed appropriate to consider some aspects of the dichotomy between natural history and art museums, and to ask why some objects are put into one and some into the other. If at first the answers seem obvious, there will, I think, prove to be enough difficulties to lead us into interesting if obscure regions of the human mind.

Initially the objects in a collection were assembled to be looked at. They are to arouse admiration and pleasure in their beauty, wonder at their strangeness or history, envy or awe at their costliness or rarity. The simple reactions of the unlearned to the strange or marvellous give some idea of the primary reactions to objects in a collection, reactions which most of us have forgotten. A peasant woman enquires if the *pala d'oro*, the great gold and enamelled Byzantine altar frontal in San Marco in Venice, is really made of gold. Napoleon or his officials are said to have been persuaded that it was much too big to be really golden, and so left it unconfiscated and unmolested.

In the crypt of the Basilica of Sant' Ambrogio in Milan, the great Saint Ambrose lies between two somewhat undocumented martyrs, San Gervasio and San Protasio; another peasant woman exclaims " che nomi " at hearing the unfamiliar names attached to venerable skeletons in a sacred place. In a secular context many people entering a natural history museum for the first time must wonder, if only for a moment, whether a pterodactyl or a dinosaur really could have lived, and how they got their names.

We can begin to get some insight into our problem by considering a group of rare and strange objects, that achieved their greatest popularity during the period just about the time that collections were beginning to become differentiated, in which the properties of some natural object play a very great part in the decorative qualities of an *objet d'art*. Some of these composite objects, such as richly mounted bezoar stones or the nuts of the mysterious coco-de-mer *Lodoicea maldivica* (Gmelin) Pers, were treasured for their fancied alexipharmic properties. Most however are purely decorative and we can divide these into two more or less discrete classes. In the one which we will call *self-theorising objects* the natural structure that provides

FIGURE 2. **Left:** *Nautilus* cup, Augsberg 17th century; **center:** Ostrich-egg goblet, Leipzig 1560–80; **right:** *Turbo marmoratus* cup, ?Nürnberg, 16th century (Vienna, Kunsthistorisches Museum no. 62, 95, and 116; E. Kris Publ. Kunsthist. Samml. Wien: Goldschmiedearbeiten I, Tafeln 5 and 67). By kind permission.

the decorative form, also displays, or would if we fully understood it, the deterministic laws by which it came into being. Here we have an example of the effects of rotation during translation down the oviduct, in the form of an ostrich egg, mounted as a goblet (Figure 2, center). On a much smaller scale rotation of pearls against some more resistant part of a mollusk's foot or mantle can make an acorn pearl. This probably most often happens in mobile pearl-producing mollusks, such as the freshwater *Margaritifera* (Figure 3, top right). Crystals, which proclaim at least part of their atomic structure in their macroscopic shape, can be mounted in their natural condition to make jewels (Figure 3, top center). The banding of an agate, pre-

FIGURE 3. **Above:** Sah Oved: contemporary English; topaz crystal mounted as a pin, between earrings, Mermaid in her Vanity, and Pelican in her Piety, the former with acorn pearl. **Below:** Ring, 16th century (Yale Univ. Art Gallery 1959–43–24; *ex. coll.* Margaret Hutchinson, Sir Francis Cook, Marlborough and quite possibly Thomas Howard, 2nd Earl of Arundel, the greatest collector in 17th century England).

sumably exemplifying Liesegang phenomena of diffusion in a colloid, can be used in conjunction with the form of the bezel (Figure 3, below) of a ring or any other mounting. The example illustrated, a ring dating from the 16th century is of interest in that the concentrically circular agate was often regarded as a toadstone and as such protective or magical properties were ascribed to it; this ring moreover also came through Sir Francis Cook from the Marlborough Collection, which was largely formed from the jewellery of the great 17th century Arundell collection. This identical ring may therefore have been in a cabinet of a great aristocratic virtuoso at the time when J. van Kessel painted the picture we have just examined. Horns, very early made into drinking vessels, exhibit a form clearly dependent on the mode of their growth, and shells of *Nautilus* (Figure 2, left) and *Turbo* (Figure 2, right), used to form magnificent cups, specifically display the logarithmic spiral characteristic of the growth processes of many mollusks and some other animals.

At least in the use of shells and possibly also of more perishable materials, the employment of actual organic objects for their decorative properties and of copies of them in various workable media must have grown up side by side giving the huge range of phytomorphic and zoomorphic forms known to art. At a more recent time wax flowers were doubtless valued for looking natural before they were valued for looking artificial.

In direct antithesis to the *self-theorising object,* may be set the *elegant ink blot,* the baroque pearl mounted to bring out its resemblance to the torso of a mythological figure (Figure 4), or on a less princely scale the driftwood or other *objets trouvés* of the surrealists and some later schools. In these, stochastic processes dominate the form, the selection and appreciation of which obviously involves some sort of psychological projection.

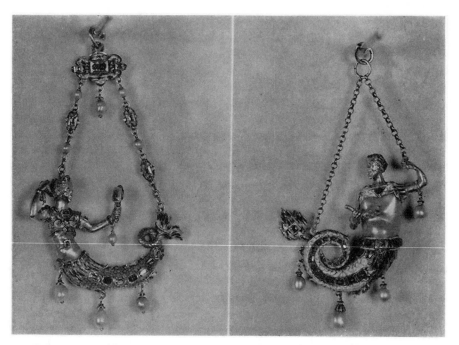

FIGURE 4. **Left:** Mermaid; **right:** Triton, baroque pearls, jewelled and enamelled gold, Italian 16th century (Widener Coll., National Gallery of Art, Washington, D. C.). By kind permission.

In the recent history of museums the various attitudes expected to arise in the minds of the observers have been sorted out, purified or perhaps merely divided into categories convenient in administration. We see the mixed objects that I have just described primarily as artifacts; if we happen to be interested in the pure natural history of the natural part, the artificial part is regarded as an excrescence that gets in the way of scientific vision.

Today we enter an art gallery expecting to be delighted by the beauty of certain works of man; we enter a natural history museum expecting to

104 PROC. ACAD. NAT. SCI., PHILA. [VOL. 115

be instructed in the workings of nature. There are also museums in which archaeological or ethnographical material is displayed to illustrate something about man that is akin to natural history, and indeed the same point of view is apparent in the grouping of works of art in any modern art gallery, where the pictures are placed by schools and periods, i.e. geographically and chronologically, just like fossils in a palaeontological collection. What seems often to be lacking, at least explicitly on the part of the intelligent public, is the realization that a number of objects in the natural history museum are of extraordinary natural beauty and that they should be valued quite simply as such, as well as for their scientific connotations. In practice in any good museum such as this one, the public displays are largely implicitly based on such aesthetic considerations. The question however of the nature of the beauty of the natural world and its relation to human art deserves more consideration than it is customarily given, and deserves such consideration quite specifically in the context of the natural history museum.

If we enquire why we make a distinction between the work of art and the object of natural beauty, which enquiry is a partial rephrasing of our original question, I suppose that at the present time the essential difference would usually be described in terms of communication or expression. What is valued in the work of art is supposedly not the sort of intrinsic beauty that we find in nature, but some evidence of a message from, or expression of, the personality of another human being, the artist who made the work. This concept however leads us into very considerable difficulties. The late Bernard Berenson said at the end of his life, of which seventy odd years had been largely spent in problems of attribution, that it did not matter who painted a picture as long as it was a real picture. This obvious truth, coming from him, carries non-obvious overtones. In the more limited modern vocabulary that we are using, it may be rephrased that it does not matter who painted a picture as long as the picture is a genuine expression. In the light of such a statement let us look at an oil study of a mulatto lady (Figure 5) wearing only a red and green turban and holding a long bamboo cane, first deciding, without asking about its history, if it is a real picture genuinely expressing something to the observer.

The painting is known to have been among the effects of Eugène Delacroix, sold at auction in Paris in 1864.[4] It was apparently included in a miscellaneous lot of seventeen studies supposedly by the painter himself, which were not described individually in the sale catalogue. It passed into the Cheramy collection, and was sold, as by Delacroix, in 1908, though not

---

[4] Goodison, J. W. and Denys Sutton in *Fitzwilliam Museum Catalogue of Paintings.* Volume I, French, German and Spanish: pp. 172–174 for full discussion; also *Art News,* New York 53, p. 47, 1954; *Connoisseur,* London, 133, p. 260, 1954; *Fitzwilliam Museum Annual Report,* 1954. Pl. IV, pp. 5–6. 1955. Mr. Goodison kindly writes that there can be little doubt that the attribution to Delacroix was mistaken, but that at present the evidence is quite insufficient that the painting is by Auguste.

FIGURE 5. Study of a Mulatto Woman, French School 1820–25. (Fitzwilliam Museum, Cambridge.) By kind permission.

listed as by him in the Cheramy catalogue published in that year. In 1954 it appeared, as by Delacroix, in an exhibition of 19th and 20th century French painting at the Lefèvre Gallery in London. At this time it evidently generated much excitement; it was reproduced in two art journals; the *Art News*, published in New York, wrote of it as " one of the chief pleasures " of the exhibition, " amazingly forceful though only 22 inches high." It was bought from the exhibition by a leading English museum, with a subvention from the National Art Collections Fund, and at the time of the purchase was hailed as of outstanding quality, and was praised for its distinction of vision and surety and sensitivity of handling. It evidently gave great satisfaction to all concerned, as it did to me when I saw it in 1958 and again in 1963. Later, however, the painting was regarded, as compared to certainly authenticated works of Delacroix's early years, " as mannered and timid in character and superficial in draughtsmanship and anatomical structure." It has indeed been suggested that it was probably the production of a dilettante called Jules-Robert Auguste, who for a time knew Delacroix; on at least one occasion they both worked from the same model. Auguste's known works, in pastel rather than oil paint, are said to possess a " preciosity of vision and meticulousness of style quite in conformity " with the picture we are considering. It is very hard to avoid the feeling that great subjectivity is involved in the appreciation of such a work; so long as it comes from the brush of M. Delacroix its virtues are emphasized, when his authorship is suspected all the faults suddenly become apparent, perhaps indeed overapparent. This leaves us in a very difficult, though admittedly honest, position in the face of the majority of the works of art in the world, whose makers are unknown. In the case of the painting we have been considering, without being able to express any real expert opinion, I have no difficulty in believing that it is not by Delacroix, but if, as seems rather unlikely, it is by a really weak painter, as Auguste seems to have been, he must have been so much under the influence of a better painter when he painted the work, that some of the virtues of the greater artist could be borrowed and incorporated into the work of the lesser man. It is worth noting that sometimes supposedly most characteristic works of major masters have turned out, as scholarship progresses, to be copies, studio pieces or even works of fairly independent pupils. Since in some cases only the more obvious qualities of the master may be caught and transferred to the derivative work, the latter may become a sort of elementary introduction to the subtilties of the master, with an immediate appeal leading in the right direction. The painting which first gave me insight into Zurburan, for instance, is a Santa Rufina now believed to emanate from his studio but not from his hand. At any rate all of us who have frequented art collections for any length of time must realize that we have almost certainly got what seems to be pleasure of the very highest order out of works of suspect attribution. The exact origin of the message conveyed is perhaps of less importance than is often believed.

We may now as naturalists raise a still more awkward problem, one that was adumbrated by the *singerie* painters of the 17th century,[5] the problem of the ape as artist. Unfortunately a certain amount of inevitable commercialism and humor has tended to obscure the extraordinary significance of the work that was started fifty years ago by Kohts in Russia and which has been recently greatly developed by many workers, of whose studies Desmond Morris has provided an illuminating synthesis.[6]

The great apes and some other primates, notably capuchin monkeys, when put into an experimental environment in which they can exhibit it, have a sense of symmetry in design, which is most easily demonstrated by giving the animal a paper, blank except for a square set eccentrically. In a highly significant number of cases the animal will tend to mark the paper in such a way as to balance the design. Rensch, moreover, in experiments in which animals can make choice of ready-made designs, finds that balanced patterns pleasing to ourselves also seem to please many other vertebrates. When the animal is given more elaborate opportunities for artistic expression, Morris concludes that in all cases, there is, as well as compositional control of balance, an attempt at calligraphic differentiation of line; thematic variation within an individual style appears when paintings of the same animal are compared. There is also an attempt to achieve a degree of optimum heterogeneity giving a sense that a painting is complete. At least in young apes the activity is highly self-rewarding or autotelic. Any intrusion is resented more than if the animal had for instance been disturbed when eating. Providing the young apes with paints, brushes and canvas gives it, for the first time in its life, something very important to do.

The general level of achievement, though compared to action painting or abstract expressionism by some critics, appears to be that of a three year old child just prior to the development of diagrammatic representation of the human face.

What these studies show clearly is that the desire and capacity to engage in some sort of self-expressive autotelic activity exists in animals that have diverged from the human line many millions of years ago, and do not have the intellectual capacity to invent the mechanisms to provide the sort of satisfaction that is within their intellectual range.

Other examples of animals being able to gain satisfaction from far more complicated types of behavior than they can invent in nature could be multiplied, though none I think are more interesting than the artistic activities of primates. The capacity of seals to learn to perform on musical instruments and in some cases to get enjoyment from doing so, is perhaps another example; here we may suspect that an interest in the rhythmical

---

[5] For a series of 17th and 18th century satiric paintings of the ape as artist see H. W. Janson, *op. cit.*, chap. X.

[6] Morris, Desmond. The Biology of Art. London, 1961. 176 pp.

sounds of breaking waves on the rocks or beaches of a shoreline has some initial adaptive value. It is evident that in a sense the more highly developed mammals are preadapted to inventions that for most of them have not become available.

It is reasonably certain that a large part of human intellectual evolution must have consisted in the rare invention of such activities, painting, dancing, music, games, counting and elaboration of language, which once they had been achieved accidentally or by exceptional insight of a genius, caught on with a large part of, if not the whole, population.

Whatever the expressiveness that is required to put an object in an art gallery may be, it is clearly not quite confined to the genus *Homo;* as the evolutionist would expect, it has a history and this history can be traced outside our own genus or family.

If we are prepared to grant that at least some of the qualities present in a human painting are also present in a very rudimentary form in those of the great apes, we may legitimately inquire about certain other kinds of animal activity which seem to us to have aesthetic properties. Most conspicuous are the songs and displays of many birds, the latter perhaps culminating in the extraordinary activities of bower birds in collecting and arranging decorative objects.

We may in the present state of knowledge make the following statements about such activities.

They are all parts of adaptive behavior directed to ends that are significant in the life of the animal, notably the holding of territory, retention of interest in a mate and the like. The significance always implies some sort of social interaction. Though usually both innate and learnt behavior are involved, in many cases the behavior has a stereotyped innate component that is largely lacking when a human being sings, dances or paints, and for that matter when an ape is given the chance to do the last named. There is often a great discharge of neuromuscular activity which is reasonably regarded as comparable to what we know subjectively as emotion.

In a very large number of cases structures or activities used as social signals of a visual or auditory kind are found to be aesthetically significant to human beings. Apart from the fact that in many cases the activities involved are largely innate, which allies them perhaps more to elaborately grown structure than to learnt activity, the social and emotional aspects of animal display and the activities involved in the production of much so-called primitive art appear to be comparable. In both cases the aesthetic elements which we value are originally secondary to the social functions subsumed by displays or rituals. If we compare the voice of a peacock with his tail we get a clear hint that what is needed to produce the secondary aesthetic effect, is a considerable degree of elaboration. In all structures used in display, the elaboration is no doubt correlated with the need for quite specific signals different from anything else. The greater the elaboration of two structures the less the probability that they will resemble each other. Moreover, if we look at all the organisms which we at first sight would regard as strikingly beautiful in a decorative rather than a purely functional way, or for that matter inanimate structures which give the same sort of impression, we find that nearly all the extraordinary cases are the product of some sort of differentiation in a relatively free environment, in water, or growing up into the air or at least moving about above the ground, rather than burrowing in sand or mud. I give no examples on the screen; being within a major natural history museum, *si exemplum vis, circumspice.* Wherever there is a physical possibility of developing in a spacially unrestricted way in a context which either calls for or merely permits elaboration, we get natural beauty. Moreover in all cases we have more than a hint of what I initially referred to as a self-theorising property. The elaborate form tends to express deterministic laws that brought it into being, though often we may not know what they are but merely feel that the symmetry and elegance of the object before us implies a symmetry and elegance in the theory describing its genesis.

We have seen how the random irregularities of what I have called elegant ink blots are the vehicles for certain sorts of psychological projection, entirely irrelevant to the nature of the object, yet capable of giving considerable satisfaction under certain circumstances. We have seen also how in looking at an entirely conventional human work of art there can be an enormous subjective element in evaluation; a study of either forgeries or fashions in appreciation no less than overenthusiastic attributions would lead to a comparable conclusion. We have to go out to meet the work of art on some ground between it and ourselves to receive its message; the place where we stand may make all the difference.

We have further seen that there is apparently a continuum from conscious human works of art, through immensely beautiful but in purpose only secondarily artistic works of primitive art, to animal activities and structures employed socially and then to those that are not so employed, and so finally to inanimate structures which we recognize as beautiful. As we get further from the human work, we find that what we see as beautiful comes into being largely as elaboration in a relatively unrestricted space, whatever the actual mechanisms of its development. This happens because there are orderly processes occurring in nature and when they get a chance to show what they can do, they produce elaborate works in which symmetry and elegance in the external world suggest that, even if we cannot explain the process yet, and we often can, the explanation would involve elegant theory, which it often does.

Again as with human works, our viewpoint makes a considerable difference. An unforced feeling for how a form may arise can enhance its natural beauty. Some people may be willing to stop at this point, as every philosophical position, or lack of position, implies enormous difficulties. Others may want to go on further, feeling themselves in the presence of a message from nature or the external world which they go out to meet with their understanding. To be meaningful such a position would have, I think, to be theistic. It does not involve any logically compelling argument for the existence of God, but like each of the arguments on this matter, it makes its point if one is prepared to accept some of the others.

Meanwhile I think if the general trend of my line of thought makes at least partial sense, we can agree that in large measure the public exhibits in a good natural history museum are in some ways the modern counterparts of the nautilus cups and ostrich egg goblets of the Renaissance, constructed of both natural objects and a highly skilled kind of applied art. Yet they are far more important, because they are made to contain not wine, which anyway would be hard to drink from such objects, but scientific truths, made plain by the art with which the self-theorising properties of the specimens are exhibited. If the whole aspect of the work of a natural history museum is considered in this light, a taxonomically arranged set of diatom slides or drawer of insects, no less than a habitat group or your magnificent

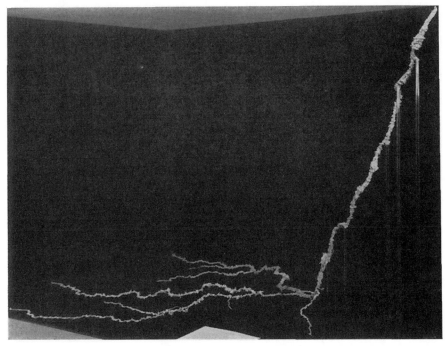

FIGURE 6. Fulgurite from Santa Rosa Island, near Fort Walton, Florida. (Academy of Natural Sciences, Philadelphia. Collected and mounted by Josephine Henry.)

fulgurite (Figure 6), are seen to have some of the properties of works of arts. Although I think there are good reasons for separating art galleries and natural history museums, they still, even after more than a century and a half of autonomous development, may have much in common.

Acknowledgments. I am much indebted to the Soprintendenza alle Gallerie, Florence, to Dr. Erwin M. Auer, Director of the Kunsthistorisches Museum, Vienna, to Mr. J. W. Goodison of the Fitzwilliam Museum, Cambridge, and the Director of the National Gallery of Art, Washington, for permission to publish the photographs reproduced in figures 1, 2, 4, and 5; to Mr. Emiddio DeCusati and the Yale Art Gallery for figure 3, and to Dr. Ruth Patrick and the Philadelphia Academy for figure 6. I am most grateful to Mrs. Sah Oved and Professor Charles Seymour, Jr., for assistance and information about several objects discussed, to Miss Yamaiel Oved for one of the ideas that I have used, and to my wife for much help.

THE CREAM IN THE GOOSEBERRY FOOL

By G. E. HUTCHINSON

THE Reverend Gilbert Henry Raynor, who was born in 1854 and died in 1929, spent a quarter of a century, and a third of his life, as rector of Hazeleigh, a small village near Malden in Essex. He is known today almost exclusively as an ardent collector of Lepidoptera, and apart from such activities, it is hard to learn much of him from what has been published. His two obituary notices in entomological journals emphasize his genial and helpful disposition, and it is by his helpfulness as much as by his skill as a lepidopterist that he achieved a small but definite place in the history of science. Fortunately, through the kindness of the Reverend C. G. Bartle, rector of the now combined parishes of Woodham Mortimer and Hazeleigh, Miss C. Evelyn Croxon, who knew Raynor well, prepared a manuscript account which she generously sent to me. She writes

"he was a kind friend to all his parishioners in their joys and sorrows; he had a ready wit and a keen sense of humor. His interests were wide and varied... His garden contained a wonderful collection of rare plants, bulbs and shrubs, and he was never tired of showing them, and explaining their origin to his friends. . . He was also a collector of old china."

"A peep into his study would reveal the life cycle of many rare butterflies and moths, which he bred with much success. As a child I was often instructed by him to collect caterpillars from various plants and shrubs, and I remember being rebuked by my mother for getting a rash on my hands from handling some of these creatures, but I felt nevertheless that it had all been very worthwile!"

How right Miss Croxon had been. She continues, "Mr. Raynor was a keen cricketer and tennis player, and he gave much help and inspiration to young people in their games." He clearly made the rectory at Hazeleigh a center of civilization for his small village during the quarter century of his incumbency.

Apart from this picture of Raynor in his prime, his friends C. R. N. Burrows (1929), and N. D. Riley (1929), record that his interest in natural history developed early; at sixteen, he was already publishing records in the *Entomologist* (Raynor 1870). He read classics at Cambridge, went to Australia for a time as a teacher, his collections made there passing to the British Museum (N. H.), and later taught classics at Kings School, Ely, and at Brentwood Grammar School.

Raynor is now mainly remembered as a collector and breeder of varieties of the magpie moth *Abraxas grossulariata*. This conspicuous white moth, somewhat unpleasant to taste according to Ford (1955), spotted with black and streaked with yellow, is common in the Palaearctic and has an even wider distribution in picture galleries, for it is one of the ancilliary subjects often introduced into still life paintings by Jan Breughel the elder, Jan van Kessel, and other 17th century artists of the

Low Countries. Eleazar Albin, who was among the 18th century Englishmen to continue the tradition of the artist-naturalist, dedicated plate XLIII of *A Natural History of English Insects*, on which the moth is depicted, to Mrs. Bovey of Flaxley, but whether she had any other connection with *Abraxas grossulariata* is unelucidated.

The species is familiar in Britain wherever gooseberries and currants are cultivated, the larvae feeding on their leaves and sometimes being a pest. The adult is, in fact, often called the currant or gooseberry moth. As with many spotted insects, there is much variation in size and disposition of spots, some of which may run together producing dark forms or become obsolete producing pale varieties. There are, in addition, three rather striking mutant genes, which have become of genetic interest and which produce in homozygous (or hemizygous) condition the aberrations *lutea* Cockerell, *varleyata* Porritt, and *dohrnii* Koenig.

The first of these, widespread but uncommon in nature, has the whitish areas of the wing suffused with yellow; when heterozygous the *lutea* gene usually produces a yellow tinge to the front wings (ab. *semi-lutea* Raynor).

The second of these aberrations, *varleyata*, has black wings with a sub-basal white stripe from costal to anal border. It was originally found by a Mr. J. Varley (1864) of Huddersfield, who bred a specimen in 1864 which was figured on the frontispiece of the first volume of *The Naturalist*, published that year, a rather fuzzy color plate. Later he is said to have obtained ten more specimens that he sold at £1 apiece. The aberration was apparently not named till much later, by Raynor's "old friend" and frequent critic, G. T. Porritt, an amateur entomologist of great knowledge who did not hesitate to get involved with difficult groups such as the caddis flies. Ab. *varleyata* has occurred very sporadically in nature in Lancashire and South Yorkshire. Near Huddersfield, collections made by two working men (Porritt 1905) gave, on one occasion, 15 specimens in 4000 pupae collected, which, for an autosomal recessive, as *varleyata* proved to be, corresponds to a gene frequency of about 1 in 16 in the population sampled.

The third of the important aberrations, *dohrnii*, better known in the genetic literature by Raynor's later name *lacticolor*, and also named *deleta* Cockerell and *flavofasciata* Huene, occurs as a rare sporadic form in various parts of England and the Continent of Europe, at least east to Esthonia. All the wild-caught specimens are females. The black markings of ab. *dohrnii* are much reduced and the ground color of the wing has a creamy tint.[1] It was perhaps this form that was noted almost two hun-

---

[1] Two forms, one due to a sex-linked recessive like *dohrnii*, reared by Poulton (Ford 1937), the other to an autosomal recessive (Woodlock 1916), are known in which a comparable but less extreme reduction of black occurs; at least in Poulton's form the ground color was whitish rather than cream.

dred years ago (1764) at Enfield Chase by Drury, who records a "Magpie (in the Eveng.) without any black spots on it scarcely an extraordinary odd Fly" (Hobby and Poulton 1934).

Raynor (1902–03) started serious rearing of the magpie moth in 1899, obtaining many larvae from different parts of England. The study of variation by amateurs had become fashionable, partly under the in-

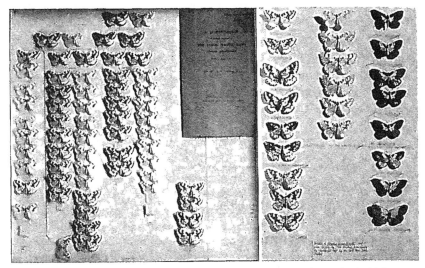

Left Hand Panel: Part of the material given by Raynor to the Zoological Museum, Cambridge University, including at the top a pair consisting of a wild type male on the left and a female of *dohrnii* on the right, the latter being a granddaughter of Raynor's original specimen of 1899, and the ancestor of most of the experimental families. The arrangement follows manuscript notes prepared by Doncaster and preserved in the drawer with a letter of Doncaster to David Sharp, and a copy of Raynor's Compendium, once belonging to Francis Jenkinson, presumably added by Hugh Scott.

Right Hand Panel: Forms of *A. grossulariata* reared by Onslow and in part derived from Raynor's breeding stocks. Left hand column; above wild type and five aberrations of the dark pattern, below three specimens of ab. *lutea:* center column; ab. *dohrnii:* right hand column; from top, two specimens ab. *varleyata*, the lower one with yellow basal band on forewing; two specimens of ab. *actinota* and one of ab. *leucosticta*, varying phenotypic expressions in the male of a gene causing white radiating markings in *varleyata;* one specimen of *exquisita* which is presumably *varleyata-dohrnii*, and two specimens of *pulchra*, with very wide basal band, genetically unanalyzed.

fluence of J. W. Tutt, who in his unfinished work on the British Lepidoptera attempted what was probably the most comprehensive and ambitious natural history of a group ever conceived. The first specimen of *dohrnii* to appear in Hazeleigh Rectory emerged on July 7, 1899, one of a large number of moths produced by larvae obtained from Lancashire. This specimen was mated to *grossulariata* and bred but, as we now would expect, its whole progeny was wild type. Fortunately, Raynor was not discouraged, and, breeding from these insects, obtained in 1901 rather over a score of the aberration, all females. When he published (Raynor 1902–03) an account of his experiments late in the next year, all the specimens of the succeeding generation were wild type *grossulariata*. It must have been about this time he came to know Leonard Doncaster. As Bateson (1921) wrote in his obituary of Doncaster, Raynor told the latter that all the specimens of the variety that he had bred were females. He also can hardly have avoided mentioning that his experience had indicated that the variety appeared in alternate generations. "At that time no example of what is now called 'sex-linked' inheritance amenable to experimentation had been studied. He [i.e., Doncaster] at once saw the extraordinary importance of the subject, and, as the result of correspondence with Mr. Raynor, matings were arranged and a critical investigation of the case was begun" ([Bateson] 1921). The results, disclosing a "Mendelian recessive of quite a new type," were first made known at an exhibit at the Cambridge meeting of the British Association in 1904 (Raynor and Doncaster 1905), and after more generations had been bred were described in a paper "On Breeding Experiments with Lepidoptera" published in the Proceedings of the Zoological Society under joint authorship (Doncaster and Raynor 1906). In this, two sets of experiments were reported, one of no great general importance, on *Angerona prunaria* "(experiments by L. Doncaster)," the other on *Abraxas grossulariata* "(experiments by the Rev. G. H. Raynor)." Raynor presented to the Zoological Museum, of Cambridge University, in the late summer or early autumn of 1907 "two cabinet drawers of *grossulariata* (chiefly ab. *lacticolor*), containing all the families that he had reared for heredity purposes and which were described by Mr. Doncaster and himself" ([Tutt] 1907). These specimens are still at Cambridge, with pedigrees and a suggested arrangement of the specimens written out by Doncaster. An ancestral pair of 1901, which were progenitors of most of the families studied, are included; the greater part of the material was bred in 1903 and 1904. Raynor therefore was certainly in touch with Doncaster between writing his account published late in 1902 and the breeding season of 1903.

In an obituary of Raynor by his friend the Rev. C. R. N. Burrows, it is suggested that part of Raynor's interest in the varieties of the Lepidoptera, and in particular of the magpie moth, was due to the recent rediscovery of Mendel's writings. It is, however, clear that he started work before this had happened. In none of his published notes or papers did he demonstrate any familiarity with genetic principles. He appeared more concerned with the beauty and strangeness of his specimens and with the extraordinary potentialities for variation that the species seemed to show. Moreover, at this time, as apparently earlier in his life, Raynor's entomological activities were not exclusively scientific and aesthetic, for, shortly after his munificent gift to the Zoological Museum at Cambridge, he put his collection up to auction, realizing £487, of which nearly £200 came from 170 specially selected specimens of *A. grossulariata*, each provided with an aberrational name. Raynor's mentor and friend J. W. Tutt ([Tutt] 1907), wrote of the Raynor sale.

> "We never saw such a jam of 'gooseberries' as at Stevens' room... when Mr. Raynor's collection was sold. Nor were there wanting samples of the gooseberry fool, mellowed though they were by some of that cream, which regards these fine aberrations as matters of scientific interest and not, as so many say, postage stamps."

Tutt felt that the sum realized was increased by the erroneous belief that some of the material was derived from the experiments done with Doncaster. The highest price of £6-10-0, was, however, paid for ab. *melanozona*, an aberration with fused spots forming a stripe across the fore-wing and for ab. *nigrolutea* in which the fore-wing is largely black and the pale ground color a fine yellow, rare and striking forms but not ones that had been the subject of any recorded genetic experiments. Higher prices, even allowing for inflation between 1907 and 1919, of up to 17½ guineas, were paid (Raynor 1920) in the latter year for 19th century specimens of the completely white ab. *candida* Raynor, which, like his counterpart in John Moore's (1946) *The Fair Field*, Raynor himself was never able to breed. Nor did he obtain the "pure black form which Mr. W. Beattie bred from Mickleham, and Mr. L. W. Newman from larvae of ab. *varleyata*. This I think may aptly be called ab. *nigra* n. ab. Should I be fortunate enough to rear either of these, I shall not say of *candida*, as Virgil did of horses 'color est deterrimus *albis*,' but in praise of *nigra*, I might be tempted to fire off the famous line 'Rara avis in terris, *nigroque* simillima cygno' " (Raynor 1909).[1]

Sales of collections of Lepidoptera were an important feature of English natural history at the time; many specimens, including the two of *candida* (Dr. Kettlewell disagrees with Raynor's judgement on their appearance, in fact in them *color est deterrimus albis*) that passed, under Mr. Stevens' hammer, from one famous collection to another, must now have found a more abiding home in the great Rothschild-Cockayne-Kettlewell collection at Tring. Sales of insect collections still occur, but the prices paid for varieties of *A. grossulariata* today are lower; a year ago a couple of ab. *nigra* fetched only five shillings (de Wurms 1963). The gooseberry fool seems rarer than at Raynor's sale in 1907 and is perhaps approaching extinction.

[1] The genetics of *nigra* are problematic and obviously peculiar; it may be, at least in some cases, an accentuated form of the black-dotted *nigrosparsata* (Porritt 1912 a,b 1914). Porritt claimed that the specimen bred by Newman is not really comparable to the other known examples of *nigra* which have nothing to do with *varleyata*. Cockayne (1915) figured a magnificent somatic mosaic, normal on the right, almost entirely black on the left, and later (1924) another extraordinary specimen, *dohrni* on the right side, normal on the other.

After the dispersal of his first collection of *A. grossulariata*, Raynor continued to breed the species almost till the end of his life, though accidents shortly before his death appear to have caused the loss of all his breeding stock. In his later years, from about 1916 onward, he became associated with the Hon. Huia Onslow, one of the most remarkable biologists of his time. Miss Croxon indicates that, in 1921, when Raynor retired, he went to live at Brampton in Huntingdonshire to be within easy distance of Cambridge, as he had many interests there. Foremost of these must have been Onslow and his experiments. Having injured his spine most seriously in a diving accident as an undergraduate, Onslow spent the remaining years of his life in a semirecumbent position. In spite of this immense handicap he achieved during the span of his scientific career, of little less than a decade, an astonishing amount of work. He died at the age of 32, a year after Raynor retired to live nearer Cambridge. Onslow was one of the founders of biochemical genetics, and also left an extraordinary investigation of the iridescent colors of butterflies. The greater part of his work, however, concerns the genetics of color pattern in moths. In the first of his seven papers on the subject, published in 1919, Onslow studied the yellow ab. *lutea* of *A. grossulariata*, proving, by an ingenious colorimetric method, that what looked like blending inheritance really involved Mendelian segregation. Again Raynor is standing in the background supplying breeding stock, information, and aberrational names. Onslow's pair of color plates illustrating the paper in fact provide the only published illustrations of some of Raynor's named aberrations. In the large pedigree that Onslow gives of his breeding experiments, the earlier part is due to Raynor who clearly kept careful notes of the descent of his specimens, even though his published remarks give the impression of quite unsystematic breeding. It is certainly incorrect to suggest, as Burrows did in his obituary, that there was no written record; rather it seems that it was destroyed posthumously (Cockayne *fide* Kettlewell, *personal information*). Onslow's work in turn led on to that of Ford (1940) who demonstrated that the incomplete dominance expressed in *semilutea* could be greatly modified in either direction by selection, so that whether the yellow color was dominant or recessive depended on the genetic constitution of the animal under observation. Onslow, in a later paper (1921), returned to *Abraxas grossulariata*, working this time with the melanic *varleyata*, obtained in part from Raynor, in part from G. T. Poritt. Onslow found *varleyata* to depend on an autosomal recessive gene, as was indeed fairly clear from the anecdotal data of the earlier breeders. He also studied a number of modifications of the *varleyata* pattern, including ab. *actinota* Raynor which seems to involve a peculiar type of perhaps sex-limited inheritance confined to the male, and ab. *exquisita* Raynor which appears to be the expression of homozygous *varleyata* in an individual that would otherwise be phenotypically *dohrnii* (*cf.* Ford 1937, 1955). Raynor, with his customary abstention from genetic interpretation, merely indicates that certain yellow forms derived from *dohrnii* had been crossed with *varleyata* in producing *exquisita;* he is more interested in expatiating on the beauties of what to the nonconnoisseur would look like an unfamiliar but by no means extraordinary insect. Onslow promised a further analysis of *exquisita*, but died before it was completed.

Though Onslow's work on *A. grossulariata* produced nothing as fundamental as sex-linked inheritance, the studies on ab. *lutea* are of considerable theoretical importance as a remarkable early example of the Mendelian analysis of apparent blending inheritance. As with Doncaster, so with Onslow, it was Raynor's breeding experience which made the work possible. In the case of Doncaster's work, the results represented an immense contribution to the study of heredity, which contribution led toward the chromosomal theory developed in America in the next decade, and so to all that has happened subsequently in genetics. Raynor's skill in breeding and his obvious delight in the protean beauty of the magpie moth lie behind this and so are one of the small roots of modern science—*exaltavit humiles.*

Acknowledgments

Apart from my very great debt to Miss Croxon and the Reverend C. G. Bartle, I should like also to express my thanks to Dr. H. B. D. Kettlewell of Oxford, Dr. John Smart of Cambridge, and Dr. C. R. Remington of Yale University who have helped in various ways in the preparation of this account.

REFERENCES

[BATESON] W., 1921, Leonard Doncaster 1877–1920, *Proc. Roy. Soc. Lond., 92* B, xli–xlvi.

BURROWS, C. R. N., 1907, Sale of the "Raynor" Collection of Lepidoptera, *Entom. Rec., 19,* 293–297.

[BURROWS] C. R. N., 1929, Obituary, The Rev. Gilbert Henry Raynor, M.A., *Entom. Rec., 41,* 139–140.

COCKAYNE, E. A., 1915, "Gynandromorphism" and kindred problems, with descriptions and figures of some hitherto undescribed examples, *J. Genet., 5,* 75–131.

———, 1924, A somatic mosaic or mutation in *Abraxas grossulariata, Entom. Rec., 36,* 17–20.

de WURMS, C. G. M., 1963, The "Canon Watkinson" sale of Lepidoptera, *Entom. Rec., 92,* 22–24.

DONCASTER, L. and RAYNOR, G. H., 1906, Breeding experiments with Lepidoptera, *Proc. Zool. Soc. Lond.,* 1906 *1,* 125–133.

FORD, E. B., 1937, Problems of heredity in the Lepidoptera, *Biol. Rev., 12,* 461–503.

———, 1940, Genetic research in the Lepidoptera, *Ann. Eugen.* 10, 227–252.

———, 1955, Moths (New Naturalist) London.

HOBBY, B. M. and POULTON, E. B., 1934, William Jones as a student of the British Lepidoptera, *Trans. Soc. Brit. Entom., 1,* 149–155.

MOORE, J., 1946, *The Fair Field,* New York, Simon and Schuster XIV, 240 pp.

ONSLOW, H., 1919, The inheritance of wing colour in Lepidoptera I., *Abraxas grossulariata* var. *lutea* (Cockerell), *J. Genet., 8,* 209–258.

———, 1921, The inheritance of wing colour in Lepidoptera, V, Melanism in *Abraxas grossulariata* (var. *varleyata* Porritt), *J. Genet., 11,* 123–139.

PORRITT, G. T., 1905, Abraxas grossulariata var. varleyata at Huddersfield, *Entom. Mont. Mag.*, *41*, 211.

PORRITT, G. T., 1912a, Melanism in Abraxas grossulariata, *Entom. Month Mag.*, *48* 214.

PORRITT, G. T., 1912b, Abraxas grossulariata, var. nigra, *Entom. Month Maq.*, *48*, 215.

———, 1941, On the breeding of the variety nigrosparsata of Abraxas grossulariata, *Entom. Rec.*, *21*, 270–272.

RAYNOR, G. H., 1870, Early appearance of Platypteryx lacertula, *Entom.*, *5*, 147.

———, 1902–03, Notes on Abraxas grossulariata and how to rear it, *Entom. Rec.*, *14*, 321–325, *15*, 8–11.

———, 1909, Further notes on Abraxas grossulariata, *Entom. Rec.*, *21*, 270–272.

———, 1920, A compendium of named varieties of the large magpie moth *Abraxas grossulariata* with label list. . . obtainable only of the author, Hazeleigh Rectory, Maldon, Essex.

RAYNOR, G. H. and DONCASTER, L., 1905, Experiments on heredity and sex determination in *Abraxas grossulariata*, *Rep. Brit. Assoc. Adv. Sci.*, 1904 (Cambridge), 594–595.

R[ILEY] N. D., 1929, Rev. G. H. Raynor, *Entom.*, *62*, 239–240.

[TUTT, J. W.] 1907, Current Notes, *Entom. Rec.*, *19*, 304.

VARLEY, J., 1864, Remarkable varieties of *Abraxas grossulariata* and *Arctia caja*, *Naturalist*, *1*, 136–137.

WOODLOCK, J. M., 1916, Some experiments in heredity with *Abraxas grossulariata* and two of its varieties, *J. Genet.*, *5*, 183–187.

On Being a Meter
and a Half Long

G. EVELYN HUTCHINSON

I AM PARTICULARLY PLEASED to be contributing to the same symposium as Sir Kenneth Clark, known to many readers as the author of, among other books, *Landscape into Art*,[1] because the subject of much of my professional activity has consisted of what may perhaps be called *Landscape into Science*. I would begin by reminding you of what Sir Kenneth says about the Predella of Gentile da Fabriano's *Adoration of the Magi* in the Uffizi, namely, that it has often been noted that here for the first time the sun shone in a picture. This "great gold sun that gardener spring has brought into perfection," in Edith Sitwell's beautiful words,[2] is admittedly stylized. It is shining as the result of the work of an artist who has a great and rather conservative medieval tradition behind him, and it illuminates a landscape, vernal as it may now seem to us, which was the scene, as Huizinga[3] has so clearly demonstrated, of an age of anxiety. At the present time the sun shines much less in pictures. The anxiety that our age feels has been transfused throughout our art. It is only in the works of scientists, who seem wisely or unwisely to be happier, less anxious, and more simple-minded people than do our contemporary writers and artists, that the twentieth century can allow a little sunshine to appear.

I hope that my landscape is sufficiently full of light to be recognizable; it is a landscape with figures some of which we learn to recognize as ourselves. In my remarks about this landscape with figures—plant, animal, or human—I shall emphasize those kinds of study most

G. EVELYN HUTCHINSON

appropriate to natural-history museums, because a major function of the Smithsonian Institution is to provide a home for the greatest of such museums in the New World.

There are, I think, two rather different ways of looking at nature, which may be termed *extensive* and *intensive*. The greatest investigators doubtless combine both methods, but in any one discovery or group of discoveries we can usually recognize the predominance of one point of view. The final Newtonian triumph of celestial mechanics was an intensive triumph, though not without some extensive background, notably in the then existing body of astronomical observations used by Newton's predecessors. The theory of natural selection of Darwin and Wallace was an extensive vision, though obviously not without intensive elements of theory. The same is probably true of Freud's exploration of the unconscious, which is usually regarded as having produced the third revolution in the intellectual life of the past half millennium. The dichotomy is perhaps basically less a methodological one than a psychological distinction of the kinds of things people like thinking about. The extensive worker prefers sets of examples as subjects for his initial speculations and keeps these or comparable examples in mind throughout. The intensive starts perhaps with a single hint in building a new deductive theory and may not bother to go back to look at nature until many steps later in the mathematical development of the initial intuition.

It is commonly believed that the first steps in any science will be extensive, while the culminating developments will be intensive. Most working investigators probably do not bother very much about this unless they are forced to do so for economic reasons. Many publicists and professional administrators worry very much too much about the matter because they are concerned about making sure that really contemporary developments are considered in their editorials or supported by their institutions. One can easily be told that major academic institutions should support taxonomy only if it is numerical or non-Linnaean or whatever happens at the moment to be the most recent approach. It is considered far more important to be up-to-date than to be interesting and useful. The fact is, however, that except for treatments of the flowering plants, birds, and butterflies of Western Europe and North America, no major area possesses adequate taxo-

ON BEING A METER AND A HALF LONG

nomic treatises on any large group, in which all the significant known complexities are clarified. The production of such works, if done really well, using all modern methods when they are appropriate but not as ends in themselves, would be extremely useful to both pure and applied biologists—a fact that should be sufficient justification for really large-scale support of what is often called old-fashioned. This means, besides adequate institutional aid, which in this country is most fortunately often forthcoming, an active policy of not discouraging people who like such work from doing it. The social and financial pressures that can be put on students and young extensive-minded investigators who might be first-rate taxonomists to become second-rate experimentalists are considerable and in some branches of biology definitely injurious. To a person of the requisite temperament, the construction of a good key to a genus, as readers of Elizabeth Sewell's *Orphic Voice*[4] may suspect, is an activity more clearly allied to the writing of poetry than is any other branch of science. A taxonomic key is in fact a special kind of poem which happens to be of great practical utility. I have heard of people who, knowing little or nothing about plants, have read Bentham and Hooker's *British Flora*[5] for pleasure, just as on a higher level of poetic achievement one can read Shakespeare's sonnets without worrying about the identity of Mr. W. H. or whether the initials may not be, as has recently been suggested, the anagrammatic inversion of those of someone called Harold Wilson.

There is a correlative aspect of taxonomic research which is of some interest in relation to any organization such as the Smithsonian Institution that is directly responsible for collections. From a strictly scientific point of view the results of an investigation constitute a set of propositions, of less or greater generality, that are stated in some more or less formal language. The maintenance of the collection in a museum is, from this point of view, merely something preserved in case more propositions can be based on it. The question arises, however, whether propositions about material objects based on nothing that now materially exists would be in many cases of any great interest. A very good case can be made for the most careful preservation of samples of the things about which we have knowledge, which otherwise might seem remote and unreal. This is true whether or not

85

G. EVELYN HUTCHINSON

the object in question has any immediate practical relevance. Dinosaurs and dodos, apes and elephants, all serve to give us a rich time dimension and help us to avoid the all too common triviality of living in the moment as a continuous prelude to rushing thoughtlessly into the future. The whole beauty of nature and of man's work is needed to tell us what the world can be and what there really is to enjoy if we look carefully. The provision of this sort of enjoyment, which is becoming more and more difficult as there are more and more people to be satisfied, needs very careful consideration. A world of boxlike apartment houses provides little that seems marvelous and beautiful. It may be necessary for the majority of people to be engaged in producing and maintaining such a world, but it is also necessary for certain people to make sure that there are an adequate number of emeralds, giraffes, *Welwitschia* plants, birds-of-paradise, *Sequoia* trees, swallowtail butterflies, and giant tortoises, as well as music, painting, and sculpture, to provide the marvelous and the beautiful. This is by no means an insignificant function of the zoological gardens and the national museum. As I have pointed out elsewhere, these activities grade insensibly into those of art galleries.

What I have just been saying is related to an even broader problem that is very seldom considered. I was thinking about part of this address in a botanic garden, founded in the seventeenth century, in which there are a number of rectangular beds set in a lawn, each devoted to a family of flowering plants. Many of these plants were in flower at the time I was there, and the similarity and diversity of the members of each family were obviously apparent and attractive to the eye and the mind. As I was told that certain biologists regarded this display as of little or no educational significance, I began to think about such matters in terms of what was before my eyes. Much of the information set forth in the labeling, grouping, and choice of specimens has been available for two centuries, some of it for much longer. The details of classification and most notably its phylogenetic interpretation have greatly developed in the past century, but what Linnaeus or even Ray wrote can be read comprehendingly today. Much of the information provided by the display in the garden is indeed old-fashioned, in the sense that the propositions embodying it were fashioned some centuries ago. No one denies that if one happens to be interested in the biogenesis of a particular alkaloid it is desirable to

86

ON BEING A METER AND A HALF LONG

have available expert knowledge as to the systematics of the Papaveraceae so that the right plant for the investigation can be obtained. The question that seems to be raised is whether this sort of information is so esoteric and specialized that it is not worth while implanting it in the minds of students of biology or the general public. In other words, are we living in a world that is inevitably so artificial that the rich diversity of natural objects is an irrelevance of no great educational significance?

Actually the current trend is to go even further, so that the only process that is regarded as of any significance is the obtaining of new information which is stored in libraries or other nonhuman memory stores, from which it can be obtained by taking a book off the shelf or by retrieval from whatever inanimate memory store may be used. Only information acquired in the past five years is often now regarded as significant in some branches of science. The period during which new knowledge retains its bouquet moreover seems to decrease steadily, so that if the present attitude persists we might expect the content of science to lose its significance in five months, five minutes, and ultimately five milliseconds or whatever period is needed to get it from recorder to storage unit. In the end no one would know anything except how to keep the apparatus growing, and the learned man would have been automated out of existence. At present he is indeed often condemned to the position of quarryman or miner, encouraged to produce with ever-increasing rapidity more and more information about nature. When a quarry or mine is exhausted the worker must find a new one, if he is young enough. The pleasure of the chase is no doubt great, but it easily becomes obsessive, producing no real satisfaction. This seems to be the situation developing in our universities, the persistent argument between research and teaching being largely an argument as to whether it is more important to get new knowledge or to enjoy the whole form, insofar as we can see it, of what we have. An obsessional attitude to the chase and an obsessional attitude to possession—the attitudes of the hunter or of the miser—are equally inappropriate in their psychopathic form in either learning or teaching. Because Ray or Linnaeus loved what they had found out does not mean that we cannot enjoy it; we merely are more fortunate in having more to enjoy.

It is not, however, only with taxonomy, where almost the whole

87

G. EVELYN HUTCHINSON

effort is extensive, that I am concerned, but rather with the complete science of landscape with figures, or what is variously called synecology, community ecology, or biocoenology. Here the significant feature is that we are called on to consider the extensive comparative relationships as the subject of intensive study. At the present time this sort of approach is developing a strong evolutionary complexion; in fact at the moment scientific natural history, the proper subject of natural history museums, includes as its central activity what has been called evolutionary ecology.

Since, as I have indicated, we are inevitably among the figures in the landscape and, being roughly 1.5 to 2.0 m. in length, fall into an intermediate size range, though in the larger part of that range, of living organisms, we can only with difficulty get outside the landscape to look at it, and even then we have to allow for the disturbing effect of our activities. The result is that it is rather hard to get a clear idea of the nature of the subject that we are trying to approach. I believe this is the fundamental difficulty in making clear to people outside the field what it is that ecologists actually are trying to do. Perhaps this is best illustrated by a concrete example.

The example that I shall use is the elegant recent work of MacArthur and Levins,[6] who showed, by a simple but quite deep mathematical approach, that given a minimum set of reasonable and very general postulates about possible food habits in competing animal species, two extreme types of evolutionary path are possible. Taking one, the animals tend to develop behavioral mechanisms decreasing the probability of their paths crossing, so emphasizing local ecological allopatry, or alloecism as we may call it; taking the other they tend to develop increasing structural diversity while living in such a way that their paths continually cross, in ecological sympatry or synoecism. In both cases specialization will have occurred allopatrically, but the way in which the species build up communities will be different. Once the theory is developed the dichotomy appears obvious, and once we begin to look for examples, the higher more mobile metazoans provide them in numbers.

A very large amount of work is in fact accumulating that suggests that in most families of insects, in which there are often a number of sympatric species, these species tend to be separated by their choice of specific parts of the biotope in which to live, even though there may be little or no difference in the kind of food resources that are used. In other cases, as in monophagous insects, the behavior will in fact involve a particular choice of food, but, as seems to happen so often in the true sucking bugs of the family Miridae, once the bug is on the right plant it can be either herbivorous or feed on other insects already there. In general, the first case of MacArthur and Levins is likely in small animals in biotopes providing a great deal of mosaic diversity, large enough for each kind of diverse element to be the habitat of one of the species present.

In contrast to this we have in larger more mobile and generally carnivorous animals a marked tendency to synoecism, which is possible only if the species already differ in some way permitting rather different utilization of the varied resources of a habitat in any part of which there is an equal chance of all the species occurring. In general we should expect it in animals whose dimensions are greater than the elements of the mosaic pattern of the habitat, so that as they move about, though the different elements of the mosaic satisfy different needs of the animals, they do not supply specific habitats to different species. Cases, particularly in rather undifferentiated habitats such as the open water of lakes and ponds, can easily be found of structural divergence of allied synoecic animals of very small size, while much larger animals such as rodents and insectivores may show much alloecism, yet on the whole it is reasonable to suppose that the synoecic evolution of communities will mainly occur in rather large animals.

The two extreme kinds of evolution which seem indicated by the MacArthur and Levins[6] approach will have very different paleontological results. In the vertebrates, and most notably in the mammals, sympatric species will probably always be capable of some crossing of paths and as such, if they tend to feed on the same general sort of food, will be subject to evolution by character displacement. At least in mammals what is fossilized is nearly always bone, that is to say the hard structures to which trophic and locomotory muscles are attached. Any adaptive change in movement or feeding is likely to involve skeletal adjustments, either in size or form, or most often in both.

In a great many invertebrates in which the most important differentiae are pleiotropic concomitants of physiological efficiencies in slightly different habitats and of the response mechanisms by which

G. EVELYN HUTCHINSON

these habitats are found, we may expect to have far less obvious progressive evolution along adaptive lines because the evolution of a specific set of responses, maintaining an organism in a very special optimal habitat, is most unlikely to be recognizable from visible changes in the available fossils.

We thus may, at least in the Metazoa, recognize two modal types of evolutionary change. One is most likely to occur in animals relatively small compared with the mosaic structure of the environment, and involves initially the evolution of mechanisms maintaining the organism in its optimal habitat.

The other is most likely to occur in animals relatively large compared with the mosaic structure of the environment and involves initially the evolution of mechanisms favoring specialized efficient utilization of a particular but not spatially restricted part of the total resources of the habitat.

Often the two types of evolutionary change will be concomitant, but where the first is predominant it will be hard to detect in paleontological material any clear indications of the nature of the adaptations involved, as they are primarily behavioral. In the second type skeletal material should often reflect the adaptation.

Most groups of insects probably exhibit mainly the first type of evolution, most groups of mammals mainly the second. Although in some cases of moderately small animals, such as some carnivorous water bugs, there can be three or perhaps even six sympatric species living synoecically, it is probable that in large mobile animals which provide most of the cases, the synoecic groups will consist only of two or three species per genus. Where more appear, rather subtle alloecic behavioral mechanisms, such as those demonstrated by MacArthur[7] in the American warblers of the genus *Dendroica*, are operating.

In our own species, in which we are large enough to move about over enormous stretches of habitat, it is quite clear that there is no specific part of the terrestrial habitat to which we are limited by innate response mechanisms. At present there is no other species of our family Hominidae living, and so competition with a closely allied species no longer occurs. It is probable from the finds of Leakey that there were a pair of sympatric Hominidae in the early Pleistocene and that these differed in food habits and to some extent in structural

90

differences that were correlated with the difference in food. Even though the MacArthur and Levins dichotomy is not entirely exclusive and not hard and fast where it is applicable, it does at least permit us to recognize in our own nature, in being animals large enough to walk over the ranges of innumerable smaller organisms, large enough to develop an adaptable nervous system independent of inbuilt response mechanisms, and large enough to live more than a season and so have time to learn with our large brain, that we are very different organisms from insects.

Perhaps distinguishing man from insects is not a great feat, but what I want to do is to show that evolutionary processes, even though they all have a basis in natural selection leading to allopatric speciation, are nevertheless of a number, but probably a limited number, of recognizable kinds, of which the kind producing a vast number of insect species is very different from the kind producing ourselves. I want to emphasize this in the present context because it seems to me that evolutionary ecology in conjunction with taxonomy is a most significant activity for a major natural-history museum. The example that I chose led to a very striking if somewhat obvious dichotomy, but other examples of equal human interest could easily be used. The great advantage of a comparative approach, particularly if we are dealing with ourselves, is that it forces us to examine things that are so familiar and obvious that we are apt not to consider their meaning. Instead of comparing large animals and small, we could compare other major kinds of ecological or behavioral categories, diurnal animals with nocturnal, or social with nonsocial, and again find characteristic kinds of evolution. That the right point of view here can lead to the recognition of unsuspected matters of great importance is shown by the beautiful work of Allison Jolly[8] on the differences in the social life of lemurs in Madagascar, which suggest a wealth of new ideas about human evolution. At the present time there is obviously an enormous potential field opening up in evolutionary ecology; we may look forward to the various relevant branches of the Smithsonian Institution producing many "delightful truths," as Sir Thomas Browne would have called them, from this field before a third centennial of James Smithson's birth.

91

G. EVELYN HUTCHINSON

REFERENCES

1. CLARK, SIR KENNETH MACKENZIE, Landscape into art, xix + 147 pp. London: J. Murray, 1949.
2. SITWELL, EDITH, The child who saw Midas in Troy Park, 103 pp. London: Duckworth, 1925.
3. HUIZINGA, JOHAN, The waning of the middle ages, viii + 328 pp. Translated by F. Hopman. London: E. Arnold, 1924.
4. SEWELL, ELIZABETH, The orphic voice: poetry and natural history, 463 pp. New Haven: Yale University Press, 1960.
5. BENTHAM, GEORGE, Handbook of the British flora, 6th ed., lxxx + 584 pp. Revised by SIR J. D. Hooker. London: L. Reeve, 1892.
6. MACARTHUR, ROBERT, and LEVINS, RICHARD, Competition, habitat selection, and character displacement in a patchy environment. *Proc. Nat. Acad. Sci.,* vol. 51, no. 6, pp. 1207–1210, June 1964.
7. MACARTHUR, ROBERT H., Population ecology of some warblers of northeastern coniferous forests. *Ecology,* vol. 39, no. 4, pp. 599–619, Oct. 1958.
8. JOLLY, ALLISON, Two social lemurs. Chicago: University of Chicago Press. (In press.)

CHAPTER TWO

AYSTHORPE

WHEN I WAS NINE I was enrolled in one of the three private schools for boys, in local parlance, preparatory schools, in Cambridge, this one being Saint Faith's. We had moved from Belvoir Terrace to a larger house set in a garden of just under an acre, and the school was about three minutes walk from this house that we called Aysthorpe. The house was built on land leased from Trinity College. The road on which it stood, Newton Road, was the first of several in a small development and was called after Trinity's greatest alumnus. Next came Bentley Road, and after my time Barrow Road followed.

The house looked out on meadows, over which kestrels (*Falco tinnunculus*) hawked and skylarks (*Alauda arvensis*) sang. The yellow and black warningly colored caterpillars of the cinnabar moth (*Hypocrita jacobaeae*) fed on numerous plants of ragwort (*Senecio jacobaea*) among the grasses, while in the swampier places ghost moths (*Hepialus humuli*) could be seen appearing and disappearing as they fluttered in the summer dusk. A flowering almond stood by the front gate, and I remember one Sunday evening in May, when the tree was in full bloom, a nightingale (*Luscinia megarhynchos*) chose it as a singing place.

23

24   THE KINDLY FRUITS OF THE EARTH

Aysthorpe was a name synthesized by my parents from
Ayscough, Sir Isaac's mother's maiden name, and Wools-
thorpe, where he was born. The house was built for my
parents by an architect called Alan Munby, an old friend
of my father, who had spent most of his professional career
designing laboratories, on which subject he wrote a book.
His son was later librarian of King's and a very respected
bibliographic scholar. Munby secured from somewhere a
set of plaster zodiacal signs that he set, to our great delight,
in a circle in the ceiling of the dining room. My parents
must have become rather sick of hearing us chant at
mealtimes:

> The Ram, the Bull, the Heavenly Twins
> And next the Crab, the Lion shines
> The Virgin and the Scales
> The Scorpion, Archer and He-goat
> The Man who holds the watering pot
> And the Fish with glittering tails.

This room was the scene of the disgraceful incident
I described in "Cambridge Remembered," where my
brother and I locked in all the guests, including Sir George
and Lady Darwin, at a dinner party with my parents, and
then turned out the lights at the main.[1] Being presumably
now the only living person to have incarcerated one of
Charles Darwin's sons, has often provided a useful link
with the past when giving a class on evolution.

Saint Faith's, at the corner of Newton Road and
Trumpington Road, taught everything that a young gentle-
man was supposed to know: English poetry, at first
Macaulay's *Horatius* and *The Battle of Lake Regillus*, which

---

1. G. E. Hutchinson, *The Enchanted Voyage* (New Haven and
London: Yale University Press, 1962), (reprinted., Greenwood Press,
1978) see pp. 149–56.

didn't have enough about the lake for me, but later, to my lasting delight, we did Chaucer's *Prologue to the Canterbury Tales*, Gray's *Elegy*, Milton's *L'Allegro*, *Il Penseroso*, and *Lycidas*, and Coleridge's *Ancient Mariner*. English history, geography—in retrospect largely lists of the products of the British possessions in Africa, each list ending in pignuts —arithmetic, algebra, and geometry, French and Latin constituted the rest of the curriculum, with, for all but me, Greek. My father insisted that I do more mathematics instead, which was his only educational mistake. His similar insistence that I was also to do algebra instead of the ridiculous exercise of writing unspeakably bad Latin verses was obviously wise. It did, however, keep me from the unconscious English riches of some of the textbooks on that infertile subject, which, among the examples to be done into Latin, rise to the heights, so I understand, of:

> Grinder, jocund-hearted grinder
> By whom Barbary's agile son,
> Deftly poised upon his hinder
> Paw, accepts the pro-offered bun.

Organ grinders went round the streets of English towns, pushing barrel organs which they played at intervals. The better ones each had a monkey in a red jacket with a mug in which to collect the "penny to play in the next street." As children we loved them. They all came from Cassino in central Italy, but, alas for the veracity of the poet, the monkeys I believe were mainly *Cebus* from South America.

The place of Greek in education had for long played a large role in university politics. The language was required for entrance to the older universities, but at least at Cambridge a growing number of dons were opposed to its retention in what was officially the Previous Examination, and colloquially the Little-Go, which gave entrance to all higher studies. A few Hellenists were violently

26   THE KINDLY FRUITS OF THE EARTH

opposed to this change, and among them none more so
than T. R. Glover.

My sister relates in a letter of 12 October 1977 that
"Mother had sat next to Glover at a dinner party on a
Saturday night and had had a heated discussion on the
value of compulsory Greek in schools, he defending and
she attacking it. Next morning he was to give one of a
series of Ecumenical sermons at St. Edward's Church, so
we abandoned St. Botolph's and trooped down to St.
Edward's. Arriving (as usual) rather late, we were shown
into the front pew. When the time came for the sermon
Glover leaned forward from the pulpit and announced
his text 'And the centurion said unto Paul, canst thou
speak Greek?' Collapse of Hutchinson family, as mother
had regaled us with an account of the discussion all through
breakfast." I have no recollection of this, but if it occurred
in term time with more than one child at home it must have
been during my time at Saint Faith's. Greek was finally
made optional in the Little-Go shortly after the First
World War.

A very important aspect of going to a boys' school
involved changing one's name. In the largely feminine
world of childhood I was Evelyn; now I had become, at
school, Hutchinson. Of course, I already understood about
Christian names and surnames, but the latter were empirically unimportant. My cousin who was very close to me
was a Stewart, my uncle who was important to me, a
Shipley. Becoming "Hutchinson" was in a sense a process
of initiation. It meant belonging in a large world in which
a few other people whom I learned to feel were relatives
had the name, but, more importantly, belonging in a small
world, the school, where it happened that initially no one
else bore the name. It was, moreover, a process that only
happened to boys. In some cases it would emphasize primo-
geniture. My brother had had rheumatic fever when he

was about eight years old; in Cambridge it was commonly, if unconsciously, believed that Hippocrates would have disapproved of the local air, much preferring that of Surrey. Accordingly, it was felt wiser for Leslie to go to a boarding school at Hindhead, so he escaped being Hutchinson 2 at Saint Faith's after a year. Later he was Hutchinson minor at Gresham's School, Holt, which must inevitably have been somewhat galling. Moreover, it could happen quite accidentally. Robinson tertius was no relative of Robinson major and minor. Not being well liked, he soon became Robinson dirt and was still less liked.

Though most adult men called each other by their surnames ("My dear Darwin" and "My dear Huxley" would be typical Victorian salutations), there was some use of nicknames, often derived from the surname in a way parallel to derivation within a family from a Christian name. Thus E. A. Wilson of the Terranova Expedition seems to have been Dr. Bill on the Antarctic continent. There was another usage denoting intimacy, namely the use of Christian-name initials. This was very common in some circles in late Victorian and Edwardian times. My father never called my uncle anything but A. E. Moreover, he used to call me G., which I think was an elision of G. E. and denoted a desire to think of me as a close contemporary friend. Among people of my age the practice must have been virtually extinct, though one American who became a fellow of Saint John's in the 1930s always addressed me as G. E. when we met. By the time that I was an undergraduate, men were beginning to use Christian names for their close friends. This involved, I think, some rejection of the arrogant masculinity of the schoolboy and an acceptance of more mature values in which tenderness was not out of place. While adolescent, the boys I knew, when we went to young people's dances, would always call the girls by their Christian names, Diana, Barbara, and so on. When one got to college, every new acquaintance was, in ordinary mixed contexts, Mr. and Miss, until one was ready for kissing, of which there was far less than today. The Christian name was not merely adult, but it had a definite sexual connotation; thinking of its use to someone attractive thus became a mild but very charming erotic fantasy.

The full sequence of names used for one's close associates in late Victorian times would then have been,

childhood	Christian name
boyhood	surname, often with nickname
early manhood	surname, initial name, or nickname
courtship, love, and marriage	Christian name.

By my time the Christian name was beginning to play both the last two roles.

About women's names I know little, but I suspect an almost universal use of Christian names in private, with Miss becoming more used publicly as the girl matured. There was, of course, a movement by "advanced" young women in the 1920s to imitate men, of which movement poor Carrington is perhaps the only significant memorial. Insofar as the practice was not a meaningless imitation of a rather tiresome male world, it obviously denoted faith in one's powers to do something as well as a man could. Unfortunately, in Carrington's case, everyone knows that she wanted to be called Carrington and that she died for the love of someone who was no doubt personally much inferior to her, however eminent a Georgian he may have been. Between name and death practically nobody till recently considered the fate of her pictures and whether she had been a good painter.

The mandatory change in all this was the adoption of the surname as the main appellation of a boy going to school. Whether later a nickname or an initial name were used, and when a young woman was no longer Miss, depended on personal judgments. One could always be wrong and so any situation involving names initially might be embarrassing. It generally was not, but the fear that it might be was real.

The headmaster of Saint Faith's School was Henry Lower; he was assisted by his sister and a small staff. Mr. Lower taught the two upper classes side by side and in retrospect he gives the impression of having been able to teach Greek, Latin, and algebra simultaneously. I can remember nothing much about the other teachers, save that I was first taught arithmetic by a Miss Chrystal, who I think recently had been, or was about to become, a student at the university. I hope, and indeed suspect, that she was the daughter of Professor George Chrystal of Edinburgh, the great investigator of the seiches or periodic water movements of lakes.

The curriculum was completed by cricket, in which I excelled negatively, in spite of the beauty of the game, and by soccer, which on one occasion I played very well. My subjective experience of adrenalin is ultimately based, so I imagine, on that one match. Almost everything I learned at Saint Faith's has proved of great value, but while I was there I was also conducting my own education in other ways.

The country around Cambridge lies on Cretaceous rocks and in my boyhood there were exposures of the whole sequence from the Lower Greensand to the Upper Chalk in cuttings, clay pits, and more or less abandoned chalk pits, from some of which, with tenacity, very nice fossils could be obtained. Further north at Upware on the edge of the fen country a lime pit exposed late Jurassic coral rock with fragments of tropical-looking sea-urchin spines, an unbelievably romantic circumstance to a small boy. In

30   THE KINDLY FRUITS OF THE EARTH

geological studies I had of course help from my father, who
always traveled on our holidays with a geological atlas
of the British Isles, so that we could appreciate from the
window of the train the ultimate nature of the countryside
through which we passed. The geological structure of
Britain, with the usually great dip of the beds, permitting
one to pass one formation after another in a distance of a
few tens of miles, was just what one needed to learn about
rocks or, in my case, where to look for fossils. I was also
greatly helped by two other older friends. One was Dr.
Cowper Reed, a very eminent professional paleontologist
and curator of the Sedgwick Museum of the Department
of Geology in the university. The other was the Reverend
J. W. E. Conybeare, a local antiquarian and fossil collector
who had been vicar of Barrington, famous for a gravel pit
that had yielded many *Hippopotamus* bones. He had, however, become converted to Roman Catholicism and now
lived privately in Cambridge. He had a large collection of
specimens from the Cambridge Greensand, a very narrow
bed full of phosphatic nodules which had been extensively
worked in the nineteenth century and had produced an
extraordinary fauna, including a few bird bones. Only one
exposure still existed, at the top of a very deep clay pit;
I managed to make friends with the workmen and visited
the locality, but I cannot remember what I found. Fortunately I presented the one or two really good fossils
I had collected near Cambridge to the Sedgwick Museum,
for the rest were given away, without my knowledge, when
I went to South Africa.

The churches in the country around Cambridge contain a number of medieval brasses, and brass rubbing, now
so popular that its devotees flock from Bengal and Oklahoma to Oxford and Westley Waterless, was a favorite
avocation of a few of us. I had learned the art, rather imper-
fectly, from an older boy called George Anderson who

was technically a superb brass-rubber and who had traveled in Belgium and Germany rubbing some of the extraordinary brasses in the Continental or "Flemish" style.
I hope his collection survives, as some of the originals, after
two wars, well may not, and his rubbings could hardly
have been excelled. I am told that I passed on the art to
Michael Ramsey, later one-hundredth archbishop of Canterbury, though I greatly regret not having any memories
of this; the Ramseys were valued friends. I have written
elsewhere of Frank Ramsey mystifying a children's party
by choosing the "left horn of a dilemma" as the object to
be guessed in a game of Animal, Vegetable, or Mineral,
which we called "clumps."

Three miles south of Cambridge lay Sir Roger de
Trumpington, a magnificent knight and the second oldest
survivor in Britain.[2] A short bicycle ride to Fulbourn
brought one to William de Fulburne in a splendid cope,
under a canopy, and on the way home one could speculate
on the meaning of lunacy when passing the county asylum.
At Hildersham a skeleton in a shroud yielded a rubbing
useful for scaring the housemaid. Brasses, moreover, led
to heraldry, which when once recognized greeted one from
every building. "Barry argent and azure, an orle of martlets

2. Sir Roger de Trumpington, whose effigy must adorn walls
all over the world, died in 1289; his armor fits this date, but the figure
was originally surrounded by a fillet, of which only the indent remains,
bearing an inscription. No other brass or indent before 1300 has such
a fillet, the earliest inscriptions always consisting of letters set separately
in the stone matrix. If the figure and fillet are coeval, the brass probably
dates from c. 1300. However, it is set on an altar tomb under an ogival
arch, which looks even later. There is some indication that the heraldry
of the figure has been altered. Sir Roger's brass may well have been
reused on the tomb of his son and successor Sir Giles, who died in 1327.
If this happened the inscription on the fillet would be later than the
effigy (see *Inventory of the Historical Monuments in the City of Cambridge*, part 1, pp. cvii–cviii).

32   THE KINDLY FRUITS OF THE EARTH

gules," which describes the blue and white bars and ring
of little red birds of one half of the Pembroke arms, is
surely one of the most beautiful abstract lines in any
language. I began a book on heraldry, of which three
completed chapters still exist. It was to have been very
learned and exhaustive, so it had to mention all the very
rare tinctures or heraldic colors used occasionally by Continental heralds but regarded in England as foreign bar-
barisms. I remember trying to make *cendré* paint out of
ashes from a fireplace to illustrate a coat of arms with this
ash-gray field.

The loss of brasses at the Reformation and after was
appalling, initially because they asked in Norman French,
Latin, or English, "Dieu de sa alme eyt mercie," and later
because, broken and outmoded, they reached a good price
as scrap metal. Other remarkable things, however, might
be found in Cambridgeshire churches. Ickleton, which
takes its name from the pre-Roman Icknield Way, has a
set of Roman monolithic pillars capped by Norman arches
and is said by Pevsner to be "far too little known."[3] I won
a prize for a highly unoriginal essay on its architectual
history when I was at Saint Faith's, thirty-five or forty
years before Pevsner wrote.

Another favorite was Hauxton, a small church with a
beautiful Norman south door by which, in the wall, is a
scratch dial, a very primitive sundial with a hole for the
gnomon, for which the user of the dial supplied his own
piece of stick. Inside, at the end of the south aisle of the
church, is one of the very few surviving medieval paintings
of Saint Thomas of Canterbury, a bête noire of Henry VIII,
who attempted as far as possible to blot out the archbishop's
memory. Sundials were a great part of my childhood, as

3. Nikolaus Pevsner, *The Buildings of England*, B. E. 10 Cambridgeshire (London: Penguin Books, 1954), see p. 331.

my father had a passion for them. He not only possessed some very good sixteenth- and seventeenth-century pocket dials, now in the Whipple collection of the Department of the History and Philosophy of Science in Freeschool Lane in Cambridge, but also loved regraduating, for the contemporary epoch, any painted wall-dial in Cambridge that had become illegible.

Fortunately, for some inexplicable reason I did not visit Whittlesford Church, very close to Hauxton, until I was an adult; it would either have been lost on me or would have worried me greatly. On the tower is a sheela-na-gig,[4] a squat, intensely erotic female figure, being approached by a zoomorphic man (figure 3, above). Such female figures, of which there are over thirty conspicuously set on British churches, more than twice that number in Ireland, and some in Normandy, raise unsolved questions of medieval belief. As a great contrast, in a glass case in the same church, among a number of broken fifteenth-century alabaster figures found bricked up in a wall, a lovely fragment of a *virgo lactans* seems, by her very low gown, a late medieval symbol of the unmarried state, exposing her breasts, to convey with great elegance both the virginity and maternity of Mary (figure 3, below).[5]

4. For the most recent account of sheela-na-gigs, see J. Andersen, *The Witch on the Wall* (London: Allen & Unwin, 1977). See also G. E. and A. L. Hutchinson, The "Idol" or Sheela-na-gig at Binstead with Remarks on the Distribution of Such Figures, *Proc. Isle of Wight Nat. Hist. Archaeol. Soc.* 6:237–81, 1970.

5. The Whittlesford alabaster fragments are described in J. H. Middleton, On Fragments of Alabaster Retables from Milton and Whittlesford, Cambridge, *Proc. Cambridge Antiquar. Soc.* 7:106–11, 1893. H. J. E. Burrell and G. M. Benton, The English Alabaster Carvings of Cambridgeshire, with Special Reference to Fragmentary Examples at Wood Ditton Church, *Proc. Cambridge Antiquar. Soc.* 34:77–83, 1924.

The Whittlesford *virgo lactans* is clearly very close to a large (c. 90

34   THE KINDLY FRUITS OF THE EARTH

In the winter of 1917–18, the weather was abnormally
cold and many birds died. I remember just after Christmas
finding six redwings (*Turdus iliacus*) dead in a ditch over
a stretch of about fifty feet along the Trumpington Road.
One would have thought that so boreal a bird would have
known how to look after itself. I tried to skin one; my
father mentioned this to Dr. F. H. H. Guillemard,[6] possibly at a meeting of the vestry of Saint Botolph's Church,[7]
where they were both sidesmen. Guillemard had been a
remarkable traveler and collector in early life, was the
author of a book on the birds of Cyprus, and had for a
short time been lecturer in geography at Cambridge,
though he never lectured. He was eccentrically conservative
and eschewed gas or electricity in the Old Mill House in
Trumpington where he lived. He immediately asked my
father if I was aware of the importance of paper in skinning

---

cm.) and perfect figure of Saint Catherine illustrated in an advertisement
in *Burlington Magazine*, Dec. 1968, p. xxxix; the two pieces may well
be by the same hand. The occurrence of broken images embedded in
walls may suggest that such images were, in spite of Puritan desecrators,
believed to have some numinous significance that they could continue
to exercise even though fragmented and unseen.

6. For F. H. H. Guillemard see S. C. Roberts, *Adventures with
Authors* (Cambridge: Cambridge Univ. Press, 1966), pp. 26–29; D. R.
Stoddart, The RGS and the Foundations of Geography in Cambridge,
Geogr. J. 141:216–39, 1975, where two fantastic photographs of
Guillemard are published.

As I indicated in *The Enchanted Voyage*, Guillemard, till he was
in his seventies, made it a practice to attend every auction of a great auk
(*Alca impennis*) skin or egg held in Britain.

7. The rector of Saint Botolph's, the Reverend Canon A. W.
Goodman, was, I believe, the last examiner in Paley's *Evidences of
Christianity*, which could be taken instead of chemistry or logic as a
subject in the university entrance or previous examination, usually
called the Little-Go. Very few students can have elected Paley in my
day, but the existence of the subject is of some interest in view of the
influence of the book on Darwin.

FIGURE 3   Whittlesford Church. *Above*, sheela-na-gig on the tower, early twelfth century; *below*, fragment of *virgo lactans*, now on north wall of nave, c. 1480.

36   THE KINDLY FRUITS OF THE EARTH

a bird. My father must have answered no, because next
morning I rode out on my bicycle to Trumpington with
a dead redwing, scissors, and scalpel. The secret of preparing a good study skin was to have a supply of small
pieces of toilet paper about one inch square, with which
every piece of flesh exposed was immediately covered. No
blood or exudates ever soiled the feathers. Many years
later, my wife pointed out to me that Mary Kingsley, the
great African explorer, had had lessons from Guillemard,
about a quarter of a century before I did, in the preparation
of natural history specimens. I am delighted to have this
link with her.

Before my last year at Saint Faith's I had come to
regard the Lepidoptera, quite wrongly and indeed snob-
bishly, as too commonplace a subject for study. Later I
realized that they carry all the major problems of evolutionary biology set out in colored two-dimensional diagrams on their wings. At thirteen I had decided that they
were too much collected, by too uninteresting people, to
be worthy of my attention.

The most remarkable butterfly in the area was the
swallowtail, *Papilio machaon britannicus*, which occurred
only at Wicken Fen and in the Norfolk Broads. The British
race has heavier black markings than the Continental, is
usually univoltine, or with one brood each year, and feeds
as a larva on milk parsnip, *Peucedanum palustre*, while the
paler form in France is bivoltine and feeds on cultivated
carrot and some other plants. French individuals often
blow over and many have been taken in southeastern England, but the insect cannot now establish itself.[8]

8. Most of the information available on the distribution of the
British and Continental swallowtails in England is summarized in E. B.
Ford, *Butterflies* (London: Collins, 1945), pp. 302–03; T. G. Howarth,
South's British Butterflies (London and New York: Frederick Warne,
1973), p. 38; and R. L. H. Dennis, *The British Butterflies* (Faringdon,

AYSTHORPE 37

About the time I was born, the swallowtail occurred on Burwell Fen near Cambridge as well as on Wicken Fen. During my boyhood it was common at the latter locality, a large stretch of undrained fenland covered with sedge and in places with buckthorn. In the 1940s it died out at Wicken but has now, after considerable efforts, been introduced from Norfolk. It was hard to catch. On my first trip to the fen I was riding on my bicycle along the lode or ditch that separated the fen from the lower drained pasture to the north. Along came a swallowtail. I swept at it with my net, more or less falling off my bicycle. On getting up I picked up the net, and out flew the butterfly. At least I have both caught a *P. machaon britannicus* riding a bicycle and in retrospect need not feel I have contributed to the decline of an endangered species. I think this incident may have contributed to a temporary disillusionment with the Lepidoptera.

Rather later in my school days my interest in the Lepidoptera revived, partly as the result of seeing, at teatime on one or two Sundays, the splendid collection of European butterflies made by Dr. J. Neville Keynes, Registrary of the university and father of Lord Keynes, the economist, and Sir Geoffrey Keynes, the surgeon and incomparable bibliographer. My interest in variation early led me to the magpie or currant moth, *Abraxas grossulariata*, and I have published papers on this protean organism in recent years.

The two elder sons of William Bateson, John and Martin, both to die tragically, to the world's great loss, overlapped me at Saint Faith's and turned me momentarily to beetles. I was also failing to learn to play the piano, and

Oxon.: E. W. Classey, 1977), pp. 120–21, 140. It is quite possible that the Continental *P.m. bigeneratus* was a regular inhabitant of southern England until the early nineteenth century.

after my lessons Francis Jenkinson, the University Librarian and husband of my teacher, often showed me, through his binocular microscope, the flies that he had been collecting. If I had been told about counterpoint I might have combined music and entomology successfully. As it was, the flies proved more fascinating but hopelessly difficult. There were too many species and no handy guide to their identification. I felt, moreover, that I should look into something not otherwise studied at Cambridge. The true bugs or Hemiptera seemed appealing. Good books existed on them and my father could borrow them from the university library. Moreover, my friend Jim Pearce, who had become very interested in pond life, was a devotee of beetles, so that on our expeditions we could divide the spoils without contention. We formed a Cambridge Junior Natural History Society. I think the members were Jim Pearce, David Pollock, who I suspect went into law, Audrey Lloyd Jones, whose medical father was an amateur paleontologist, but of whom I know nothing more, and probably my brother Leslie, who was drawn in by Jim and myself. For a time he collected caddis flies, though largely through my prodding. As the group was so little known and as he had superb manual dexterity in mounting such things, his specimens were ultimately added to the national collection, along with many of my bugs. My sister, whom I persuaded to collect shells, tells me she also occasionally attended our meetings, which were held in the dining room of the lodge of Corpus Christi College, where Jim's father was the master. Later I was to discover that there had been a similar society more than a century before in Bottisham, east of Cambridge, brought into existence by Leonard Jenyns, whose greatest contribution to science was arranging for Darwin to sail on the *Beagle*.

Jim Pearce, now an Anglican monk at Mirfield, became an authority on the Haliplidae, and later Pselaphidae and Scydmaeidae. The coleopterist will realize that he had a passion for the minute, allied I think to the Blessed Dame Julian of Norwich seeing the universe as a hazelnut placed in her hand and to other medieval attitudes that appear in miniature painting. Alas, he is now too blind to see his beloved little beetles.

Through David Pollock, I met a woman, I imagine of about thirty, called Dorothy Elizabeth Thursby-Pelham. She had been a postgraduate student of or assistant to Richard Assheton, a mammalian embryologist of great distinction who died relatively young. Assheton had started some work on the placentation of the coney or dassie, that curious little relative of the elephant, scientifically then known as *Hyrax*. After Assheton's death, Miss Pelham completed the work. She used to take me to her laboratory to show me what she was doing, with an extremely discreet flirtatiousness, which was just right for my age and insured that for the appropriate time I should be in love with her. Later she got involved with defenses against gas warfare under Barcroft and then went into fisheries biology at Lowestoft. I owe her a great deal.

At about this time, through Jim Pearce, I also got to know E. J. Bles, whose name is to be found in the bibliographies of all the early biological work on *Xenopus*, which he was the first to breed in captivity. He was a man of private means who had built a beautiful laboratory behind his house. Initially his interests were in the life histories of Amphibia, but later he turned to cell physiology and did a remarkable study on the production of gas vacuoles in *Arcella*, a work continually interrupted by illness and most tenaciously completed.[9] Bles remained my friend through the rest of his life and was a continual source of exciting biological conversation.

In the winter months, when little was stirring in the

9. E. J. Bles, Arcella. A Study in Cell Physiology, *Quart. J. Microsc. Sci.* 72:527–48, 1929.

field, the various university museums in Cambridge were a constant source of delight. I probably knew the specimens shown in the old University Museum of Zoology by heart by the time I was fourteen. Most of them were displayed in a large hall on a site that had been the University Botanic Gardens until the middle of the nineteenth century. The ground floor of this hall was occupied by stuffed mammals and even more numerous skeletons. A gallery ran round the hall and was devoted to lower vertebrates.

Hanging from the roof and easily studied from the gallery was a magnificent skeleton of a rorqual or finwhale, *Balaenoptera musculus*. I remember hearing a most romantic story of this whale being stranded on the east coast sometime in the last century. Its body was said to have been brought to Cambridge by train, on a line of flatcars, by the Great Eastern Railway. When it arrived, during the Long Vacation, everyone connected with the museum was away. The decomposing whale, lying near the station, caused more and more distress, until the vicar of Little Saint Mary's, who held a position once occupied by George Herbert, valiantly assembled a group of osteological volunteers to prepare the skeleton. Unhappily this elaborate and beautiful story is completely untrue. The animal was beached dead, at Pevensey Bay, Sussex, on 13 November 1865. Claimed by the Crown, it was sold at auction and ultimately was exhibited by a speculator who had roughly prepared a skeleton. Later the skeleton was auctioned again, purchased by the university, and properly mounted. The romantic tale is an interesting example of a myth generated from the flimsiest factual foundations.[10]

In two galleries at the side of the main hall invertebrates and specimens to illustrate the principles of zoology were installed. In the museum I was able to learn

10. The true story is given in A. E. Shipley, "*J.*" *A Memoir of John Willis Clark* (see chap. 1, n. 10), pp. 265–66.

of Mendelism and of sex-linked inheritance in the magpie moth, *Abraxas grossulariata*, of cryptic coloration and mimicry, of the sequestering of carotinoids from chilis in the plumage of orange "color-fed" canaries, of Abbott Thayer's discovery of countershading, giving meaning to the dark upper and paler lower surfaces of most vertebrates, as well as of a host of strange beasts such as the Surinam toad and its compatriot with an enormous tadpole and small frog, which fooled Maria Sibylla Merian, or one of her daughters, into thinking she had found a frog that grew up into a tadpole. Upstairs in a building which mostly housed the department of zoology and the study collection of insects, the bird room contained skeletons of the dodo and of both male and female solitaires, among the best testimonies of these extinct beings in existence. All these things and doubtless many more were essential ingredients of my life.

I was first taken to the Sedgwick Museum across the street from the Museum of Zoology, by my father, on my birthday, probably my eighth. We went specifically to see the skull of an aurochs, the old wild ox of Europe, in the forehead of which a Neolithic ax was embedded. It was supposed, I imagine, that the beast had been killed by a Neolithic hunter but had fallen into a water-filled pit in the fen country and been covered with peat. In later times the skull was certainly discovered in a peat cutting, but I believe that there is some doubt about the authenticity of its association with the ax. In the museum, however, the specimen fitted in perfectly with the bones of *Hippopotamus* from Barrington nearby and with all the marine fossils from the Chalk. Not only in space was an extraordinary variety of creatures to be found but the geological collections showed that my home itself had once been on the seafloor and that later it was inhabited by large exotic animals and wild men.

The Museum of Ethnology and Archaeology close by

contained a number of superb casts of Maya stelae and great collections of artifacts from Malaya and the Pacific Islands. On another birthday, either my tenth or eleventh, we were shown some of its special treasures by the curator, Baron Anatole von Hügel, the brother of the well-known theologian. I had for a minute or so a magnificent Hawaiian feather cloak draped over my shoulders, the sartorial high point of my existence. I remember also on the same occasion Baron von Hügel telling my mother of the extraordinary personal cleanliness of the Fijians. She received this information with great relief and obvious skepticism.

The last and greatest museum in Cambridge was the Fitzwilliam; with its beautiful vases of flowers, it is, perhaps, the most welcoming art gallery in the world. The style that has characterized the Fitzwilliam throughout most of this century is fundamentally due to its former director Sir Sidney Cockerell. He had a reputation in Cambridge for being impossible; in Oxford *to cockerell* an object meant to acquire it by any means, fair or foul. I distinctly remember him, when I was eleven or twelve, going out of his way to show and explain to me a number of Dürer prints.

When I first knew the Fitzwilliam it retained the crowded appearance of a nineteenth-century gallery, picture ranged above picture, closely covering the whole wall. I was very small, probably about five years old, when I was taken there to see an ivory model of the Taj Mahal. This object, no doubt an embarrassing gift, was later removed, but it was a source of real delight to children. Throughout my early life it gave me a haunting hope that one day I should see the building itself. When I did see it, it did not let me down. Of another, more adult work in the collection, I shall have a little to say on a later page. Revisiting the Fitzwilliam has always been one of the great pleasures of returning to Cambridge. Over the years my favorite painting there, or indeed perhaps anywhere, has come to be a small annunciation by Domenico Veneziano, a predella panel originally associated with an altarpiece now in the Uffizi. In this painting the two figures are in a cortile opening into a garden with a path leading to a wooden double door, which must be that *ex qua mundo lux est orta.*

Opposite the Fitzwilliam was a confectioner's shop called Mason's which produced a special large sponge cake; officially called a Fitzwilliam Cake, everyone called it a Fitzbilly. A legend grew up that the recipe had been invented by Sir Ellis Minns, a great authority on the Scythians and my father's dearest friend in Cambridge. He was also supposed to have blended a tea, before he was knighted, known as "Mr. Minns' Mixture," which could be obtained many years after his death at a grocer's shop called G. P. Jones. It was, I think, the best tea that I have drunk. Alas, G. P. Jones has closed and all efforts to learn the nature of the blend have failed.

Shortly before his death I asked Sir Ellis why Michael Rostovtzeff, who became the most significant member of the department of classics in my early days at Yale, had left Oxford, where he had come from Berlin in the early 1930s. Sir Ellis replied: "In Oxford if you disagree radically with someone, you say to him, 'My dear fellow, don't you think it is *just* possible that there is an alternative explanation, somewhat along the following lines?' But Rostovtzeff said, 'You are wrong, and I will tell you why you are wrong.'" Henry Andrews suspected that the words had been said to Sir Maurice Bowra. At Yale, however, Rostovtzeff, in spite of a rather leonine appearance, had a reputation for his very mild treatment of examination and term papers; attending the course rated a C, handing in some sort of an exercise a B, knowing anything at all an A.

During the time when I was at Saint Faith's, real

44   THE KINDLY FRUITS OF THE EARTH

tragedy first touched my life. E. A. Wilson, the medical officer of the Terranova Expedition led by Captain Robert F. Scott to explore Antarctica, was a friend of my uncle. I can well remember the horror that struck us when we learned that Scott and four of his companions, including Wilson, had perished on their return from the South Pole. Wilson was an admirable ornithologist, a good landscape painter, and a most saintly man.

Later I discovered a curious link with the expedition. I had, as a boy, known that a Cambridge undergraduate called D. G. Lillie, at the end of the first decade of the present century, had been an extraordinary caricaturist.[11] He had painted William Bateson holding up two dark chickens labeled F_1 while a white F_2 emerged from his pocket. He had also done Sir Sidney Harmer, later the meticulous director of the Natural History Museum in South Kensington, with a red-tape worm in a museum jar.

About 1933 my father, who had found four of Lillie's works among my uncle's papers, gave them to me. They hung on my study wall until 1972, when I gave them to the National Portrait Gallery in London. Mr. Kingsley Adams, late director of the gallery, and I engaged in considerable research on the artist. It turned out that Lillie had been marine biologist on the Terranova, had done a

11. All the documentary evidence relating to Lillie that Mr. Adams, Lillie's goddaughter Miss Muriel Arber, or I have been able to discover is now in the National Portrait Gallery in London, with the four caricatures that I gave to that institution. There are four more Lillie originals in the Scott Institute for Polar Research at Cambridge, including the one of Wilson here reproduced. Besides those of William Bateson (see *The Enchanted Voyage*, frontispiece), and of Harmer, my four included one of Sir William Ridgeway (see A Cambridge Caricaturist, *Country Life* 144:644–45, 1968) and one of Tansley and Blackman (see G. E. Hutchinson, *An Introduction to Population Ecology* [New Haven and London: Yale University Press, 1978], figure 83).

FIGURE 4 D. G. Lillie, Caricature of E. A. Wilson. Scott Institute for Polar Research, Cambridge University.

46   THE KINDLY FRUITS OF THE EARTH

wonderful caricature of Wilson as a penguin with a human
head (figure 4), and had spent some time after the other
survivors returned to England studying whales off New
Zealand. He became involved in medical research in the
First World War, but he had a mental breakdown in 1919
from which he never recovered, dying in 1963.

The loss of Scott's party seems in retrospect to have
been the beginning of the carnage of 1914. Any day one
might learn that someone close had been killed. Christina
lost both her brothers. John Bateson was killed. Every
boy of my age must have felt that if the stalemate in the
trenches went on indefinitely, his turn one day would come.

Early in the war a camp was established in the University Polo Ground south of Aysthorpe, and my father
decided that the soldiers there lacked all facilities for
recreation. He got permission to set up two marquees,
one as a writing and reading room and one as a sort of
café. The camp was very temporary, but I remember
serving ginger ale in the café and selling Woodbine ciga-
rettes throughout one summer vacation.

Later, when gas warfare began, my father was much
worried by the insensitivity of the navy to the dangers
of a gas attack by sea. With a high naval medical officer,
Fleet-Surgeon Hewitt by name, he managed to get invited
to Scapa Flow to convince the high command. He did
this by letting out a little brombenzene downwind, which
immediately sent Admiral of the Fleet Lord Jellicoe below,
weeping. This earned my father an O.B.E.

Apart from selling gingerpop and Woodbines, my
wartime career consisted of being a boy scout orderly at
the First Eastern General Hospital, erected on the King's
College cricket ground, now the site of the Cambridge
University Library.

The matron did not like me; she was a good authori-
tarian and much preferred subservient working-class boy

scouts. My duties at the hospital were mainly sorting out good fruit from very mixed barrels of apples. I think, however, that at this point I began to realize that the world in which I really lived consisted of people who were passionately devoted to the fascinating and beautiful wonders provided naturally or artificially by the world, though the troubles of the times might force them to defer their passion. Insofar as we shared interests there was love between us, even though I was a small boy and they might be grown and eminent men and women. The Cambridge philosopher J. M. Ellis McTaggart believed that reality consisted of selves loving each other. People complained that he modeled the universe on an ideal conception of the fellowship of Trinity College. Certainly not all fellows at Trinity loved one another, but looking back on my own boyhood I see a little of what he meant, even though he insisted that he derived his ideas from first principles and not merely from the best that Cambridge could then offer empirically.

What I have written will I hope indicate the extraordinary richness of the environment in which I grew up. My first thought as a teacher has been to try to pass on as much of that richness as possible. Quite recently, in studying the career of Maria Sibylla Merian, the remarkable entomological artist who worked in Surinam about 1700, I have been impressed with the artistic and intellectual quality of her family environment, ranging from her father's connections with the De Bry publishing firm to her granddaughter marrying Leonhard Euler.[12] Through her stepfather and his student Abraham Mignon, who

12. G. E. Hutchinson, The Influence of the New World on Natural History, in *The Changing Scenes in Natural Science, 1776–1976,* ed. Clyde E. Goulden. Academy of Natural Sciences, Philadelphia Special Publ. 12, 1977, pp. 14–22.

48   THE KINDLY FRUITS OF THE EARTH

taught Maria Sibylla as a girl, she inherited the tradition of Dutch flower and insect painting that goes back to the late Middle Ages. Wherever she went she seems to have been helped by other naturalists, as I was. The role of social structure in providing a really rich mental and artistic environment might well be a most significant subject of study for an intellectual historian. I suspect that quite unexpected parts of society may be involved.

CONJECTURES ARISING IN A QUIET MUSEUM

G. Evelyn Hutchinson, Yale University, New Haven, Connecticut, U.S.A.

These speculations arose largely from visits to Tring to gaze at the huge series of *Abraxas grossulariata* in the Rothschild-Cockayne-Kettlewell collection, in those days still housed there. The fantastic variation exhibited by a small minority of the specimens of the species, an aristocratic variability remote from the daily life of the average currant moth, is always beautiful but also in turn puzzling, frustrating and challenging. Now that I think I see what some of it may mean, the meaning tends to be related to other organisms, watersnails, stoats and even man himself. It may be that my conclusions are unoriginal; this does not worry me, because it seems as though in evolutionary ecology an idea only achieves its full usefulness after it has been stated a number of times by completely independent investigators.

One of the very few dominant mutants known in magpie moth populations (Stovin 1940) is ab. *nigrosparsata* Raynor, in which the white areas of the wings are peppered with small black dots. Porritt (1921, 1926) to whom we owe most of our information about the aberration, believe it to be the "real" i.e. the industrial, melanic of *A. grossulariata*. It appeared in South Wales before 1900 (Barrett 1910) and in Yorkshire about 1905; in the area around Huddersfield towards 10% of the population of the moth belonged to this form in 1917. A comparable though quantitatively undocumented rise in *nigrosparsata* certainly occurred in the environs of the city of York. At Huddersfield the species became very scarce after 1917 and when its numbers increased again several years later, *nigrosparsata* was present only casually.

The ordinary pattern of *A. grossulariata* is presumably an aposematic one, adapted to be conspicuous at dusk or by moonlight as well as by day. The history of ab. *nigrosparsata* is in accord with Porritt's speculation that it was an industrial melanic and so cryptic rather than, or perhaps as well as, being (Rothschild 1963) aposematic.

Stovin's breeding experiments suggested to Cockayne that the *nigrosparsata* gene was dominant but of very low penetrance. If we suppose that the optimal visual environments of *grossulariata* and *nigrosparsata* are rather different, but that their incidence is irregular and unpredictable in any given locality, a dominant mutation towards any new type of adaptation would have a greater chance of survival if its penetrance was not complete. In any period when no environment favouring the mutant existed, the individuals carrying the gene would include some in which it was not expressed phenotypically and these would to some extent carry the gene over until natural selection again could operate in its favour.

De Laramberque (1939) has described populations of the African watersnail *Bulinus contortus* in which there are two morphs, one euphallic with a full male system capable of cross-fertilizing other individuals acting as females, the other aphallic without an intromittent organ and capable only of self-fertilizing hermaphroditism or of acting in cross-fertilization as a female. A number of similar cases are known in various slugs and snails. In *B. contortus*, populations consisting of nearly all euphallic or nearly all aphallic individuals are known in nature. Intense selection within laboratory populations, either almost completely euphallic or almost completely aphallic, has no effect on the composition of the colony under treatment; the most euphallic lines still produce a few aphallic individuals and vice-versa. The genetic constitution must control the probability of aphallism or euphallism in any individual rather than determining strictly the structure developed. Presumably aphallic individuals save some resources in not having a penis, while they can reproduce if isolated alone; euphallic individuals gain whatever advantage recombination actually may ensure. In a varying environment a probabilistic mechanism permits the snail as a population to make the best of whichever of two possible worlds is presented by nature.

Richards (1963, 1968) has reported another case in freshwater gastropod molluscs. The tropical American planorbid *Biomphalaria glabrata* has two forms, one without lamellae at the aperture, which matures rapidly and does not aestivate, the other, with lamellae just inside the aperture, which delays reproduction and aestivates. The lamellae provide a place of attachment for the margin of the epiphragm which is secreted to cover the aperture in aestivation. In perennially wet habitats the non-lamellate form, which can begin breeding earlier, is clearly the fitter of the two. In seasonally wet habitats the lamellate form is obviously the only one that can survive. Purely non-lamellate strains are known, but it is apparently impossible to get, even with prolonged selection, a strain that produces nothing but lamellate individuals. Richards points out that in an uncertain environment some specimens are always able to tide over a dry year and others to take advantage of the higher reproductive potential of the non-lamellate form in a wet year. He believes that the realisation of the lamellate condition depends on the satisfaction of critical thresholds dependent on unstable environmental factors. A similar explanation of comparable uncertainties in alary polymorphism in insects is advanced by Harrison (1980).

A comparable possibility, though with a somewhat

(Continued on p. 97)

(Continued from p. 92)

different genetic background, has been suggested (Hutchinson 1978 p. 175, note 49) in relation to human ability. Since the coding of a complex nervous system is likely to be difficult, it is supposed that it is achieved within a certain tolerance level that ordinarily ensures adequate if not really outstanding performance. Mechanisms may exist to prevent random variation from producing too often systems below the tolerated limit, but not from producing better ones, which might occur fortuitously. The occasional very remarkable people then represent a kind of unrepeatable random variation that is not really genetic since it is not inherited nor can it be attributed deterministically to the environment. The New Zealand statesman who is said to have announced sententiously that what was not produced by inheritance was due to the environment may perhaps have been, in practice, wrong.

The type of adaptive mechanism suggested by the occurrence of far from complete penetrance in nature is comparable to another suggested possibility (Hutchinson 1974) in the case of a gene that is dominant in one sex but recessive in the other. Though the gene is more vulnerable to adverse selection than it would be if a full recessive, if the environment changes frequently enough the fact that there is always phenotypic expression of both alleles in some individuals permits selection to get to work without waiting for rare homozygous recessives to appear. Though a supposed example of this in *Philaenus spumarius* may not (cf. Halkka, Halkka, Hovinen, Raatikainen and Vasarainen 1975) be valid, Hutchinson and Parker (1978) have suggested that the sexual dimorphism in winter whitening in *Mustela erminea* and *M. frenata*, in zones in which whitening is incomplete, has an adaptive meaning along these lines.

Finally, it is obvious that insofar as recessive genes, of no value in an old environment, are to become useful in producing new adaptations when the environment changes, potential utility will be greater when homozygous recessives turn up in fair frequency even if they are doomed to perish in the old environment. For autosomal genes the proportion of such homozygotes will ordinarily be the square of the gene frequency, so that rare genes show up phenotypically very rarely indeed in a randomly mating population. If the gene at times is useful, anything that keeps heterozygotes around in the intervening times will be selected for, be it a high mutation rate or heterosis. Every time a major shift occurs there will be some adjustment in the genetic environment of the genes that are basic to the new adaptation. After a long period of time when several adaptations have been evolved, discarded and perhaps reemployed, it would not be unreasonable to expect the population to contain a fair number of potentially useful recessives that have been selected, used and partly abandoned but are well integrated into their genetic milieu (cf. Chetverikov [1926] 1961; Ford 1975) and are carried along at a usable frequency, so that with the return of past times they may be taken out of the closet and get their chance again. Such a low grade balanced polymorphism built up as the genome wanders around for millions of years in its evolutionary space would fit a case like that of *A. grossulariata* where there are probably over sixty wing-colour mutants, many quite spectacular, known. Nearly all of them are likely to be autosomal recessives involving a gene frequency (Hutchinson 1969, 1975) of not less than 0.001, and often much higher. Most of them produce harmonious symmetrical patterns that would not look out of place as specific characters. The colouration being generally aposematic, selection against the double recessive is likely to be less severe than if the pattern were cryptic and this may explain why the numerous rare mutants beloved of collectors are not so rare that they are never collected.

The question of the damage caused to aposematic patterns by random alterations may possibly involve a further little considered problem, as to whether an aposematic pattern may not also be a means of species recognition in mating and so be stabilised to a degree not really necessary to the warning function of the pattern. In the waterbugs of the genus *Notonecta*, there is a tantalising case that requires further study. Any kind of bright colouration, epigamic or aposematic, is extremely rare in aquatic insects. In *Notonecta*, however, there are a number of cases of reddish brown or pinkish colours as the background tint of the hemelytra. In one small group, placed in the subgenus *Erythronecta*, inhabiting south-western North Amercia and Central America, the females may exhibit a conspicuous red and black colouration while the male is usually less brilliant and darker. In the cabinet, or on plate 3 of Hungerford's (1933) monograph, the red and black insects look aposematic. Puncture by the beak of such waterbugs is certainly painful. Recognition of females by males is known to be largely by sight in the genus (Clark 1928). Against the attractive view that this is indeed a case of dual purpose colouration may be set the fact that the hemelytra are covered by an air film and the colour may not be visible in the water from below as a swimming male approaches a female at rest at the surface. A further difficulty may arise from the fact that the red colour matures slowly and is at first grey. In the laboratory grey females do not breed but evidently under some circumstances may do so in the field (Fox 1975). Clearly we need much more study of *Erythronecta*, including direct observation from below in the field. Meanwhile I am assured that artificial epigamic female colouration in our own species is used all the time, both to attract and as a warning.

I present these conjectures in an entomological context because they began as the result of moth gazing, and since most varying eucaryotes are insects, anything of value in my thoughts is likely to be developed by students of the Hexapoda. I particularly would pay tribute to the three men who built up the great collection that inspired me. My friend Jeffrey Powell most kindly read this note; to him I owe my knowledge of Chetverikov's remarkable paper. I am also indebted to Nancy Knowlton and Marjorie Garber for insights about aposematic and epigamic colouration.

References

BARRETT, G. G. 1901. *The Lepidoptera of the British Isles. II. Heterocera Geometrina.* London 335 pp.

CHETVERIKOV, S. S. [1926] 1951. On certain aspects of the evolutionary process, from the standpoint of modern genetics (trans. M. Barker, ed. I. M. Lerner). *Proc. Amer. Philos. Soc.* **105:** 167-195 [Russian original *Zhurn. Eksper. Biol.* A2: 3-54].

CLARK, L. D. 1928. Seasonal distribution and life-history of *Notonecta undulata* in the Winnipeg region, Canada, *Ecology* 9: 383-403.

DE LARAMBERQUE M. 1939. Etude de l'autofécondation chez les gastéropodes pulmonés recherches sur l'aphallie et la fécondation chez *Bulinus (Isodora) contortus* Michaud. *Bull. Biol. Fr. Belg.* 73: 19-231.

FORD. E. B. 1975. *Ecological Genetics* 4th ed. London, Chapman and Hall. 442 pp.

FOX, L. R. 1975. Some demographic consequences of food shortage for the predator *Notonecta hoffmanni. Ecology* 56: 868-880.

HALKKA, O., HALKKA, L., HOVINEN, R., RAATIKAINEN, M. and VASARAINEN, A. 1975. Genetics of *Philaenus* colour polymophism: the 28 genotypes. *Hereditas* 79: 308-310.

HARRISON, R. G. 1980. Dispersal polymorphism in insects. *Ann. Rev. Ecol. Syst.* 11 (in press).

HUNGERFORD, H. B. 1933. The genus *Notonecta* of the world. *Univ. Kansas Sci. Bull.* 21: 5-195.

HUTCHINSON, G. E. 1969. Some Continental European aberrations of *Abraxas grossulariata* Linn. (Lepidoptera) with a note on the theoretical significance of the variation observed in this species. *Trans. Conn. Acad. Arts Sci.* 43: 1-24.

1974. New and inadequately described aberrations of *Abraxas grossulariata* (Linn.) (Lep.: Geometridae). *Entom. Rec.* 86: 199-206.

1975. Variations on a theme by Robert MacArthur *in* M. L. Cody and J. M. Diamond eds. *Ecology and Evolution of Communites.* Cambridge, Mass. and London. Belknap Press of Harvard University. pp. 492-521.

1978. *Introduction to Population Ecology.* New Haven and London. Yale University Press. 260 pp.

HUTCHINSON, G. E. and PARKER, P. 1978. Sexual dimorphism in the winter whitening of the stoat, *Mustela erminea. J. Zool., Lond.* 180: 560-563.

PORRITT, G. T. 1921. The Huddersfield varieties of *Abraxas grossulariata,* with description of a new variety. *Entom. Month. Mag.* 57: 128-135.

1926. The induction of melanism in the Lepidoptera, and the subsequent inheritance. *Entom. Month. Mag.* 62: 107-111.

RICHARDS, C. S. 1963. Apertural lamellae, epiphragms and aestivation of planorbid molluscs. *Amer. J. Trop. Med. Hyg.* 12: 254-263.

1963. Aestivation of *Biomphalaria glabrata* (Basommatophosa Planorbidae) genetic studies. *Malacologia* 7: 109-116.

ROTHSCHILD, M. 1963. An aposematic moth, the small magpie (*Eurrhypara horticulata* [L.]) (Lep.: Pyraustidae). *Entom. Month. Mag.* 98: 203-204.

STOVIN, G. 1940. Some breeding experiments with *Abraxas grossulariata* (Lep.: Geometridae). *Entom.* 73: 265-267.

APPENDIX   *Publications of G. Evelyn Hutchinson*

Listed in chronological order.

Hutchinson, G. E. 1918. A swimming grasshopper. *Entomological Record and Journal of Varia-tion* 30:138.

Hutchinson, G. E. 1919. *Gerris asper* Fab., in Norfolk. *Entomological Monthly Magazine* 55:33.

Hutchinson, G. E. 1919. *Notonecta halophila* Edw. in Cornwall. *Entomological Monthly Maga-zine* 55:261.

Hutchinson, G. E. 1921. *Nabis boops,* Schodte, in Wiltshire, etc. *Entomological Monthly Maga-zine* 57:18.

Hutchinson, G. E. 1921. Two records of Hemiptera. *Entomological Monthly Magazine* 57:39.

Hutchinson, G. E. 1922. Localities for *Notonecta halophila* J. Edw. *Entomological Monthly Magazine* 58:255.

Hutchinson, G. E. 1923. A preliminary account of Hemiptera-Heteroptera. In *The Natural History of Wicken Fen,* ed. J. S. Gardiner, pp. 100–103. Cambridge: Bowes and Bowes.

Hutchinson, G. E. 1923. Miscellaneous notes on stridulation in Corixidae. Incorporated into E. A. Butler, *Biology of the British Hemiptera-Heteroptera,* pp. 575–576, 601. London: Witherby.

Hutchinson, G. E. 1924. Contributions towards a list of the insect fauna of the South Ebudes. IV. The Hemiptera (Rhynchota). *Scottish Naturalist* 1924:21–27.

Hutchinson, G. E. 1926. Hemiptera-Heteroptera. Pt. I. Hydrobiotica and Sandaliorhyncha. In *Natural History of Wicken Fen,* pp. 234–252.

Hutchinson, G. E. 1927. On new or little known Notonectidae (Hemiptera-Heteroptera). *Annals and Magazine of Natural History,* Series 9, 19:375–379.

Hutchinson, G. E. 1927. Psychological dissociation as a biological process. *Nature* 120:695.

Hutchinson, G. E. 1928. The branchial gland of the Cephalopoda: a possible endocrine organ. *Nature* 121:674–675.

Hutchinson, G. E. 1928. On the temperature characteristics of two biological processes. *South African Journal of Science* 25:338–339.

Hutchinson, G. E. 1928. On Notonectidae from central Africa (Hemiptera-Heteroptera). *Annals and Magazine of Natural History,* Series 10, 1:155–166.

Hutchinson, G. E. 1928. Notes on certain African and Madagascan waterbugs (Notonectidae and Corixidae). *Annals and Magazine of Natural History,* Series 10, 1:302–306.

Hutchinson, G. E. 1928. On certain palearctic species of *Notonecta* Linn. *Entomological Monthly Magazine* 64:35–37.

Hutchinson, G. E. 1928. Another Cumberland locality for *Corixa dentipes* Th. *Entomological Monthly Magazine* 64:236.

Hutchinson, G. E. 1928. Notes on Westmoreland Corixidae. *Entomological Monthly Magazine* 64:13–14.

Hutchinson, G. E. 1929. Observations on South African Onychophora. *Annals of the South African Museum* 25:337–340.

Hutchinson, G. E. 1929. A revision of the Notonectidae and Corixidae of South Africa. *Annals of the South African Museum* 25:359–474 and Pl. 27–41.

Hutchinson, G. E., G. E. Pickford, and J. F. M. Schuurman. 1929. The inland waters of South Africa. *Nature* 123:832–833.

Hutchinson, G. E. 1930. Fisheries survey of Lakes Albert and Kioga. Report on Notonectidae and Corixidae. *Annals and Magazine of Natural History*, Series 10, 6:57–65.

Hutchinson, G. E. 1930. Restudy of some Burgess Shale fossils. *Proceedings of the US Natural History Museum* 78(2854): Art. 11, pp. 1–24 and 1 Pl.

Hutchinson, G. E. 1930. On the chemical ecology of Lake Tanganyika. *Science* 71:616.

Hutchinson, G. E. 1930. Report on Notonectidae, Pleidae, and Corixidae (Hemiptera). Mr. Omer Cooper's investigation of the Abyssinian fresh-waters. (Dr. Hugh Scott's expedition). *Proceedings of the Zoological Society of London* 1930:437–466.

Hutchinson, G. E. 1930. Two biological aspects of psycho-analytic theory. *Journal of Psychoanalysis* 1:83–86.

Hutchinson, G. E. 1931. On the occurrence of *Trichocorixa* Kirkaldy (corixidae, hemiptera-heteroptera) in salt water and its zoo-geographical significance. *The American Naturalist* 65:573–574.

Hutchinson, G. E. 1931. New and little known Rotatoria from South Africa. *Annals and Magazine of Natural History*, Series 10, 7:561–568.

Hutchinson, G. E. 1931. The hydrobiology of arid regions. *Yale Scientific Magazine* 5(2):2.

Hutchinson, G. E. 1932. On Corixidae from Uganda. *Stylops* 1:37–40.

Bond, R. M., M. K. Cary, and G. E. Hutchinson. 1932. A note on the blood of the hag-fish *Polistotrema stouti* (Lockington). *Journal of Experimental Biology* 9:12–14.

Hutchinson, G. E. 1932. Experimental studies in ecology. I. The magnesium tolerance of Daphniidae and its ecological significance. *Internationale Revue der gesamten Hydrobiologie* 28:90–108.

Hutchinson, G. E., G. E. Pickford, and J. F. M. Schuurman. 1932. A contribution to the hydrobiology of pans and other inland waters of South Africa. *Archiv für Hydrobiologie* 24:1–154.

Hutchinson, G. E. 1932. Supplementary report on Notonectidae, Pleidae, and Corixidae (Hemiptera). Mr. Omer Cooper's investigation of the Abyssinian fresh-waters. (Dr. Hugh Scott's expedition). *Proceedings of the Zoological Society of London* 1932:125–130.

Hutchinson, G. E. 1932. Reports on the Percy Sladen expedition to some Rift Valley lakes in Kenya in 1929. II. Notonectidae, Pleidae, and Corixidae from the Rift Valley lakes in Kenya. *Annals and Magazine of Natural History*, Series 10, 9:323–329.

Hutchinson, G. E., and G. E. Pickford. 1932. Limnological observations on Mountain Lake, Virginia. *Internationale Revue der gesamten Hydrobiologie* 27:252–264.

Hutchinson, G. E., G. E. Pickford, and J. F. M. Schuurman. 1932. A contribution to the hydrobiology of pans and other inland waters of South Africa. *Archiv für Hydrobiologie* 24:1–154.

Hutchinson, G. E. 1933. Limnological studies at high altitudes in Ladak. *Nature* 132:136.

Hutchinson, G. E. 1933. The zoo-geography of the African aquatic hemiptera in relation to past climatic change. *Internationale Revue der gesamten Hydrobiologie* 28:436–468.

Hutchinson, G. E. 1933. A revision of the Distantian and Palvaian types of Notonectidae and Corixidae in the Indian Museum. *Records of the Indian Museum* 35:393–408.

Hutchinson, G. E. 1934. Yale North India Expedition. *Nature* 134:87–88.

Hutchinson, G. E. 1934. Yale North India Expedition—Report on Amphipod Crustacea of the genus Grammarus by Masuzo Ueno. Note by the Biologist of the Expedition. *Memoirs of the Connecticut Academy of Arts and Sciences* 10:72–73.

Edmondson, W. T., and G. E. Hutchinson. 1934. Yale North India Expedition—Report on Rotatoria. *Memoirs of the Connecticut Academy of Arts and Sciences* 10:153–186.

Hutchinson, G. E. 1934. Yale North India Expedition—Report on terrestrial families of Hemiptera-Heteroptera. *Memoirs of the Connecticut Academy of Arts and Sciences* 10:119–146.

de Terra, H., and G. E. Hutchinson. 1934. Evidence of recent climatic changes shown by Tibetan Highland lakes. *Geographical Journal* 84:311–320.

Hutchinson, G. E. 1936. *The clear mirror. A pattern of life in Goa and in Indian Tibet.* Cambridge: Cambridge University Press.

Hutchinson, G. E. 1936. Alkali deficiency and fish mortality. *Science* 84:18.

Hutchinson, G. E. 1937. Limnological studies in Indian Tibet. *Internationale Revue der gesamten Hydrobiologie* 35:134–177.

Hutchinson, G. E. 1937. A contribution to the limnology of arid regions. *Transactions of the Connecticut Academy of Arts and Sciences* 33:47–132.

Hutchinson, G. E. 1938. Chemical stratification and lake morphology. *Proceedings of the National Academy of Sciences USA* 24:63–69.

Hutchinson, G. E. 1938. On the relation between the oxygen deficit and the productivity and typology of lakes. *Internationale Revue der gesamten Hydrobiologie* 36:336–355.

Hutchinson, G. E. 1939. Freilebende Ruderfusskrebse (Crustacea Copepoda), Addendum, Scientific Results of the Yale North India Expedition, Biological Report No. 19. *Memoirs of the Indian Museum* 13:199–200.

Hutchinson, G. E. 1939. Ecological observations on the fishes of Kashmir and Indian Tibet. *Ecological Monographs* 9:142–182.

Hutchinson, G. E., E. S. Deevey, and A. Wollack. 1939. The oxidation reduction potentials of lake waters and their ecological significance. *Proceedings of the National Academy of Sciences USA* 25:87–90.

Hutchinson, G. E. 1940. Bio-ecology. *Ecology* 21:267–268.

Hutchinson, G. E. 1940. A revision of the Corixidae of India and adjacent regions. *Transactions of the Connecticut Academy of Arts and Sciences* 33:339–476.

Hutchinson, G. E., and A. Wollack. 1940. Studies on Connecticut lake sediments. II. Chemical analyses of a core from Linsley Pond, North Branford. *American Journal of Science* 238:493–517.

Hutchinson, G. E. 1941. Ecological aspects of succession in natural populations. *The American Naturalist* 75:406–418.

Hutchinson, G. E. 1941. Limnological studies in Connecticut. IV. Mechanism of intermediary metabolism in stratified lakes. *Ecological Monographs* 11:21–60.

Hutchinson, G. E. 1942. Note on the occurrence of *Buenoa elegans* (Fieb.) (Notonectidae, Hemiptera-Heteroptera) in the early postglacial sediment of Lyd Hyt Pond. *American Journal of Science* 240:335–338.

Hutchinson, G. E. 1942. The history of a lake. *Yale Scientific Magazine* 16:13–15, 22.

Hutchinson, G. E. 1942. Noti sunt mures, et facta est confusion (Review of *Voles, mice and lemmings: problems in population dynamics* by C. Elton). *Quarterly Review of Biology* 17:354–357.

Hutchinson, G. E. 1942. Addendum (on the death of Raymond J. Lindeman). *Ecology* 23:417–418.

Hutchinson, G. E. 1943. Marginalia. *American Scientist* 31:270–278.

Hutchinson, G. E. 1943. Marginalia. *American Scientist* 31:346–355.

Hutchinson, G. E. 1943. The biogeochemistry of aluminum and of certain related elements. 4 parts. *Quarterly Review of Biology* 18:1–29, 128–153, 242–262, 331–363.

Kubler, G. A., and G. E. Hutchinson. 1943. Shepherd rockeries. *Notes and Queries* 185:65–66.

Hutchinson, G. E. 1943. Thiamin in lake waters and aquatic organisms. *Archives of Biochemistry* 2:143–150.

Hutchinson, G. E., A. Wollack, and J. K. Setlow. 1943. The chemistry of lake sediments from Indian Tibet. *American Journal of Science* 241:533–542.

Hutchinson, G. E., and A. Wollack. 1943. Biological accumulators of aluminium. *Transactions of the Connecticut Academy of Arts and Sciences* 35:73–128.

Hutchinson, G. E. 1943. Food, time, and culture. *Transactions of the New York Academy of Sciences* 5:152–154.

Hutchinson, G. E. 1944. Marginalia. *American Scientist* 32:78–83.

Hutchinson, G. E. 1944. Marginalia. *American Scientist* 32:150–157.

Hutchinson, G. E. 1944. Marginalia. *American Scientist* 32:288–293.

Kubler, G. A., and G. E. Hutchinson. 1944. David Forbes and guano archaeology. *Nature* 154:773.

Kubler, G. A., and G. E. Hutchinson. 1944. Guano archaeology. *Notes and Queries* 147:212.

Hutchinson, G. E. 1944. Limnological studies in Connecticut. VII. A critical examination of the supposed relationship between phytoplankton periodicity and chemical changes in lake waters. *Ecology* 25:3–26.

Hutchinson, G. E. 1944. Conway on the ocean—an appreciation and a criticism. *American Journal of Science* 242:272–280.

Hutchinson, G. E. 1944. Nitrogen in the biogeochemistry of the atmosphere. *American Scientist* 32:178–195.

Hutchinson, G. E. 1944. Notes on Hemiptera. *Annals and Magazine of Natural History,* Series 11, 11:769–778.

Hutchinson, G. E. 1944. A century of atmospheric biogeochemistry, I. Introduction: the carbon metabolism and photosynthetic efficiency of the earth as a whole (G. A. Riley). *American Scientist* 32:129–132.

Hutchinson, G. E. 1944. Problems of biogeochemistry, II. Introduction: the fundamental matter-energy difference between the living and inert natural bodies of the biosphere (W. L. Verdansky). *Transactions of the Connecticut Academy of Arts and Sciences* 35:483–517.

Hutchinson, G. E. 1945. Marginalia. *American Scientist* 33:55–58.

Hutchinson, G. E. 1945. Marginalia. *American Scientist* 33:120–125.

Hutchinson, G. E. 1945. Marginalia. *American Scientist* 33:194–201.

Hutchinson, G. E. 1945. Marginalia. *American Scientist* 33:262–269.

Hutchinson, G. E. 1945. Aluminum in soils, plants, and animals. *Science* 60:29–40.

Hutchinson, G. E. 1945. Alexander Petrunkevitch—an appreciation of his scientific works and a list of his published writings. *Transactions of the Connecticut Academy of Arts and Sciences* 36:9–24.

Hutchinson, G. E. 1945. On the species of *Notonecta* (Hemiptera-Heteroptera) inhabiting New England. *Transactions of the Connecticut Academy of Arts and Sciences* 36:599–605.

Hutchinson, G. E. 1945. Note on a specimen of silk from Dura. In R. Pfisler and L. Bellinger, *The excavations of Dura-Europos,* Final Rep. 4 Pt. 2., p. 54. New Haven: Yale University Press.

Hutchinson, G. E., and J. K. Setlow. 1946. Limnological studies in Connecticut. VIII. The niacin cycle in a small inland lake. *Ecology* 27:13–22.

Hutchinson, G. E. 1946. Social theory and social engineering. *Science* 104:166–167.

Hutchinson, G. E. 1946. Marginalia. *American Scientist* 34:115–120.

Hutchinson, G. E. 1946. Marginalia. *American Scientist* 34:264–266, 280, 282, 284–286.

Hutchinson, G. E. 1946. Marginalia. *American Scientist* 34:477–482, 506, 508, 510.

Hutchinson, G. E. 1946. Marginalia. *American Scientist* 34:650–653, 672–677.

Hutchinson, G. E., J. K. Setlow, and R. L. Brooks. 1946. Biochemical observations on *Asterias forbesi. Bulletin of the Bingham Oceanographic Collection* 9:44–58.

Hutchinson, G. E. 1947. Marginalia. *American Scientist* 35:117–119, 136, 138, 140, 142.

Hutchinson, G. E. 1947. Marginalia. *American Scientist* 35:249–257.

Hutchinson, G. E. 1947. Marginalia. *American Scientist* 35:404–405, 426, 428, 430–431.

Hutchinson, G. E. 1947. Marginalia. *American Scientist* 35:544–549, 582.

Hutchinson, G. E. 1947. The problems of oceanic geochemistry. *Ecological Monographs* 17:299–307.

Hutchinson, G. E., and V. T. Bowen. 1947. A direct demonstration of the phosphorous cycle in a small lake. *Proceedings of the National Academy of Sciences USA* 33:148–153.

Hutchinson, G. E. 1947. A note on the theory of competition between 2 social species. *Ecology* 28:319–321.

Hutchinson, G. E. 1947. The problems of oceanic geochemistry. *Ecological Monographs* 17:299–307.

Gannister, F. A., and G. E. Hutchinson. 1947. The identity of minervite and palmerite with taranakite. *Mineralogical Magazine* 28:29–35.

Hutchinson, G. E. 1948. Appendix B in G. Kubler, "Toward absolute time: guano archaeology." In *A reappraisal of Peruvian archaeology,* Memoirs of the Society for American Archaeology, No. 4, comp. W. C. Bennett, pp. 29–50. Menasha, Wisconsin: Society for American Archaeology and the Institute of Andean Research.

Hutchinson, G. E. 1948. Marginalia—Comments on *The religious man* by Robert G. Aitken. *American Scientist* 36:147–149.

Hutchinson, G. E. 1948. Marginalia. *American Scientist* 36:155–160.

Hutchinson, G. E. 1948. Marginalia. *American Scientist* 36:208, 210, 212, 214, 216, 218, 220, 222, 291–294.

Hutchinson, G. E. 1948. Marginalia. *American Scientist* 36:356, 358, 428–430.

Hutchinson, G. E. 1948. Marginalia—Comments on *Card-guessing experiments* by B. F. Skinner. *American Scientist* 36:461–462.

Hutchinson, G. E. 1948. Marginalia—In Memoriam D'Arcy Wentworth Thompson. *American Scientist* 36:577–581, 601–606.

Hutchinson, G. E. 1948. Circular causal systems in ecology. *Annals of the New York Academy of Sciences* 50:221–246.

Hutchinson, G. E. 1948. On living in the biosphere. *Science* 108:587.

Hutchinson, G. E. 1948. Biology. *Encyclopedia Brittanica,* v. 3, pp. 598–609.

Hutchinson, G. E. 1949. Marginalia. *American Scientist* 37:266–272.

Hutchinson, G. E. 1949. Marginalia. *American Scientist* 37:593–597, 614, 616, 618.

Hutchinson, G. E. 1949. A note on two aspects of the geochemistry of carbon. *American Journal of Science* 247:27–32.

Hutchinson, G. E. 1949. Lakes in the desert. *American Scientist* 37:385–409.

Hutchinson, G. E. 1949. The guano islands ecologically and climatically and paleogeographically considered (Discussed by Lt Col R. B. Seymour Sewell, C.I.E., F.R.S., Dr. Radcliffe N. Salaman, F.R.S., and the President; Prof. Hutchinson replied). *Proceedings of the Linnean Society of London* 162:2.

Hutchinson, G. E., and E. S. Deevey. 1949. Ecological studies on populations. In *Survey of biological progress,* v. 1, ed. G. S. Avery, pp. 325–329. New York: Academic Press.

Hutchinson, G. E. 1950. Marginalia. *American Scientist* 38:282–289.

Hutchinson, G. E. 1950. Marginalia. *American Scientist* 38:612–619.

Hutchinson, G. E. 1950. Survey of contemporary knowledge of biogeochemistry III. The biogeochemistry of vertebrate excretion. *Bulletin of the American Museum of Natural History* 96:1–554.

Hutchinson, G. E. 1950. Limnological studies of Connecticut. IX. A quantitative radiochemical study of the phosphorus cycle in Linsley Pond. *Ecology* 31:194–203.

Brooks, J. L., and G. E. Hutchinson. 1950. On the rate of passive sinking in *Daphnia. Proceedings of the National Academy of Sciences USA* 36:272–277.

Hutchinson, G. E., and V. T. Bowen. 1950. Limnological studies of Connecticut. IX. A quantitative radiochemical study of the phosphorous cycle in Linsley Pond. *Ecology* 31:194–203.

Hutchinson, G. E. 1950. Notes on the functions of a university. *American Scientist* 38:127–131.

Hutchinson, G. E. 1951. Marginalia—Tuba mirum spargens sonum per sepulcra regionum. *American Scientist* 39:145–150, 181.

Hutchinson, G. E. 1951. Marginalia. *American Scientist* 39:306–314.

Hutchinson, G. E. 1951. Marginalia. *American Scientist* 39:473–479.

Hutchinson, G. E. 1951. Marginalia. *American Scientist* 39:717–724.

Hutchinson, G. E. 1951. Review of *Die Binnengewasser Bd. 18* by A. Thienneman. *Science* 114:190.

Hutchinson, G. E. 1951. Copepodology for the ornithologist. *Ecology* 32:571–577.

Hutchinson, G. E. 1951. Review of *The sensory world of the bee* by K. von Frisch. *Yale Review* 40:546–548.

Hutchinson, G. E. 1951. The loves of the card index cards (Review of *Patterns of sexual behavior* by C. S. Ford and F. A. Beach). *Ecology* 32:569–571.

Hutchinson, G. E., and E. M. Low. 1951. The possible role of sexual selection in the etiology of endemic goiter. *Science* 114:482.

Hutchinson, G. E. 1952. Methodology and value in the natural sciences in relation to certain religious concepts. *Journal of Religion* 32:175–187.

Hutchinson, G. E. 1952. The biogeochemistry of phosphorous. In *The biology of phosphorous,* ed. L. F. Wolterink, pp. 1–35. Lansing: Michigan State College Press.

Hutchinson, G. E. 1952. Marginalia. *American Scientist* 40:146–153.

Hutchinson, G. E. 1952. Marginalia. *American Scientist* 40:336–341.

Hutchinson, G. E. 1952. Marginalia. *American Scientist* 40:509–517.

Hutchinson, G. E. 1952. Marginalia. *American Scientist* 40:685–689.

Hutchinson, G. E. 1953. Marginalia. *American Scientist* 41:117–124.

Hutchinson, G. E. 1953. Marginalia. *American Scientist* 41:303–308.

Hutchinson, G. E. 1953. Marginalia. *American Scientist* 41:464–467.

Hutchinson, G. E. 1953. Marginalia. *American Scientist* 41:627, 628–634.

Hutchinson, G. E. 1953. The concept of pattern in ecology. *Proceedings of the Academy of Natural Sciences of Philadelphia* 105:1–12.

Hutchinson, G. E. 1953. *The itinerant ivory tower.* New Haven: Yale University Press.

Hutchinson, G. E. 1953. The natural sciences in Seventy-five. A study of a generation in transition. *Yale Daily News,* New Haven.

Hutchinson, G. E. 1953. Review of *The origin and history of the British fauna* by B. P. Beirne. *American Journal of Science* 251:837–838.

Hutchinson, G. E. 1953. Review of *Reports of the Swedish deep-sea expedition, 1947–48* edited by H. Pettersson. *American Journal of Science* 251:907.

Hutchinson, G. E. 1954. Review of *The medusae of the British Isles* by F. S. Russell. *American Journal of Science* 252:511–512.

Hutchinson, G. E. 1954. *Scientific language and religious faith.* New York: National Council of the Episcopal Church, 25 pp.

Hutchinson, G. E. 1954. Review of *Historical aspects of organic evolution* by P. G. Fothergill. *American Journal of Science* 252:317–318.

Deevey, E. S., M. S. Gross, G. E. Hutchinson, and H. L. Kraybill. 1954. The natural C_{14} contents of materials from hard water lakes. *Proceedings of the National Academy of Sciences USA* 40:285–288.

Hutchinson, G. E. 1954. Marginalia. *American Scientist* 42:136–142.

Hutchinson, G. E. 1954. Marginalia. *American Scientist* 42:300–308.

Hutchinson, G. E. 1954. Marginalia. *American Scientist* 42:500–504.

Hutchinson, G. E. 1954. Marginalia. *American Scientist* 42:661–666.

Hutchinson, G. E., and S. D. Ripley. 1954. Gene dispersal and the ethology of the Rhinocerotidae. *Evolution* 8:178–179.

Hutchinson, G. E. 1954. Theoretical notes on oscillatory populations. *Journal of Wildlife Management* 18:107–109.

Hutchinson, G. E. 1954. The threat of overpopulation (Review of *The challenge of man's future* by H. Brown). *Yale Review* 43:611–613.

Hutchinson, G. E., and J. R. W. Vallentyne. 1955. New approaches to the study of lake sediments. *International Association of Theoretical and Applied Limnology* 12:669–670.

Hutchinson, G. E. 1955. Marginalia—Coda. *American Scientist* 43:144–147.

Hutchinson, G. E. 1955. The enchanted voyage—a study of the effects of the ocean on some aspects of human culture. *Journal of Marine Research* 14:276–283.

Hutchinson, G. E., R. J. Benoit, W. B. Cotter, and P. J. Wangersky. 1955. On the Nickel,

Cobalt, and Copper sediments of deep-sea sediments. *Proceedings of the National Academy of Sciences USA* 41:160–162.

Hutchinson, G. E. 1956. Ice ages—past and future. *New York Herald Tribune,* 29 July.

Hutchinson, G. E., and H. Loffler. 1956. The thermal classification of lakes. *Proceedings of the National Academy of Sciences USA* 42:84–86.

Hutchinson, G. E., R. Patrick, and E. S. Deevey. 1956. Sediments of Lake Patzcuaro, Michoacan, Mexico. *Bulletin of the Geological Society of America* 67:1491–1504.

Hutchinson, G. E. 1957. *A treatise on limnology, v. 1. Geography, physics and chemistry.* New York: Wiley.

Hutchinson, G. E. 1957. Future of marine paleoecology. In "Treatise on marine geology and paleoecology," ed. H. S. Ladd. Special issue, *Geological Society of America Memoir* 67(2): 683–689.

Hutchinson, G. E. 1957. On a new species of *Anisops* (Hemiptera, Notonectidae) from the Moluccas. *Postilla (Peabody Museum of Natural History, Yale University)* 33, 4 pp.

Hutchinson, G. E. 1957. *A preliminary list of the writings of Rebecca West 1912–1951.* New Haven: Yale University Library.

Hutchinson, G. E. 1957. Review of *Roots of scientific thought* edited by P. Wiener and A. Noland. *Library Sci Book News* 4.

Hutchinson, G. E. 1957. Marginalia. *American Scientist* 45:88–96.

Hutchinson, G. E. 1957. Marginalia—Letter to Editor. *American Scientist* 45:120A, 122A.

Hutchinson, G. E. 1957. Population studies—animal ecology and demography—concluding remarks. *Cold Spring Harbor Symposia on Quantitative Biology* 22:415–427.

Hutchinson, G. E. 1958. Religion and the natural sciences. In *Religion and the state university,* ed. E. A. Walters, pp. 156–171. Ann Arbor: University of Michigan Press.

Wangersky, P. J., and G. E. Hutchinson. 1958. Deposition and deep water movements in the Caribbean. *Nature* 181:108–109.

Hutchinson, G. E. 1959. Homage to Santa Rosalia, or Why are there so many kinds of animals? *The American Naturalist* 93:145–159.

Hutchinson, G. E. 1959. A speculative consideration of certain possible forms of sexual selection in man. *The American Naturalist* 93:81–91.

Hutchinson, G. E. 1959. Il concetto moderno di niccia ecologica. *Memorie dell'Instituto Italiano di Idrobiologia* 11:9–22.

Hutchinson, G. E. 1959. Anybody here seen the Abominable Snowman lately? (Review of *On the track of unknown animals* by B. Heuvelmans). *New York Herald Tribune Book Review,* 7 June.

Hutchinson, G. E. 1959. The role of the museum in teaching and research. *Yale Alumni Magazine* 22:17–19.

Hutchinson, G. E. 1959. The pageant of the animal kingdom (Review of *The road to man* by H. Wendt). *New York Herald Tribune Book Review,* 12 July.

Hutchinson, G. E., and R. H. MacArthur. 1959. A theoretical model of size distributions among species of animals. *The American Naturalist* 93:117–125.

Hutchinson, G. E., and R. H. MacArthur. 1959. On the theoretical significance of aggressive neglect in interspecific competition—appendix. *The American Naturalist* 93:133–134.

Hutchinson, G. E. 1960. On evolutionary euryhalinity. *American Journal of Science* 258:98–103 Suppl. S.

Hutchinson, G. E. 1960. John Farquhar Fulton. *Yearbook of the American Philosophical Society 1960*:140–142.

Hutchinson, G. E. 1960. Landscape with animals—zoolog's romantic story (Review of *Out of Noah's ark* by H. Wendt). *New York Herald Tribune Book Review,* 21 February.

Hutchinson, G. E. 1960. A set of informed guesses (Review of *The story of early man* by H. E. L. Mellersh). *New York Herald Tribune Book Review,* 12 August.

Hutchinson, G. E. 1961. Review of *Evolution above the species level* by B. Rensch. *American Anthropologist* 63:880–881.

Hutchinson, G. E. 1961. The paradox of the plankton. *The American Naturalist* 95:137–145.

Hutchinson, G. E. 1961. Fifty years of man in the zoo. *Yale Review* (50th anniversary issue): 56–65.

Hutchinson, G. E. 1961. Contribution: the challenge of graduate education. *Yale Alumni Magazine* 24:17 (also published in *Ventures* 1:13–15).

Hutchinson, G. E. 1961. Review of *Darwin's biological work—some aspects reconsidered* edited by P. R. Bell. *American Journal of Science* 259:558–559.

Hutchinson, G. E. 1962. *The enchanted voyage and other studies.* New Haven and London: Yale University Press.

Tsukada, M., U. M. Cowgill, and G. E. Hutchinson. 1962. The History of Lake Petenxil, Departamento de El Petén, Guatemala [abstract]. *Science* 136:329.

Cowgill, U. M., R. J. Andrew, A. Bishop, et al. 1962. An apparent lunar periodicity in sexual cycle of certain prosimians. *Proceedings of the National Academy of Sciences USA* 48:238–241.

Hutchinson, G. E. 1962. Review of *Survival of the free* edited by W. Engelhardt. *New York Herald Tribune Book Review,* 14 October.

Hutchinson, G. E. 1963. Natural-selection, social organization, hairlessness, and the Australopithecine canine. *Evolution* 17:588–589.

Hutchinson, G. E., and U. M. Cowgill. 1963. Chemical examination of a core from Lake Zeribar, Iran. *Science* 140:67–69.

Hutchinson, G. E., and U. M. Cowgill. 1963. El Bajo de Santa Fe. *Transactions of the American Philosophical Society* 53:1–51.

Cowgill, U. M., and G. E. Hutchinson. 1963. Differential mortality among sexes in childhood and its possible significance in human evolution. *Proceedings of the National Academy of Sciences USA* 49:425–429.

Hutchinson, G. E. 1963. A note on the polymorphism of *Philaenus spumarius* (L.) (Homopt. Cercopidae) in Britain. *Entomological Monthly Magazine* 99:175–178.

Cowgill, U. M., and G. E. Hutchinson. 1963. Sex-ratio in childhood and depopulation of Petén, Guatemala. *Human Biology* 35:90–103.

Hutchinson, G. E. 1963. The cream in the gooseberry fool. *American Scientist* 51:446–453.

Cowgill, U. M., O. Joensuu, and G. E. Hutchinson. 1963. An apparently triclinic dimorph of crandallite from a tropical swamp sediment in El Petén, Guatemala. *American Mineralogist* 48:1144–1153.

Hutchinson, G. E. 1963. Is an electron smaller than a dream? *Journal of Parapsychology* 27:301–306.

Hutchinson, G. E. 1963. The prospect before us. In *Limnology in North America,* ed. D. G. Frey, pp. 683–690. Madison: University of Wisconsin Press.

Hutchinson, G. E. 1963. The naturalist as an art critic. *Proceedings of the Academy of Natural*

Sciences, Philadelphia 115:99–111 (also appeared in *Association of Southeastern Biologists Bulletin* 10:47–54).

Cowgill, U. M., and G. E. Hutchinson. 1963. Ecological and geochemical archaeology in the southern Maya lowlands. *Southwestern Journal of Anthropology* 19:267–286.

Hutchinson, G. E. 1964. The influence of the environment. *Proceedings of the National Academy of Sciences USA* 51:930–934.

Cowgill, U. M., and G. E. Hutchinson. 1964. Archaeological significance of a stratigraphic study of El Bajo de Santa Fe. *Actas (Mexico City, 1964), 35th International Congress of Americanists, Mexico* 1:603–613.

Hutchinson, G. E. 1964. On *Filinia terminalis* (plate) and *F. pejleri sp. n.* (Rotatoria: family Testudinellidae). *Postilla (Peabody Museum of Natural History, Yale University)* 81, 8 pp.

Hutchinson, G. E. 1964. Notes on ecological principles easily seen in the Hall of Southern New England Natural History. *Peabody Museum Special Publication* 8, 7 pp.

Cowgill, U. M., and G. E. Hutchinson. 1964. Cultural eutrophication in Lago di Monterosi during Roman antiquity. *International Association of Theoretical and Applied Limnology Proceedings* 15:644–645.

Ferguson, Jr., E., G. E. Hutchinson, and C. E. Goulden. 1964. *Cypria petensis,* a new name for the ostracod, *Cypria pelagic* Brehm 1932. *Postilla (Peabody Museum of Natural History, Yale University)* 80, 4 pp.

Hutchinson, G. E. 1964. The lacustrine microcosm reconsidered. *American Scientist* 52:334–341.

Hutchinson, G. E. 1965. *The ecological theater and the evolutionary play.* London and New Haven: Yale University Press.

Hutchinson, G. E., and B. Chance. 1965. Discussion of Dr. Urey's paper. *Proceedings of the National Academy of Sciences USA* 53:1172–1173.

Hutchinson, G. E., D. I. Arnon, A. G. Fischer, B. Commoner, W. M. Elsasser, P. H. Abelson, and P. E. Cloud. 1965. Discussion of Fischer, A. G. *Proceedings of the National Academy of Sciences USA* 53:1213–1215.

Hutchinson, G. E. 1965. *Eretmia* Gosse 1886 (? Rotatoria) proposed suppression of this generic name under the plenary powers. *Bulletin of Zoological Nomenclature* 22:60–62.

Cowgill, U. M., and G. E. Hutchinson. 1965. A middle 19th century glass bottomed box. *Entomological Record* 77:264.

Hutchinson, G. E. 1966. On being a meter and a half long. Publication of the Smithsonian 200th anniversary celebration of the birth of James Smithson.

Hutchinson, G. E. 1966. To save Baikal (letter to editor), *New York Times,* 23 June.

Hutchinson, G. E. 1966. Sensory aspects of taxonomy pleiotropism and kinds of manifest evolution. *The American Naturalist* 100:533–539.

Hutchinson, G. E., and U. M. Cowgill. 1966. A general account of the basin and the chemistry and mineralogy of the sediment core. *Memoirs of the Connecticut Academy of Arts and Sciences* 17:7–62.

Cowgill, U. M., and G. E. Hutchinson. 1966. The history of the Petenxil basin. *Memoirs of the Connecticut Academy of Arts and Sciences* 17:121–126.

Cowgill, U. M., and G. E. Hutchinson. 1966. La aguadade Santa Ana Vieja—the history of a pond in Guatemala. *Archiv für Hydrobiologie* 62:335–372.

Hutchinson, G. E. 1967. Ecological biology in relation to the maintenance and improvement of the human environment. In *Applied Science and Technological Progress* (A report to the Committee on Science and Astronautics, U.S. House of Representatives, by the National Academy of Sciences), pp. 71–185.

Hutchinson, G. E. 1967. A note on the genetics of *Abruxas grossuluriatu ab. fulvapicata* Raynor. *Entomological Record* 79:80–81.

Hutchinson, G. E. 1967. *A treatise on limnology, v. 2. Introduction to lake biology and the limnoplankton.* New York: Wiley.

Hutchinson, G. E. 1968. Vittorio Tonolli 1913–1967. *Archiv für Hydrobiologie* 64:491–495.

Cowgill, U. M., G. E. Hutchinson, and H. C. W. Skinner. 1968. Elementary composition of *Latimeria chalumnae* Smith. *Proceedings of the National Academy of Sciences USA* 60:456–463.

Hutchinson, G. E. 1968. A Cambridge caricaturist. *Country Life* 144:644–645.

Hutchinson, G. E. 1969. *Aysheaia* and general morphology of Onychophora. *American Journal of Science* 267:1062–1066.

Hutchinson, G. E. 1969. Some continental European aberrations of *Abraxas grossuluriata* Linn. (Lepidoptera) with a note on the theoretical significance of the variation observed in the species. *Transactions of the Connecticut Academy of Arts and Sciences* 43:1–24.

Hutchinson, G. E. 1969. Eutrophication—past and present. In *Eutrophication: Causes, consequences, correctives; proceedings of a symposium,* pp. 17–26. Washington, D.C.: National Academy of Sciences.

Cowgill, U. M., and G. E. Hutchinson. 1969. Mineralogical examination of ceramic sequence from Tikal, El Petén, Guatemala. *American Journal of Science* 267:465–477.

Hutchinson, G. E. 1970. Marginalia. *American Scientist* 58:17.

Hutchinson, G. E. 1970. Attitudes towards nature in medieval England—Alphonso and bird psalters. *Science* 170:891.

Hutchinson, G. E. 1970. Biosphere. *Scientific American* 223:44–53.

Bormann, F. H., A. W. Galston, G. E. Hutchinson, et al. 1970. Etzioni's view of environment. *Science* 169:529.

Hutchinson, G. E. 1970. Chemical ecology of 3 species of *Myriophyllum* (Angiospermae, Haloragaceae). *Limnology and Oceanography* 15:1–5.

Hutchinson, G. E. 1970. Threatened Berkshires (letter to editor). *New York Times,* 26 April.

Hutchinson, G. E. 1970. Letter to editor. *Canadian Research and Development* 3:52.

Hutchinson, G. E., ed. 1970. Ianula: An account of the history and development of the Lago di Monterosi, Latium, Italy. Special issue, *Transactions of the American Philosophical Society* 60(4), 178 pp.

Hutchinson, G. E. 1970. Introductory account of the basin. In *Ianula: An account of the history and development of the Lago di Monterosi, Latium, Italy,* ed. G. E. Hutchinson. *Transactions of the American Philosophical Society* 60(4): 5–9.

Cowgill, U. M., and G. E. Hutchinson. 1970. Chemistry and mineralogy of the sediments and their source materials. In *Ianula: An account of the history and development of the Lago di Monterosi, Latium, Italy,* ed. G. E. Hutchinson. *Transactions of the American Philosophical Society* 60(4): 37–101.

Hutchinson, G. E., and U. M. Cowgill. 1970. The history of the lake—a synthesis. In *Ianula: An account of the history and development of the Lago di Monterosi, Latium, Italy,* ed. G. E. Hutchinson. *Transactions of the American Philosophical Society* 60(4): 163–170.

Hutchinson, G. E., and A. L. Hutchinson. 1970. The "Idol" or Sheela-na-gig at Binstead.

Proceedings of the Isle of Wight Natural History and Archaeological Society 1969:237–251.

Hutchinson, G. E. 1970. Marginalia—Wisdom is justified of all her children. *American Scientist* 58:17–20.

Hutchinson, G. E. 1970. Marginalia. *American Scientist* 58:528–535.

Hutchinson, G. E. 1971. Scale effects in ecology. In *Statistical ecology,* v. 1, ed. G. P. Patil, E. C. Pielou, and W. E. Waters, pp. xvii–xxvi. State College: Pennsylvania State University Press.

Hutchinson, G. E. 1972. Review of *Nutrients and eutrophication* edited by G. E. Likens. *Limnology and Oceanography* 17:965–968.

Hutchinson, G. E. 1972. Marginalia—Long Meg reconsidered. *American Scientist* 60:24–31.

Hutchinson, G. E. 1972. Marginalia—Long Meg reconsidered, Part 2. *American Scientist* 60:210–219.

Hutchinson, G. E. 1972. Moon and megaliths—reply. *American Scientist* 60:412.

Hutchinson, G. E., and U. M. Cowgill. 1973. Waters of Merom—study of Lake Huleh. 3. Chemical constituents of a 54 M core. *Archiv für Hydrobiologie* 72:145–185.

Hutchinson, G. E. 1973. Contribution: population and behavior, pp. 120–122; and Problems of population control, pp. 151–152, in *Ethical Issues in Biology and Medicine,* ed. Preston Williams. Boston: Schenkman.

Botkin, D. B., P. A. Jordan, A. S. Dominski, H. S. Lowendorf, and G. E. Hutchinson. 1973. Sodium dynamics in a northern ecosystem. *Proceedings of the National Academy of Sciences USA* 70:2745–2748.

Hutchinson, G. E. 1973. Marginalia—Eutrophication. *American Scientist* 61:269–279.

Hutchinson, G. E. 1974. De rebus planctonicus. *Limnology and Oceanography* 19:360–361.

Hutchinson, G. E. 1974. Attitudes toward nature in medieval England—Alphonso and Bird Psalters. *Isis* 65:5–37.

Hutchinson, G. E. 1974. Marginalia—aposematic insects and master of Brussels initials. *American Scientist* 62:161–171.

Hutchinson, G. E. 1974. New and inadequately described aberrations of *Abraxas grossulariata* (Linn.) (Lep. Geometridae). *Entomological Record* 86:199–206.

Hutchinson, G. E. 1975. Variations on a theme by Robert MacArthur. In *Ecology and Evolution of Communities,* ed. M. L. Cody and J. L. Diamond, pp. 492–521. Cambridge: Belknap Press of Harvard University Press.

Hutchinson, G. E. 1975. Mammalia in the hunting scene from the Dura Mithraeum, Appendix 13.2. In *Mithraic studies,* Proceeding of the First International Conference of Mithraic Studies, v. 1, ed. J. R. Hinnells, pp. 210–214. Manchester: Manchester University Press.

Hutchinson, G. E. 1975. For what sort of life should mankind aim? In *Environmental quality and society,* ed. R. A. Tybout, pp. 15–25. Columbus: Ohio State University Press.

Hutchinson, G. E. 1975. *A treatise on limnology, v. 3. Limnological botany.* New York: Wiley.

Hutchinson, G. E. 1975. Some biological analogies. In *Beyond growth—essays on alternative futures,* Yale School of Forestry and Environmental Studies Bulletin 88, ed. W. R. Burch, Jr., and F. H. Bormann, pp. 24–40.

Hutchinson, G. E. 1976. Marginalia—Man talking or thinking. *American Scientist* 64:22–28.

Hutchinson, G. E. 1976. A taste for learning. *Yale Alumni Magazine and Journal* 40:15.

Ohlhurst, S., and G. E. Hutchinson. 1977. Waters of Merom—study of Lake Huleh. 5. Temporal changes in Molluscan fauna. *Archiv für Hydrobiologie* 80:1–19.

Hutchinson, G. E. 1977. Review of *Island biology illustrated by the land birds of Jamaica* by David Lack. *American Scientist* 65:492.

Hutchinson, G. E. 1977. The influence of the New World on the study of natural history. In *The changing scenes in natural sciences, 1776–1976,* Academy of Natural Sciences of Philadelphia Special Publication 12, ed. C. E. Goulden, pp. 13–34.

Hutchinson, G. E. 1977. Memorial: George Alfred Baitsell, 1885–1971. *Anatomical Record* 189:299–300.

Hutchinson, G. E. 1977. Science has been liberal handed. . . . In *The Campaign for Yale,* promotional publication, March, pp. 43–44.

Hutchinson, G. E. 1977. Thoughts on $370 million. In *The Campaign for Yale,* promotional publication, June, p. 3.

Hutchinson, G. E. 1978. Review of *Animals and men* by K. Clark. *American Scientist* 66:81–82.

Hutchinson, G. E. 1978. *An introduction to population ecology.* New Haven and London: Yale University Press.

Hutchinson, G. E., and P. J. Parker. 1978. Sexual dimorphism in winter whitening of stoat *Mustela erminea. Journal of Zoology* 186:560–563.

Hutchinson, G. E. 1978. Zoological iconography in the West after AD 1200. *American Scientist* 66:675–684.

Furth, D., R. Albrecht, K. Muraszko, and G. E. Hutchinson. 1978. Scanning electron-microscope study of palar pegs of 3 species of Corixidae (Hemiptera). *Systematic Entomology* 3:147–152.

Hutchinson, G. E. 1979. *The kindly fruits of the earth—recollections of an embryo ecologist.* New Haven and London: Yale University Press.

Hutchinson, G. E. 1979. Review of *Survivals of Greek zoological illuminations in Byzantine manuscripts* by Z. Kadar. *Isis* 70:452–453.

Hutchinson, G. E. 1979. Review of *Prehistoric Avebury* by A. Burl, *Megalithic remains in Britain and Bretony* by A. Thom and A. S. Thom, and *Megaliths and masterminds* by P. L. Brown. *American Scientist* 67:728–729.

Hutchinson, G. E. 1979. The most useful, the most ornamental. *The Franklin Institute News* 44:2–3, 7.

Hutchinson, G. E. 1979. Memories of the Corixid water bugs. *Discovery (Peabody Museum of Natural History, Yale University)* 14:11–19.

Hutchinson, G. E., and S. Rachootin. 1979. Historical introduction. In *Problems of genetics,* ed. W. Bateson, pp. vii–xx. New Haven and London: Yale University Press.

Stiller, M., and G. E. Hutchinson. 1980. The waters of Merom—a study of Lake Huleh. 6. Stable isotopic composition of carbonates of a 54 M core—paleoclimatic and paleotrophic implications. *Archiv für Hydrobiologie* 89:275–302.

Hutchinson, G. E. 1980. Conjectures arising in a quiet museum. *Antenna, Bulletin of the Royal Entomological Society of London* 4:92, 97–98.

Hutchinson, G. E. 1980. Lorande Loss Woodruff, 1879–1947. *Biogeographical Memoirs of the National Academy of Sciences* 52:470–485.

Hutchinson, G. E. 1981. Thoughts on aquatic insects. *BioScience* 31:495–500.

Hutchinson, G. E. 1981. Review of *The gardens of Pompeii. Herculaneum and the villas destroyed by Vesuvius* by W. F. Jashemski. *American Scientist* 69:85–86.

Hutchinson, G. E. 1981. Review of *A bestiary of Saint Jerome—animal symbolism in European religious art* by Herbert Friedmann. *American Scientist* 69:552.

Hutchinson, G. E. 1981. Letter to the editor—Where is everybody? *American Scientist* 69:258, 260.

Hutchinson, G. E. 1981. Foreword to L. Margulis, *Symbiosis in cell evolution.* San Francisco: W. H. Freeman.

Hutchinson, G. E. 1981. Random adaptation and imitation in human evolution. *The American Naturalist* 69:161–165.

Ohlhorst, S., A. Schmida, M. M. Poulson, and G. E. Hutchinson. 1982. The waters of Merom—a study of Lake Huleh. 8. Non-siliceous plant remains, with appendices on some animal fossils. *Archiv für Hydrobiologie* 94:441–459.

Hutchinson, G. E. 1982. Review of *The monstrous races in medieval art and thought* by J. B. Friedman. *American Scientist* 70:533–534.

Hutchinson, G. E. 1982. Review of *Birds in medieval manuscripts* by B. Yapp. *Isis* 73:598.

Hutchinson, G. E. 1982. The harp that once . . . a note on the discovery of stridulation in the Corixid water bugs. *Irish Naturalists' Journal* 20:457–466.

Hutchinson, G. E. 1982. On the "Return of the soldier" by Rebecca West. *Yale University Library Gazette* 57:66–71.

Hutchinson, G. E. 1982. Life in air and water. *Discovery (Peabody Museum of Natural History, Yale University)* 16:3–9.

Hutchinson, G. E. 1982. Reminiscences and notes on some otherwise undiscussed papers. In *Selected works of Gordon A. Riley,* ed. J. Wroblewski, pp. vii–ix. Halifax: Dalhousie University.

Hutchinson, G. E. 1983. Marginalia—What is science for? *American Scientist* 71:639–644.

Hutchinson, G. E., and N. Tongring. 1984. The possible adaptive significance of the Brooks-Dyar rule. *Journal of Theoretical Ecology* 106:437–439.

Hutchinson, G. E. 1984. Rebecca West: A tribute from G. E. Hutchinson. In *Dictionary of Literary Biography,* ed. J. Bruccoli and R. Layman, pp. 143–144. Ann Arbor, Michigan: Edward Bors.

Hutchinson, G. E. 1984. Letter to the editor: Unexplained dogma. *New York Times,* 17 August.

Hutchinson, G. E. 1986. Riley, Gordon A., 1911–1985—In Memoriam. *Limnology and Oceanography* 31:233.

Hutchinson, G. E. 1987. The ecological niche. *Physiology and Ecology Japan* 24:s03–s07.

Hutchinson, G. E. 1987. Keep walking—the lecture for the Kyoto Prize 1986. *Physiology and Ecology Japan* 24:s81–s87.

Hutchinson, G. E. 1987. West, Rebecca—An account of the genesis of a complicated relationship. *Yale Review* 76:203–205.

Hutchinson, G. E. 1987. Introduction. In R. West, *The strange necessity.* London: Virago.

Hutchinson, G. E. 1988. This issue dedicated to Edmondson, W. Thomas—Introduction. *Limnology and Oceanography* 33:1231–1233.

Hutchinson, G. E. 1988. Review of *Angels fear—towards an epistemology of the sacred* by G. Bateson and M. C. Bateson. *American Scientist* 76:285–286.

Hutchinson, G. E., and N. A. Formozov. 1989. The role of A. N. Formozov in the development of ecological theory. *Archives of Natural History* 16:143–145.

Wilson, E. O., and G. E. Hutchinson. 1989. Robert Helmer MacArthur (1930–1972). *Biographical Memoirs of the National Academy of Sciences* 60:103–114.

Hutchinson, G. E. 1991. Libbie Henrietta Hyman, 1888–1969. *Biographical Memoirs of the National Academy of Sciences* 60:103–114.

Hutchinson, G. E. 1991. Alexander Petrunkevitch, 1875–1964. *Biographical Memoirs of the National Academy of Sciences* 60:235–248.

Hutchinson, G. E. 1991. Population Studies—animal ecology and demography (reprinted from 1957, *Cold Spring Harbor Symposia on Quantitative Biology* 22:415–427). *Bulletin of Mathematical Biology* 53:193–213.

Hutchinson, G. E. 1992. Correction. *Bulletin of Mathematical Biology* 54:695.

Hutchinson, G. E. 1993. *A treatise on limnology, v. 4. The zoobenthos.* New York: Wiley.

NOTES

1 Introduction. The Beauty of the World: Evelyn Hutchinson's Vision of Science

1. Jeffreys (1937).
2. Hutchinson (1953b); Hutchinson (1962). See the essay "The electronic antichrist," pp. 44–49.
3. Hutchinson (1953b). See the essay "Methodology and value in the natural sciences in relation to certain religious concepts," pp. 220–240.
4. Hutchinson (1953b). See the essay "In memoriam: D'Arcy Wentworth Thompson (1860–1948)," pp. 169–185.
5. Hutchinson (1978), pp. 237–241. The criticisms to which Hutchinson was responding were in Peters (1976).
6. Darwin (1964).
7. Slobodkin and Slack (1999).
8. Humboldt (1997). See introduction by Nicolaas A. Rupke, pp. vii–x.
9. Nicolson (1996).
10. Dettelbach (1996).
11. Humboldt (1850).
12. Humboldt (1850), pp. 21–22.
13. Hutchinson (1953b). See the essay "Methodology and value."
14. Hutchinson (1953b). See the essay "Methodology and value," p. 226.
15. Hutchinson (1953b). See the essay "Methodology and value," p. 227.
16. West (1987).
17. Bush (1945), p. 1.
18. Platt and Wolfe (1964).
19. Carson (1962).
20. Bocking (1997).
21. Hagen (1992); Golley (1993).
22. Kingsland (1995).
23. Hutchinson (1962). See the essay "The electronic antichrist," p. 49.
24. Mumford (1956), pp. 1141–1152, quoted at p. 1152.
25. Hutchinson (1953b). See the essay "The history of a lake," pp. 212–219, quoted at p. 219.
26. Hutchinson (1979), p. 252.
27. Hutchinson (1943), quoted at p. 271.
28. Hutchinson (1983), quoted p. 643.
29. Hutchinson (1953b). See the essay "In memoriam: D'Arcy Wentworth Thompson," pp. 169–185.
30. Thompson (1942).
31. Hutchinson (1953b). See p. 182.

32. Gould (1987). See the essay "Exultation and explanation," pp. 180–188.
33. Hutchinson (1979).
34. Darwin (1958). See p. 44.
35. Strong (1979).
36. Hutchinson (1983). See p. 640.
37. Hutchinson (1979). See p. 247.
38. Lindeman (1942); Cook (1977).
39. Hutchinson (1979), p. 248.

    2   Biography. From English Schoolboy to America's Foremost Ecologist

1. Sulloway (2003).
2. Sulloway (2003).
3. Hutchinson (1979).
4. Hutchinson (1918).
5. Hutchinson, letter to Jean H. Langenheim, April 1, 1986. Hutchinson Papers, Manuscripts and Archives, Yale University Library.
6. Hutchinson (1979).
7. Hutchinson (1979).
8. Hutchinson (1928).
9. Thienemann (1925).
10. Elton (1927).
11. Hutchinson (1979).
12. Hutchinson (1979).
13. Hutchinson (1979).
14. Ball (1987); Slack (1995).
15. de Terra (1932a).
16. Hutchinson (1932a).
17. Hutchinson (1932b).
18. de Terra (1932b).
19. Hutchinson (1932c).
20. Hutchinson (1932d).
21. Hutchinson (1933).
22. Hutchinson, Pickford, and Schuurman (1932).
23. Nicholas (1945).
24. Hutchinson (1936).
25. Hutchinson (1936).
26. Hutchinson (1936).
27. Hutchinson (1936).
28. Bradley (1948).
29. Hutchinson (1948a).
30. Slack (2003).
31. Cook (1977).
32. Edmondson (1971).
33. Riley (1984).
34. Hutchinson (1982).

35. Riley (1984).
36. Riley (1984).
37. Riley (1980).
38. Riley (1984).
39. Deevey (1986).
40. Hutchinson (1979).
41. Hutchinson (1984).
42. Deevey (1937).
43. Edmondson (1991).
44. Hutchinson (1979).
45. Edmondson (1991).
46. Edmondson and Hutchinson (1934).
47. Edmondson (1989).
48. Edmondson (1991).
49. Edmondson (1991).
50. Rappaport (1999).
51. Edmondson (1991).
52. C. Goulden, interviewed by N. G. Slack at the Philadelphia Academy of Sciences, September 26, 1991 (tape).
53. Dunbar (1993).
54. Dunbar (1993).
55. Dunbar (1993).
56. Hutchinson (1932e).
57. Hutchinson (1948b).
58. Hutchinson (1948a).
59. Slack (2010); Hutchinson (1943).
60. Hutchinson (1957a).
61. Hutchinson and Edmondson (1993).
62. Hutchinson (1957b).
63. Hutchinson (1959); Hutchinson (1961).
64. McIntosh (1985).
65. Platil and Rosenzweig (1979).
66. Hutchinson (1957b).
67. Kingsland (1985).
68. MacArthur (1958).

3   Limnology. Astonishing Microcosms

1. Hutchinson (1979), p. 206.
2. Hutchinson and Bowen (1947); Hutchinson (1950b); Hutchinson (1964a); Hutchinson (1918); Hutchinson (1951); Hutchinson (1959).
3. Hutchinson (1963).
4. Hutchinson (1957b).
5. Hutchinson (1933).
6. Hutchinson (1938).
7. Hutchinson (1941b).

8. Schindler et al. (1986).
9. Hutchinson and Bowen (1947).
10. Hudson, Taylor, and Schindler (2000).
11. Hutchinson (1969); Hutchinson (1941b); Hutchinson and Bowen (1947); Hutchinson and Bowen (1950).
12. Hutchinson (1942).
13. Hutchinson (1964a).
14. Forbes (1887).
15. May (1973); Tilman et al. (1996); Hooper et al. (2005); Loreau et al. (2001).
16. Hutchinson (1957a); Hutchinson (1967).
17. Riley (1971).
18. Hutchinson (1957a).
19. Hutchinson (1967).
20. Hutchinson (1941b); Hutchinson and Bowen (1947); Hutchinson (1950b); Hutchinson and Bowen (1950); Hutchinson (1969).

4 Theory. Reflection Thereon: G. Evelyn Hutchinson and Ecological Theory

1. Hutchinson (1940).
2. Clements and Shelford (1939).
3. Hutchinson (1950a).
4. Hutchinson (1941a).
5. Gause (1935).
6. Hutchinson (1947); Hutchinson (1948b).
7. Hutchinson (1951).
8. Levins (1969); Wilson (1992).
9. Tilman et al. (1994); Leibold et al. (2004); Urban and Skelly (2006).
10. Hutchinson (1953a).
11. Grinnell (1917); Elton (1927).
12. Gause (1935).
13. Hutchinson (1957b).
14. Hutchinson (1959).
15. Horn and May (1977).
16. Simberloff and Boecklen (1981).
17. Brown and Wilson (1956).
18. Dayan and Simberloff (2005).
19. Hutchinson (1961).
20. Lawton (2000).
21. Hutchinson (1964b).
22. Hendry and Kinnison (1999); Hairston et al. (2005).
23. Hutchinson (1981).
24. E.g., Werner (1986).
25. Hutchinson (1978).
26. Gould (1979).
27. Hutchinson (1962).
28. Hutchinson (1983).
29. Gould (1979).

5   *Museums. Experiencing Green Pigeons*

We are grateful to Steve Stearns for including us in the Hutchinson symposium at Yale, and to David Skelly and the other editors for encouraging our contribution (and for their great patience in awaiting our manuscript). We are especially indebted to Sharon Kingsland for sharing her thoughts and her manuscript with us, and to the staff of the Peabody Museum, who helped us greatly with their recollections. Charles Remington and Barbara Narendra provided valuable insights and information, and Ray Pupedis (division of entomology) and Kristof Zyskowski (division of vertebrate zoology) were especially helpful in the Peabody collections. For additional help with the presentation and the figures we are grateful to Joe Jolly, Sally Pallatto, Kim Zolvik, and Laura Friedman. Although neither of us had the privilege of meeting Hutchinson in person, we are both extremely grateful to him for his wise reflections on museums and collections, which have greatly enriched our own understanding and appreciation of their significance.

1. Hutchinson (1966), p. 86. References and quotations from Hutchinson's writings are cited with page numbers from his books, which are more readily accessible than the originally published essays.
2. For example, see Hutchinson (1918) in this volume (p. 57).
3. Hutchinson (1979), pp. 39–40.
4. Hutchinson (1979), p. 41.
5. Hutchinson (1966), p. 84.
6. Hutchinson (1953b), p. 144.
7. See, for example, Conn (1998).
8. Hutchinson (1979), p. 250.
9. From a letter to Keith S. Thomson, July 9, 1981. Box 57, Series II, Hutchinson Papers, Yale University Archives.
10. In a letter to biology professor Timothy Goldsmith, on the occasion of a search for a new director of the Peabody Museum, Hutchinson referenced this new development in natural history museums: "I would call attention to the historically remarkable development that has taken place in the museum during the past twenty-five years. . . . It is significant that this improvement started at the end of a period when all studies of morphological diversity and classification were regarded as old-fashioned and even evolutionary studies, looking hopelessly unexperimental, were regarded as intellectually unsatisfactory. The development of the sort of museum work that has been done in the Peabody during the last few decades represents a real modern development, in which the old traditions of systematics and evolutionary morphology have been revivified by population genetics, evolutionary ecology and comparative physiology." Letter to Timothy H. Goldsmith, not dated. Box 57, Series II, Hutchinson Papers, Yale University Archives.
11. In 1979, Charles Remington, then curator of entomology, highlighted the significance of Hutchinson's donations: "One of the most important units in the insect collections of the Peabody Museum is the rich array of water bugs presented by Hutchinson. . . . The collection contains 3000 specimens of water bugs and about 700 other insects--including the many specimens taken by Professor Hutchinson as a member of the Yale North India Expedition in 1932." Remington (1979).
12. Edelson (1982).
13. Hutchinson (1979), p. 237.
14. Hutchinson (1962), p. viii.

15. Hutchinson (1979), p. 42.
16. Hutchinson (1965), p. 99.
17. Hutchinson (1965), p. 100.
18. Hutchinson (1965), p. 107.
19. Kiester (1997).
20. Hutchinson (1965), p. 97; Weschler (1995); Donoghue and Alverson (2000).
21. Hutchinson (1979), p. 199.
22. Hutchinson (1962), pp. 90–97.
23. Hutchinson (1962), p. 97.
24. Hutchinson (1959).
25. Hutchinson (1979), p. 36.
26. Hutchinson (1965), pp. 121–130.
27. Hutchinson (1980a), p. 92.
28. Hutchinson (1965), p. 130.
29. Hutchinson (1965), p. 130.
30. Hutchinson (1977), p. 43.
31. Hutchinson (1980a).
32. Hutchinson (1980a), p. 92.
33. Hutchinson (1962), p. 108.
34. Hutchinson (1965), pp. 76–77.
35. For example, Webb et al. (2002); Wiens and Donoghue (2004); Greene (2005).
36. Cavender-Bares et al. (2004).
37. Hutchinson (1953b), p. 228.

REFERENCES

Note: All audio tapes cited are in the Smithsonian Institution Archives, Oral History Collection, G. Evelyn Hutchinson Biographical Interviews, RU9621.

Ball, J. N. 1987. *In Memoriam,* Grace E. Pickford, 1902–1986. *Endocrinology* 65:162–165.

Bocking, S. 1997. *Ecologists and Environmental Politics: A Study of Contemporary Ecology.* New Haven and London: Yale University Press.

Bradley, E. H. 1948. Letter to G. Evelyn Hutchinson, September 7, 1948. Hutchinson Papers, Manuscripts and Archives, Yale University Library.

Brown, W. L., and E. O. Wilson. 1956. Character displacement. *Systematic Biology* 5:49–64.

Bush, V. 1945. *Science, the Endless Frontier; a Report to the President.* Washington, D.C.: United States Government Printing Office.

Carson, R. 1962. *Silent Spring.* Boston: Houghton Mifflin.

Cavender-Bares, J., D. D. Ackerly, D. A. Baum, and F. A. Bazzaz. 2004. Phylogenetic over-dispersion in Floridian oak communities. *The American Naturalist* 163:823–843.

Clements, F. E., and V. E. Shelford. 1939. *Bio-ecology.* New York: John Wiley and Sons.

Conn, S. 1998. *Museums and American Intellectual Life 1876–1926.* Chicago: University of Chicago Press.

Cook, R. E. 1977. Raymond Lindeman and the trophic-dynamic concept in ecology. *Science* 198:22–26.

Darwin, C. 1964. *On the Origin of Species; a Facsimile of the First Edition.* Cambridge, Mass., and London: Harvard University Press.

Darwin, F., ed. 1958. *The Autobiography of Charles Darwin and Selected Letters.* New York: Dover.

Dayan, T., and D. Simberloff. 2005. Ecological and community-wide character displacement: the next generation. *Ecology Letters* 8:875–894.

de Terra, H. 1932a. Letter to President James R. Angell, Yale University, from Kashmir, March 22, 1932. Hutchinson Papers, Manuscripts and Archives, Yale University Library.

de Terra, H. 1932b. "Authorized Report" of the Yale North India Expedition after five months in the Tibetan-Kashmir front region. Hutchinson Papers, Manuscripts and Archives, Yale University Library.

Deevey, D. 1986. Interviewed by N. G. Slack in Queechey Lake, N.Y. (tape).

Deevey, E. S. 1937. Pollen from interglacial beds in Panggong Valley and climatic interpretation. *American Journal of Science* 33:44–58.

Dettelbach, M. 1996. Humboldtian science. In *Cultures of Natural History,* ed. N. Jardine, J. A. Secord, and E. C. Spary, pp. 287–304. Cambridge and New York: Cambridge University Press.

Donoghue, M. J., and W. S. Alverson. 2000. A new age of discovery. *Annals of the Missouri Botanical Garden* 87:110–126.

Dunbar, M. 1993. Interviewed by N. G. Slack in Montreal, June 20, 1993 (tape).

Edelson, Z. 1982. Observances honor O. C. Marsh. *Discovery* 16:33–35.

Edmondson, W. T. 1989. Rotifer study as a way of life. *Hydrobiologia* 186/187:1–9.

Edmondson, W. T. 1991. Interviewed by N. G. Slack at the U.S. National Academy of Sciences, April 30, 1991 (tape).

Edmondson, W. T., and G. E. Hutchinson. 1934. Report on Rotatoria. Yale North India Expedition. *Memoirs of the Connecticut Academy of Arts and Sciences* 10:153–186.

Edmondson, Y. 1971. Issue dedicated to G. Evelyn Hutchinson. *Limnology and Oceanography* 16, No. 2.

Elton, C. S. 1927. *Animal Ecology.* London: Sidgwich and Jackson.

Forbes, S. A. 1887. The lake as a microcosm. *Bulletin of the Scientific Association* (Peoria, Ill.), pp. 77–87.

Gause, G. F. 1935. Experimental demonstration of Volterra's periodic oscillation in the numbers of animals. *Journal of Experimental Biology* 12:44–48.

Golley, F. B. 1993. *A History of the Ecosystem Concept: More than the Sum of the Parts.* New Haven and London: Yale University Press.

Gould, S. J. 1979. Exultation and explanation. *New York Review of Books* 26(8): 3–6.

Gould, S. J. 1987. *An Urchin in the Storm: Essays about Books and Ideas.* New York: W. W. Norton.

Greene, H. W. 2005. Historical influences on community ecology. *Proceedings of the National Academy of Sciences USA* 102:8395–8396.

Grinnell, J. 1917. The niche relationships of the California thrasher. *Auk* 34:427–433.

Hagen, J. B. 1992. *An Entangled Bank: The Origins of Ecosystem Ecology.* New Brunswick, N.J.: Rutgers University Press.

Hairston, N. G., Jr., S. P. Ellner, M. A. Geber, T. Yoshida, and J. A. Fox. 2005. Rapid evolution and the convergence of ecological and evolutionary time. *Ecology Letters* 8:1114–1127.

Hendry, A. P., and M. T. Kinnison. 1999. The pace of modern life: measuring rates of microevolution. *Evolution* 53:1637–1653.

Hooper, D. U., et al. 2005. Effects of biodiversity on ecosystem functioning: a consensus of current knowledge. *Ecological Monographs* 75:3–35.

Horn, H. S., and R. M. May. 1977. Limits to similarity among coexisting competitors. *Nature* 270:660–661.

Hudson, J. J., W. D. Taylor, and D. W. Schindler. 2000. Phosphate concentrations in lakes. *Nature* 406:54–56.

Humboldt, A. von. 1850. *Views of Nature: or, Contemplations on the Sublime Phenomena of Creation; with Scientific Illustrations,* trans. E. C. Otté and H. G. Bohn. London: Henry G. Bohn.

Humboldt, A. von. 1997. *Cosmos: A Sketch of the Physical Description of the Universe,* vol. I, trans. E. C. Otté. Baltimore and London: Johns Hopkins University Press.

Hutchinson, G. E. 1918. A swimming grasshopper. *Entomologist's Record and Journal of Variation* 30:138.

Hutchinson, G. E. 1928. Cables to L. L. Woodruff, Yale University, March 31, 1928, and May 13, 1928. Yale University Library, Ross G. Harrison archives.

Hutchinson, G. E. 1932a. Letter to Grace E. Pickford, August 25, 1932. Hutchinson Papers, Manuscripts and Archives, Yale University Library.

Hutchinson, G. E. 1932b. Letter to Grace E. Pickford, September 1, 1932. Hutchinson Papers, Manuscripts and Archives, Yale University Library.

Hutchinson, G. E. 1932c. Letter to Grace E. Pickford, September 3, 1932. Hutchinson Papers, Manuscripts and Archives, Yale University Library.

Hutchinson, G. E. 1932d. Letter to Grace E. Pickford, November 13, 1932. Hutchinson Papers, Manuscripts and Archives, Yale University Library.

Hutchinson. G. E. 1932e. Letter to Grace E. Pickford, Yale University, November 8, 1933. Hutchinson Papers, Manuscripts and Archives, Yale University Library.

Hutchinson, G. E. 1933. Limnological studies at high altitudes in Ladak. *Nature* 132:136.

Hutchinson, G. E. 1936. *The Clear Mirror. A Pattern of Life in Goa and in Indian Tibet.* Cambridge: Cambridge University Press. Reprint edition, New Haven: Leete's Island Books, 1978.

Hutchinson, G. E. 1938. Chemical stratification and lake morphology. *Proceedings of the National Academy of Sciences USA* 24:63–69.

Hutchinson, G. E. 1940. Bio-ecology. *Ecology* 21:267–268.

Hutchinson, G. E. 1941a. Ecological aspects of succession in natural populations. *The American Naturalist* 75:406–418.

Hutchinson, G. E. 1941b. Limnological studies in Connecticut. IV. The Mechanisms of intermediary metabolism in stratified lakes. *Ecological Monographs* 11:21–60.

Hutchinson, G. E. 1942. The history of a lake. *Yale Scientific Magazine* 16:13–15, 22.

Hutchinson, G. E. 1943. Marginalia. *American Scientist* 31:270–278.

Hutchinson, G. E. 1947. A note on the theory of competition between two social species. *Ecology* 28:319–321.

Hutchinson, G. E. 1948a. Letter to E. H. Bradley, U.S. Geological Survey, September 24, 1948. Hutchinson Papers, Manuscripts and Archives, Yale University Library.

Hutchinson, G. E. 1948b. Circular causal systems in ecology. *Annals of the New York Academy of Sciences* 50:221–246.

Hutchinson, G. E. 1950a. Notes on the function of a university. *American Scientist* 38:127–129.

Hutchinson, G. E., 1950b. Survey of the contemporary knowledge of biogeochemistry. III. The biogeochemistry of vertebrate excretion. *Bulletin of the American Museum of Natural History* 96:554.

Hutchinson, G. E. 1951. Copepodology for the ornithologist. *Ecology* 32:571–577.

Hutchinson, G. E. 1953a. The concept of pattern in ecology. *Proceedings of the Academy of Natural Sciences of Philadelphia* 105:1–12.

Hutchinson, G. E. 1953b. *The Itinerant Ivory Tower.* New Haven and London: Yale University Press.

Hutchinson, G. E. 1957a. *A Treatise on Limnology, v. 1. Geography, Physics and Chemistry.* New York: Wiley.

Hutchinson, G. E. 1957b. Concluding remarks. *Cold Spring Harbor Symposia on Quantitative Biology* 22:415–427.

Hutchinson, G. E. 1959. Homage to Santa Rosalia, or Why are there so many kinds of animals? *The American Naturalist* 93:145–159.

Hutchinson, G. E. 1961. The paradox of the plankton. *The American Naturalist* 95:137–145.

Hutchinson, G. E. 1962. *The Enchanted Voyage and Other Studies.* New Haven and London: Yale University Press.

Hutchinson, G. E. 1963. The prospect before us. In *Limnology in North America,* ed. D. G. Frey, pp. 683–690. Madison: University of Wisconsin Press.

Hutchinson, G. E. 1964a. The lacustrine microcosm reconsidered. *American Scientist* 52:334–341.

Hutchinson, G. E. 1964b. The influence of the environment. *Proceedings of the National Academy of Sciences USA* 51:930–934.

Hutchinson, G. E. 1965. *The Ecological Theater and the Evolutionary Play.* New Haven and London: Yale University Press.

Hutchinson, G. E. 1966. On being a meter and a half long. In *Knowledge Among Men,* ed. P. H. Oehser, pp. 83–92. Washington, D.C.: Simon and Schuster and the Smithsonian Institution.

Hutchinson, G. E. 1967. *A Treatise on Limnology, v. 2. Introduction to Lake Biology and the Limnoplankton.* New York: John Wiley and Sons.

Hutchinson, G. E. 1969. Eutrophication: past and present. In *Eutrophication: Causes, Consequences, Correctives; Proceedings of a Symposium,* pp. 17–26. Washington, D.C.: National Academy of Sciences.

Hutchinson, G. E. 1977. Science has been liberal handed. . . . In *The Campaign for Yale,* March, pp. 43–44.

Hutchinson, G. E. 1978. *An Introduction to Population Ecology.* New Haven and London: Yale University Press.

Hutchinson, G. E. 1979. *The Kindly Fruits of the Earth: Recollections of an Embryo Ecologist.* New Haven and London: Yale University Press.

Hutchinson, G. E. 1980a. Conjectures arising in a quiet museum. *Antenna, Bulletin of the Royal Entomological Society, London* 4:92, 97–98.

Hutchinson, G. E. 1980b. Letter to Sharon Kingsland, October 14, 1980. Hutchinson Papers, Manuscripts and Archives, Yale University Library.

Hutchinson, G. E. 1981. Thoughts on aquatic insects. *BioScience* 31:495–500.

Hutchinson, G. E. 1982. Preface: Reminiscences and notes on some otherwise undiscussed papers. In *Selected Works of Gordon A. Riley,* ed. J. S. Wroblewski, vii–ix. Halifax, Nova Scotia: Dalhousie Printing Centre.

Hutchinson, G. E. 1983. Marginalia: what is science for? *American Scientist* 71:639–644.

Hutchinson, G. E. 1984. Topics in historical ecology. Presentation at the 70th birthday celebration for Edward Deevey, December 7, 1984 (tape).

Hutchinson, G. E. 1990. Interviewed by N. G. Slack at Yale University, November, 1990 (tape).

Hutchinson, G. E., and V. T. Bowen. 1947. A direct demonstration of the phosphorous cycle in a small lake. *Proceedings of the National Academy of Sciences USA* 33:148–153.

Hutchinson, G. E., and V. T. Bowen. 1950. Limnological studies in Connecticut. 9. A quantitative radiochemical study of the phosphorus cycle in Linsley Pond. *Ecology* 31: 194–203.

Hutchinson, G. E., and Y. Edmondson (ed.). 1993. *A Treatise on Limnology, v. 4. The Zoobenthos.* New York: Wiley.

Hutchinson, G. E., and R. H. MacArthur. 1959. A theoretical ecological model of size distributions among species of animals. *The American Naturalist* 93:117–135.

Hutchinson, G. E., G. E. Pickford, and J. F. M. Schuurman. 1932. A contribution to the hydrobiology of pans and other inland waters of South Africa. *Archiv für Hydrobiologie* 24:1–154.

Jeffreys, H. 1937. *Scientific Inference,* 2nd ed. Cambridge: Cambridge University Press.

Kiester, A. R. 1997. Aesthetics of biological diversity. *Human Ecology Review* 3:151–157.

Kingsland, S. E. 1985. *Modeling Nature: Episodes in the History of Population Ecology.* Chicago and London: University of Chicago Press.

Kingsland, S. E. 1995. *Modeling Nature: Episodes in the History of Population Ecology,* 2nd ed. Chicago and London: University of Chicago Press.

Lawton, J. H. 2000. *Community Ecology in a Changing World.* Ecology Institute: Oldendorf/Luhe, Germany.

Leibold, M. A., M. Holyoak, N. Mouquet, P. Amarasekare, J. M. Chase, M. F. Hoopes, R. D. Holt, J. B. Shurin, R. Law, D. Tilman, M. Loreau, and A. Gonzalez. 2004. The metacommunity concept: a framework for multi-scale community ecology. *Ecology Letters* 7:601–613.

Levins, R. 1969. Some demographic and genetic consequences of environmental heterogeneity for biological control. *Bulletin of the Entomological Society of North America* 15:237–240.

Lindeman, R. 1942. The trophic-dynamic aspect of ecology. *Ecology* 23:399–417.

Loreau, M., et al. 2001. Ecology—Biodiversity and ecosystem functioning: Current knowledge and future challenges. *Science* 294:804–808.

MacArthur, R. H. 1958. Letter to Bernard Patten, Duke University, from the Institute of Marine Science, Port Aranson, Texas, February 7, 1958. In possession of N. G. Slack.

May, R. M. 1973. Stability and complexity in model ecosystems. Princeton, N.J.: Princeton University Press.

McIntosh, R. P. 1985. *The Background of Ecology. Concept and Theory.* New York: Cambridge University Press.

Mumford, L. 1956. Prospect. In *Man's Role in Changing the Face of the Earth,* ed. W. L. Thomas. Chicago: University of Chicago Press.

Nicholas, J. S. 1945. Letter to Charles Seymour, Yale University, November 22, 1945. Yale University Library, Ross G. Harrison archives.

Nicolson, M. 1996. Humboldtian plant geography after Humboldt: the link to ecology. *British Journal for the History of Science* 29:289–310.

Peters, R. H. 1976. Tautology in evolution and ecology. *The American Naturalist* 110:1–12.

Platil, G. P., and M. L. Rosenzweig. 1979. Preface. In *Contemporary Quantitative Ecology and Econometrics.* Fairland, Md.: International Cooperative Publishing House.

Platt, R. B., and J. N. Wolfe, 1964. Introduction to special issue on ecology. *BioScience* 14:9–10.

Rappaport, R. 1999. Interviewed by N. G. Slack in Bar Harbor, Maine, June 1999.

Raynor, G. H., and L. Doncaster. 1905. Experiments on heredity and sex determination in *Abraxas grossulariata.* In *Report of the British Association for the Advancement of Science 1904,* pp. 594–595. London: John Murray.

Remington, C. L. 1979. The Hutchinson collection of water bugs. *Discovery* 14:12.

Riley, G. A. 1971. Introduction. *Limnology and Oceanography* 16:177–179.

Riley, G. A. 1980. Interviewed by Eric Mills in Halifax, Nova Scotia, October 27, 1980. Tape in possession of N. G. Slack.

Riley, G. A. 1984. Reminiscences of an oceanographer. Unpublished, 166 pp. In possession of N. G. Slack.

Schindler, D. W., et al. 1986. Natural sources of acid neutralizing capacity in low alkalinity lakes of the Precambrian Shield. *Science* 232:844–847.

Simberloff, D., and W. Boecklen. 1981. Santa Rosalia reconsidered: size ratios and competition. *Evolution* 35:1206–1228.

Slack, N. G. 1995. Botanical and ecological couples, a continuum of relationships. In *Creative Couples in the Sciences,* ed. H. M. Pycior, N. G. Slack, and P. G. Abir-Am, pp. 235–253. New Brunswick, N.J.: Rutgers University Press.

Slack, N. G. 2003. Are research schools necessary? Contrasting models of 20th century research at Yale led by Ross Granville Harrison, Grace E. Pickford and G. Evelyn Hutchinson. *Journal of the History of Biology* 36:501–505.

Slack, N. G. 2010. *G. Evelyn Hutchinson and the Invention of Modern Ecology.* New Haven and London: Yale University Press.

Slobodkin, L. B., and N. G. Slack. 1999. George Evelyn Hutchinson: 20th-century ecologist. *Endeavour* 23:24–30.

Strong, D. R. 1979. Review of *An Introduction to Population Ecology,* by G. E. Hutchinson. *Journal of Biogeography* 6:201–204.

Sulloway, F. J. 2003. Darwin and his doppelgänger. A review of *Charles Darwin: The Power of Place* by Janet Browne and of *In Darwin's Shadow: The Life and Science of Alfred Russel Wallace* by M. Shermer. *The New York Review of Books,* December 18, 2003.

Thienemann, A. 1925. *Die Binnengewasser Mitteleuropas.* Die Binnengewasser. Stuttgart: Schweizerbart'sche Verlagsbuchhandlung.

Thompson, D. W. 1942. *On Growth and Form,* 2nd ed. Cambridge: Cambridge University Press.

Tilman, D., R. M. May, C. L. Lehman, and M. A. Nowak. 1994. Habitat destruction and the extinction debt. *Nature* 371:65–66.

Tilman, D., D. Wedin, and J. Knops. 1996. Productivity and sustainability influenced by biodiversity in grassland ecosystems. *Nature* 379:718–720.

Urban, M. C., and D. K. Skelly. 2006. Evolving metacommunities: toward an evolutionary perspective on metacommunities. *Ecology* 87:1616–1626.

Webb, C. O., D. D. Ackerly, M. McPeek, and M. J. Donoghue. 2002. Phylogenies and community ecology. *Annual Review of Ecology and Systematics* 33:475–505.

Werner, E. E. 1986. Amphibian metamorphosis—growth rate, predation risk, and the optimal size at transformation. *The American Naturalist* 128:319–341.

Weschler, L. 1995. *Mr. Wilson's Cabinet of Wonder.* New York: Vintage Books, Random House.

West, R. 1987. *The Strange Necessity: Essays and Reviews.* London: Virago.

Wiens, J. J., and M. J. Donoghue. 2004. Historical biogeography, ecology, and species richness. *Trends in Ecology and Evolution* 19:639–644.

Wilson, D. S. 1992. Complex interaction in metacommunities, with implications for biodiversity and higher levels of selection. *Ecology* 73:1984–2000.

CONTRIBUTORS

Michael J. Donoghue Department of Ecology and Evolutionary Biology
Yale University
New Haven, Connecticut

Sharon E. Kingsland Department of History of Science and Technology
Johns Hopkins University
Baltimore, Maryland

Jane Pickering Peabody Museum of Natural History
Yale University
New Haven, Connecticut

David M. Post Department of Ecology and Evolutionary Biology
Yale University
New Haven, Connecticut

David W. Schindler Department of Biological Sciences
University of Alberta
Calgary, Alberta

David K. Skelly School of Forestry and Environmental Studies
Yale University
New Haven, Connecticut

Nancy G. Slack Department of Biology
Russell Sage College
Troy, New York

Melinda D. Smith Department of Ecology and Evolutionary Biology
Yale University
New Haven, Connecticut

CREDITS

of London 4 (1980): 92, 97–98. © 1980 by the Royal Entomological Society. Reprinted with permission.

Illustrations

Page 3: Portrait of Alexander von Humboldt painted by Friedrich Georg Weitsch, 1806. Photograph reproduced by permission of Bildarchiv Preußischer Kulturbesitz and Art Resource, NY.

Page 67: Barnet Woolf, *Sir Frederick Gowland Hopkins, O.M., F.R.S.* Originally published in *Brighter Biochemistry,* 1925. Reproduced by permission of the Department of Biochemistry, University of Cambridge.

Page 76: D. G. Lillie, *The Cambridge Natural History Society.* Reproduced by permission of the Cambridge University Museum of Zoology.

Page 135: Dr. Edward S. Deevey, Jr., "Three Stages in the Evolution of Linsley Pond." Originally published in Deevey, "Studies on Connecticut Lake Sediments III: The Biostratonomy of Linsley Pond, Part II," *American Journal of Science* vol. 240, no. 5 (1942), pp. 313–324. Reproduced with permission of the *American Journal of Science;* permission conveyed through Copyright Clearance Center, Inc.

Page 246: G. Evelyn Hutchinson receiving the Verrill Medal at the Yale Peabody Museum, 1981. From the Yale Peabody Archives; copyright Yale University.

Page 259: Illustrations of butterflies from the Aldenham-Rabinowitz "Book of Hours" folio, late fifteenth century. Courtesy of the Beinecke Rare Book and Manuscript Library, Yale University.

Page 265: Jan van Kessel (1626–1679), *Allegoria della vista.* Galleria Palatina, Florence. Reproduced with permission of the State Museums of Florence.

Page 266: *Nautilus* cup, Ostrich-egg goblet, and *Turbo marmoratus* cup, Kunsthistorisches Museum, Vienna. Image reproduced courtesy of the Kunsthistorisches Museum.

Page 268: *Pendant with a Triton* and *Pendant with a Mermaid,* Widener Collection, National Gallery of Art, Washington, DC. Image courtesy of the Board of Trustees, National Gallery of Art.

Page 270: *Study of a Mulatto Woman,* French School 1820–25. Reproduced by permission of the Fitzwilliam Museum, Cambridge.

Page 275: Fulgurite from Santa Rosa Island, FL. Image courtesy of the Academy of Natural Sciences, Philadelphia.

Page 279: Moth specimens given by Gilbert Henry Raynor to the Cambridge University Museum of Zoology. Image reproduced with permission of the Cambridge University Museum of Zoology.

Page 317: D. G. Lillie, *Our Bill,* caricature of E. A. Wilson. Reproduced with permission of the Scott Polar Research Institute, University of Cambridge.